The Hundredth Monkey

The Hundredth Monkey

BEST OF
SKEPTICAL INQUIRER
VOLUME 3

EDITED BY
KENDRICK FRAZIER

Prometheus Books
Essex, Connecticut

Prometheus Books

An imprint of Globe Pequot, the trade division of
The Rowman & Littlefield Publishing Group, Inc.
4501 Forbes Blvd., Ste. 200
Lanham, MD 20706
www.rowman.com

Distributed by NATIONAL BOOK NETWORK

British Library Cataloguing in Publication Information Available

Library of Congress Cataloging-in-Publication Data

The hundredth monkey and other paradigms of the paranormal / edited by
    Kendrick Frazier.
        p. cm.
    Forty-three essays from the Skeptical Inquirer magazine, by Carl Sagan, Isaac
Asimov, Martin Gardner, Susan Blackmore, and others.
    Includes index.
    1. Parapsychology and science. 2. Occultism and science. I. Frazier, Kendrick.

BF1045.S33H86 1991
133—dc20   90–27044
                                                                                CIP

ISBN 9781633889668 (paper : alk. paper) | ISBN 9781633889675 (ebook)

∞™ The paper used in this publication meets the minimum requirements of American
National Standard for Information Sciences—Permanence of Paper for Printed Library
Materials, ANSI/NISO Z39.48-1992

CONTENTS

Introduction   xi

Part 1: Understanding the Human Need

CARL SAGAN
The Burden of Skepticism   1

ISAAC ASIMOV
The Perennial Fringe   10

PAUL KURTZ
Reflections on the "Transcendental Temptation"   13

L. SPRAGUE DE CAMP
The Uses of Credulity   17

PHILLIPS STEVENS, JR.
The Appeal of the Occult   20

Part 2: Encouraging Critical Thinking

JAMES W. LETT
A Field Guide to Critical Thinking   31

RAY HYMAN
Assessing Arguments and Evidence   40

Part 3: Evaluating the Anomalous Experience

BARRY BEYERSTEIN
The Brain and Consciousness: Implications for Psi Phenomena   43

ROBERT A. BAKER
The Aliens Among Us: Hypnotic Regression Revisited   54

BILL ELLIS
The Varieties of Alien Experience   70

NICHOLAS P. SPANOS
Past-Life Hypnotic Regression: A Critical View   78

SARAH G. THOMASON
Past-Tongues Remembered?   85

BARRY BEYERSTEIN
The Myth of Alpha Consciousness   95

DONALD D. JENSEN
Pathologies of Science, Precognition,
and Modern Psychophysics   111

Part 4: Considering Parapsychology

SUSAN J. BLACKMORE
The Elusive Open Mind: Ten Years of
Negative Research in Parapsychology   125

ANTONY FLEW
Parapsychology, Miracles, and Repeatability   136

KENDRICK FRAZIER
Ganzfeld Studies   143

KENDRICK FRAZIER
Improving Human Performance:
What About Parapsychology?   149

MARTIN GARDNER
Psi Researchers' Inattention to Conjuring   162

MARTIN GARDNER
The Obligation to Disclose Fraud   167

Part 5: Examining Popular Claims

RON AMUNDSON
The Hundredth Monkey Phenomenon   171

BERNARD J. LEIKIND and WILLIAM J. McCARTHY
An Investigation of Firewalking   182

JOE NICKELL and JOHN F. FISCHER
Incredible Cremations:
Investigating Spontaneous Combustion Deaths   194

ADRIAN FURNHAM
Write and Wrong: The Validity of Graphological Analysis   200

ROBERT BASIL
Graphology and Personality: "Let the Buyer Beware"   206

ARLEEN J. WATKINS and WILLIAM S. BICKEL
A Study of the Kirlian Effect   209

IVAN W. KELLY, JAMES ROTTON, and ROGER CULVER
The Moon Was Full and Nothing Happened   222

PAUL KURTZ, JAMES ALCOCK, KENDRICK FRAZIER,
BARRY KARR, PHILIP J. KLASS, and JAMES RANDI
Testing Psi Claims in China   235

Part 6: Medical Controversies

KARL SABBAGH
The Psychopathology of Fringe Medicine   247

FRANK REUTER
Folk Remedies and Human Belief-Systems   256

WILLIAM JARVIS
Chiropractic: A Skeptical View   262

STEPHEN BARRETT
Homeopathy: Is It Medicine?   271

Part 7: Astrology

GEOFFREY DEAN
Does Astrology Need to Be True?
Part 1: A Look at the Real Thing   279

GEOFFREY DEAN
Does Astrology Need to Be True?
Part 2: The Answer Is No   297

FROM 'NEWS AND COMMENT'
Double-Blind Test of Astrology Avoids Bias,
Still Refutes the Astrological Hypothesis   320

Part 8: Crashed-Saucer Claims

PHILIP J. KLASS
Crash of the Crashed-Saucer Claim   323

PHILIP J. KLASS
The MJ-12 Crashed-Saucer Documents: Part 1   327

PHILIP J. KLASS
The MJ-12 Crashed-Saucer Documents: Part 2   336

PHILIP J. KLASS
New Evidence of MJ-12 Hoax   347

Part 9: Controversies Within Science

MILTON A. ROTHMAN
   Cold Fusion: A Case Study in "Wishful Science"   353

MARTIN GARDNER
   Water With Memory? The Dilution Affair   364

MARTIN GARDNER
   Science, Mysteries, and the Quest for Evidence   372

MARTIN GARDNER
   Relativism in Science   376

Recommended Reading   381

Dates of Original Publication   383

Contributors   387

Name Index   391

The Hundredth Monkey

From earliest times people have had to cope with uncertain understanding about the forces and mysteries of the natural world that permeate and affect all life. They have applied both their imagination and fears and their reason and intelligence to this task. Religions and mythologies were born to provide structure and understanding to a cosmos that otherwise overwhelmed with its awesomeness and mystery. Later natural philosophers and scientists brought new and powerful intellectual tools that explored these mysteries in novel and fruitful ways and helped peel away some of our ignorance and misconceptions, while simultaneously revealing still more and deeper mysteries. These new tools have been productive beyond measure, and a phenomenal modern world has been shaped as a result.

At the same time, we still must continually search for meaning in an often chaotic existence. While our inner nature, our psychological makeup, our human hopes and fears are products of this long and contentious human history, organized science and modern technology in comparison have been around for only a relatively short time. Yet the rapid advances of science and technology—and the new problems that have come along in their wake— have outpaced our own human ability to adapt to a changing world. No wonder, then, that for many people the quest for understanding remains essentially personal, narrow, subjective, and inner-directed. No wonder that even as the twenty-first century looms ahead, an ancient and inevitable tension still exists within us—one where our hopes and aspirations about how we'd like the world to be often struggle and conflict with the evidence of our intellect and reason. When that happens, comforting and warm, if sometimes bizarre, beliefs often win out over cool and chancy rationality and realism.

This newest collection of essays and articles from the *Skeptical Inquirer* examines important issues and tensions at the intersection of science and popular belief. Forty-three articles by noted scientists, psychologists, philosophers, writers, and other scholars and investigators explore virtually every aspect of paranormal and fringe-science beliefs and claims. Beginning with Carl Sagan's opening essay on the burden and meaning of skepticism, the authors discuss the deepest values of science and the exquisite yet little appreciated balance between science's creative openness to all new ideas and relentless scrutiny of all new claims. In these pages you will find illuminating and scientifically responsible explanations of the alien-abduction experience, the physics of firewalking, alleged spontaneous human-combustion deaths, and the MJ-12 "crashed-flying-saucer" document hoax, as well as a classic and thorough two-part examination of astrology (which the author has here expanded and updated with much new material) that has been hailed as perhaps the best ever done. Psychologists report on our understanding and misunderstandings about the brain and consciousness, post-hypnotic regression, so-called past-lives claims, and the problems with parapsychology. Experts examine medical controversies, such as homeopathy and chiropractic, and explore two highly publicized and dramatic controversies, one within science—cold fusion—and another along the margins—"remembering water." Throughout, the authors provide constructive advice about critical thinking, assessing evidence, and encouraging people to seek rational solutions to sensationalized mysteries by using common sense and their own intelligence.

The book takes its title from one of the articles, philosopher Ron Amundson's probing exposé of the so-called Hundredth Monkey Phenomenon, a remarkable claim of "group consciousness." The claim involved wild troops of monkeys on islands in Japan. When a sufficient number of them, say the hundredth monkey, learned a certain new behavior, supposedly all of the monkeys on all the islands suddenly and spontaneously adopted the new behavior as well—leaping barriers of time and space. Was such a thing possible? Did this really happen? If so, perhaps such mass consciousness could leap other barriers as well and help solve global problems by the mere application of thought and will. It was an appealing idea to many and was widely embraced by the so-called New Age movement. Amundson went back to the original sources and found that the claim was . . . well, I think I'll let you find out for yourself. It's a fascinating story, and an instructive one. Amundson's original article is here, along with a follow-up he prepared some months afterward, and a postscript written for this volume that updates and comments about the whole matter in the light of the reaction, almost all of it positive, his article received.

The articles in this collection appeared originally in the *Skeptical Inquirer*, from 1985 to 1990. This is the third such collection. The other two are *Paranormal Borderlands of Science* (1981) and *Science Confronts the Paranormal* (1986), both still in print by Prometheus. For this collection, many of the authors have written brief postscripts. These give timely new information, perspective, and comment—and summarize and respond to criticisms. They have also provided updated references.

The *Skeptical Inquirer* is published by the Committee for the Scientific Investigation of Claims of the Paranormal. CSICOP was established in 1976 to help provide the public, educators, and the media with accurate, scientifically evaluated information about the wide variety of paranormal and fringe-science claims that everywhere bombard us. The problem is that scientists, busy with their own research, typically ignore such claims as irrelevant or unimportant to real science. That may be. Yet these claims tend to have wide appeal among students and the public, and to many they may seem superficially as "scientific" as the kinds of astonishing new discoveries that one hears of almost every day. Most people don't have access to reliable information about these claims in contrast to the misinformation that is so readily available. People are intelligent enough to make their own judgments, but they need help both in gaining access to the facts and—just as important—understanding the processes of science. It is these processes by which new concepts about the world are carefully examined, subjected to tests of evidence, revised, published in refereed journals, and repeatedly challenged and tested again—gradually, perhaps, to gain a tentative foothold, yet always subject to revision in the light of future evidence.

CSICOP has helped encourage the view that scientists and scholars should devote a certain amount of their time to this public-education role. Many now do so. They are helping us understand some of these marvelous processes of science, which are a major contribution to modern culture. They investigate bizarre claims that interest the public, provide factual information, encourage critical thinking and the application of reason and common sense to arguments, and in general help separate facts from speculation, sense from nonsense, and real science from pseudoscience.

This effort was in the past carried out typically by one or two concerned persons at a time acting more or less alone. But with the establishment of CSICOP and the *Skeptical Inquirer* as a forum for published articles and debate, it has broadened steadily throughout the years. Now, many thousands of people are involved, the scientific and scholarly communities are broadly represented, and it is quite international in scope. It is not exclusive. Anyone can take part.

One of the things we have found is that the issues and demarcations are not always as simple and easy as they may at first seem; at the extremes the differences may be quite clear-cut, but at the fuzzy boundaries between science and nonscience there are frequently gray and troublesome areas. We have also found that novel-sounding claims almost identical to ones long-ago discarded keep rising up in new costume and language. Still another lesson is that emotions run high on both sides of these issues and it is a constant struggle to minimize their effect on the arguments; evidence must be the arbiter. Something about these subjects resonates with the most deep-seated needs in our psyches, and they thus have endless manifestations and reawakenings in human culture. For science-minded people concerned about clear thinking, reason, and scientific evidence there is much that is frustrating when dealing with such claims, yet there is also much that is fascinating. The effort can be quite enjoyable as well as instructive, and we now invite you to share in that quest.

This is the fifteenth-anniversary year of CSICOP, and I dedicate this volume to CSICOP's founder and chairman, philosopher Paul Kurtz. His breadth of vision, his passion for freedom of inquiry and the use of science and reason, and his extraordinary energy and skills have made this effort something everyone can value. I am proud to be his friend and colleague. I also thank the other members of the Executive Council of CSICOP and the Editorial Board of the *Skeptical Inquirer*—James Alcock, Barry Beyerstein, Martin Gardner, Philip J. Klass, Ray Hyman, Joe Nickell, and James Randi—for their effective and dedicated efforts in the cause of reason and for their longtime support and friendship. Similar very special thanks to our colleagues Carl Sagan and Isaac Asimov. And I thank all the authors of this compendium, and the many other contributors to the *Skeptical Inquirer* over the years (including the many whose articles could not be included in this volume in order to keep it to manageable length) for their care, their public spirit, and their steadfast contributions to public understanding. And finally I thank a person whose work has been almost entirely behind the scenes but has nevertheless been crucial—Doris Doyle, the *Skeptical Inquirer*'s longtime managing editor. Every article in this collection has benefited from her editorial skills and judgment.

For more information about CSICOP and the *Skeptical Inquirer* write to Box 229, Buffalo, N.Y. 14215, or call 716-636-1425.

Kendrick Frazier

Part 1: Understanding the Human Need

CARL SAGAN

The Burden of Skepticism

What is skepticism? It's nothing very esoteric. We encounter it every day. When we buy a used car, if we are the least bit wise we will exert some residual skeptical powers—whatever our education has left to us. You could say, "Here's an honest-looking fellow. I'll just take whatever he offers me." Or you might say, "Well, I've heard that occasionally there are small deceptions involved in the sale of a used car, perhaps inadvertent on the part of the salesperson," and then you do something. You kick the tires, you open the doors, you look under the hood. (You might go through the motions even if you don't know what is supposed to be under the hood, or you might bring a mechanically inclined friend.) You know that some skepticism is required, and you understand why. It's upsetting that you might have to disagree with the used-car salesman or ask him questions that he is reluctant to answer. There is at least a small degree of interpersonal confrontation involved in the purchase of a used car and nobody claims it is especially pleasant. But there is a good reason for it—because if you don't exercise some minimal skepticism, if you have an absolutely untrammeled credulity, there is probably some price you will have to pay later. Then you'll wish you had made a small investment of skepticism early.

Now this is not something that you have to go through four years of graduate school to understand. Everybody understands this. The trouble is, a used car is one thing but television commercials or pronouncements by presidents and party leaders are another. We are skeptical in some areas but unfortunately not in others.

For example, there is a class of aspirin commercials that reveals the competing product to have only so much of the painkilling ingredient that doctors recommend most—they don't tell you what the mysterious ingredient is—whereas *their* product has a dramatically larger amount (1.2 to 2 times more per tablet). Therefore you should buy their product. But why not just take two of the competing tablets? You're not supposed to ask. Don't apply skepticism to this issue. Don't think. Buy.

Such claims in commercial advertisements constitute small deceptions. They part us from a little money, or induce us to buy a slightly inferior product. It's not so terrible. But consider this:

I have here the program of this year's Whole Life Expo in San Francisco. Twenty thousand people attended last year's program. Here are some of the presentations: "Alternative Treatments for AIDS Patients: It will rebuild one's natural defenses and prevent immune system breakdowns—learn about the latest developments that the media has thus far ignored." It seems to me that presentation could do real harm. "How Trapped Blood Proteins Produce Pain and Suffering." "Crystals, Are They Talismans or Stones'" (I have an opinion myself.) It says, "As a crystal focuses sound and light waves for radio and televison"—crystal sets are rather a long time ago—"so may it amplify spiritual vibrations for the attuned human." I'll bet very few of you are attuned. Or here's one: "Return of the Goddess, a Presentational Ritual." Another: "Synchronicity, the Recognition Experience." That one is given by "Brother Charles." Or, on the next page, "You, Saint-Germain, and Healing Through the Violet Flame." It goes on and on, with lots of ads about "opportunities"—ranging from the dubious to the spurious—that are available at the Whole Life Expo.

*   *   *

If you were to drop down on Earth at any time during the tenure of humans you would find a set of popular, more or less similar, belief systems. They change, often very quickly, often on time scales of a few years: But sometimes belief systems of this sort last for many thousands of years. At least a few are always available. I think it's fair to ask why. We are *Homo sapiens*. That's the distinguishing characteristic about us, that *sapiens* part. We're supposed to be smart. So why is this stuff always with us? Well, for one thing, a great many of these belief systems address real human needs that are not being met by our society. There are unsatisfied medical needs, spiritual needs, and needs for communion with the rest of the human community. There may be more such failings in our society than in many others in human history. And so it is reasonable for people to poke around and try on for size various belief systems, to see if they help.

For example, take a fashionable fad, channeling. It has for its fundamental premise, as does spiritualism, that when we die we don't exactly disappear, that some part of us continues. That part, we are told, can reenter the bodies of human and other beings in the future, and so death loses much of its sting for us personally. What is more, we have an opportunity, if the channeling contentions are true, to make contact with loved ones who have died.

Speaking personally, I would be delighted if reincarnation were real. I lost my parents, both of them, in the past few years, and I would love to have a little conversation with them, to tell them what the kids are doing, make sure everything is all right wherever it is they are. That touches something very deep. But at the same time, precisely for that reason, I know that there are people who will try to take advantage of the vulnerabilities of the bereaved. The spiritualists and the channelers better have a compelling case.

Or take the idea that by thinking hard at geological formations you can tell where mineral or petroleum deposits are. Uri Geller makes this claim. Now if you are an executive of a mineral exploration or petroleum company, your bread and butter depend on finding the minerals or the oil; so spending trivial amounts of money, compared with what you usually spend on geological exploration, this time to find deposits psychically, sounds not so bad. You might be tempted.

Or take UFOs, the contention that beings in spaceships from other worlds are visiting us all the time. I find that a thrilling idea. It's at least a break from the ordinary. I've spent a fair amount of time in my scientific life working on the issue of the search for extraterrestrial intelligence. Think how much effort I could save if those guys are coming here. But when we recognize some emotional vulnerability regarding a claim, that is exactly where we have to make the firmest efforts at skeptical scrutiny. That is where we can be had.

Now, let's reconsider channeling. There is a woman in the State of Washington who claims to make contact with a 35,000-year-old somebody, "Ramtha"—he, by the way, speaks English very well with what sounds to me to be an Indian accent. Suppose we had Ramtha here and just suppose Ramtha is cooperative. We could ask some questions: How do we know that Ramtha lived 35,000 years ago? Who is keeping track of the intervening millennia? How does it come to be exactly 35,000 years? That's a very round number. Thirty-five thousand plus or minus what? What were things like 35,000 years ago? What was the climate? Where on Earth did Ramtha live? (I know he speaks English with an Indian accent, but where was that?) What does Ramtha eat? (Archaeologists know something about what people ate back then.) We would have a real opportunity to find out if his claims are true. If this were really somebody from 35,000 years ago, you could learn a lot about 35,000 years ago. So, one way or another, either Ramtha really is 35,000 years old, in which case we discover something about that period—that's before the Wisconsin Ice Age, an interesting time—or he's a phony and he'll slip up. What are the indigenous languages, what is the social structure, who else does Ramtha live with—children, grandchildren—what's the life cycle, the infant mortality, what clothes does he wear, what's his life expectancy, what are the weapons, plants, and animals? Tell us. Instead, what we hear are the most banal homilies, indistinguishable from those that alleged UFO occupants tell the poor humans who claim to have been abducted by them.

Occasionally, by the way, I get a letter from someone who is in "contact" with an extraterrestrial who invites me to "ask anything." And so I have a list of questions. The extraterrestrials are very advanced, remember. So I ask things like, "Please give a short proof of Fermat's Last Theorem." Or the Goldbach Conjecture. And then I have to explain what these are, because extraterrestrials will not call it Fermat's Last Theorem, so I write out the little equation with the exponents. I never get an answer. On the other hand, if I ask something like "Should we humans be good?" I always get an answer. I think something can be deduced from this differential ability to answer questions. Anything vague they are extremely happy to respond to, but anything specific, where there is a chance to find out if they actually know anything, there is only silence.

The French scientist Henri Poincaré remarked on why credulity is rampant: "We also know how cruel the truth often is, and we wonder whether delusion is not more consoling." That's what I have tried to say with my examples. But I don't think that's the only reason credulity is rampant. Skepticism challenges established institutions. If we teach everybody, let's say schoolchildren, the habit of being skeptical, perhaps they will not restrict their skepticism to aspirin commercials and 35,000-year-old channelers (or channelees). Maybe they'll start asking awkward questions about economic, or social, or political, or religious institutions. Then where will we be?

Skepticism is dangerous. That's exactly its function, in my view. It is the business of skepticism to be dangerous. And that's why there is a great reluctance to teach it in the schools. That's why you don't find a general fluency in skepticism in the media. On the other hand, how will we negotiate a very perilous future if we don't have the elementary intellectual tools to ask searching questions of those nominally in charge, especially in a democracy?

I think this is a useful moment to reflect on the sort of national trouble that could have been avoided were skepticism more generally available in American society. The Iran/Nicaragua fiasco is so obvious an example I will not take advantage of our poor, beleaguered president by spelling it out. The Administration's resistance to a Comprehensive Test Ban Treaty and its continuing passion for blowing up nuclear weapons—one of the major drivers of the nuclear arms race—under the pretense of making us "safe" is another such issue. So is Star Wars. The habits of skeptical thought CSICOP encourages have relevance for matters of the greatest importance to the nation. There is enough nonsense promulgated by both political parties that the habit of evenhanded skepticism should be declared a national goal, essential for our survival.

*   *   *

I want to say a little more about the burden of skepticism. You can get into a habit of thought in which you enjoy making fun of all those other people who don't see things as clearly as you do. This is a potential social danger present in an organization like CSICOP. We have to guard carefully against it.

It seems to me what is called for is an exquisite balance between two conflicting needs: the most skeptical scrutiny of all hypotheses that are served up to us and at the same time a great openness to new ideas. Obviously those two modes of thought are in some conflict. But if you are able to exercise only one of these modes, whichever one it is, you're in deep trouble.

If you are only skeptical, then no new ideas make it through to you. You never learn anything new. You become a crochety old person convinced that nonsense is ruling the world. (There is, of course, much data to support you.) But every now and then, maybe once in a hundred cases, a new idea turns out to be on the mark, valid, and wonderful. If you are too much in the habit of being skeptical about everything, you are going to miss or resent it, and either way you will be standing in the way of understanding and progress.

On the other hand, if you are open to the point of gullibility and have not an ounce of skeptical sense in you, then you cannot distinguish the useful ideas from the worthless ones. If all ideas have equal validity then you are lost, because then, it seems to me, no ideas have any validity at all.

Some ideas are better than others. The machinery for distinguishing them is an essential tool in dealing with the world and especially in dealing with the future. And it is precisely the mix of these two modes of thought that is central to the success of science.

Really good scientists do both. On their own, talking to themselves, they churn up huge numbers of new ideas, and criticize them ruthlessly. Most of the ideas never make it to the outside world. Only the ideas that pass through a rigorous self-filtration make it out and are criticized by the rest of the scientific community. It sometimes happens that ideas that are accepted by everybody turn out to be wrong, or at least partially wrong, or at least superseded by ideas of greater generality. And, while there are of course some personal losses—emotional bonds to the idea that you yourself played a role in inventing—nevertheless the collective ethic is that everytime such an idea is overthrown and replaced by something better the enterprise of science has benefited. In science it often happens that scientists say, "You know that's a really good argument; my position is mistaken," and then they actually change their minds and you never hear that old view from them again. They really do it. It doesn't happen as often as it should, because scientists are human and change is sometimes painful. But it happens every day. I cannot recall the last time something like that has happened in politics or religion. It's very rare that a senator, say, replies, "That's a good argument. I will now change my political affiliation."

*   *   *

I would like to say a few things about the stimulating sessions on the Search for Extraterrestrial Intelligence (SETI) and on animal language at our CSICOP conference. In the history of science there is an instructive procession of major intellectual battles that turn out, all of them, to be about how central human beings are. We could call them battles about the anti-Copernican conceit.

Here are some of the issues:

• *We are the center of the Universe. All the planets and the stars and the Sun and the Moon go around us.* (Boy, must we be something *really* special.) That was the prevailing belief—Aristarchus aside—until the time of Copernicus. A lot of people liked it because it gave them a personally unwarranted central position in the Universe. The mere fact that you were on Earth made you privileged. That felt good. Then along came the evidence that Earth was just a planet and that those other bright moving points of light were planets too. Disappointing. Even depressing. Better when we were central and unique.

• *But at least our Sun is at the center of the Universe.* No, those other stars, they're suns too, and what's more we're out in the galactic boondocks. We are nowhere near the center of the Galaxy. Very depressing.

• *Well, at least the Milky Way galaxy is at the center of the Universe.* Then a little more progress in science. We find there isn't any such thing as the center of the Universe. What's more there are a hundred billion other galaxies. Nothing special about this one. Deep gloom.

• *Well, at least we humans, we are the pinnacle of creation. We're separate. All those other creatures, plants and animals, they're lower. We're higher. We have no connection with them. Every living thing has been created separately.* Then along comes Darwin. We find an evolutionary continuum. We're closely connected to the other beasts and vegetables. What's more, the closest biological relatives to us are chimpanzees. *Those* are our close relatives—*those* guys? It's an embarrassment. Did you ever go to the zoo and watch them? Do you know what they do? Imagine in Victorian England, when Darwin produced this insight, what an awkward truth it was.

There are other important examples—privileged reference frames in physics and the unconscious mind in psychology—that I'll pass over.

I maintain that in the tradition of this long set of debates—every one of which was won by the Copernicans, by the guys who say there is not much special about us—there was a deep emotional undercurrent in the debates in both CSICOP sessions I mentioned. The search for extraterrestrial intelligence and the analysis of possible animal "language" strike at one of the last remaining pre-Copernican belief systems:

• *At least we are the most intelligent creatures in the whole Universe.* If there are no other smart guys elsewhere, even if we *are* connected to chimpanzees, even if we *are* in the boondocks of a vast and awesome universe, at least there is still something special about us. But the moment we find extraterrestrial intelligence that last bit of conceit is gone. I think some of the resistance to the idea of extraterrestrial intelligence is due to the anti-Copernican conceit. Likewise, without taking sides in the debate on whether other animals—higher primates, especially great apes—are intelligent or have language, that's clearly, on an emotional level, the same issue. If we define humans as creatures who have language and no one else has language, at least we are unique in that regard. But if it turns out that all those dirty, repugnant, laughable chimpanzees can also, with Ameslan or otherwise, communicate ideas, then what is left that is special about us? Propelling emotional predispositions on these issues are present, often unconsciously, in scientific debates. It is important to realize that scientific debates, just like pseudoscientific debates, can be awash with emotion, for these among many different reasons.

Now, let's take a closer look at the radio search for extraterrestrial intelligence. How is this different from pseudoscience? Let me give a couple of real cases. In the early sixties, the Soviets held a press conference in Moscow in which they announced that a distant radio source, called CTA-102, was varying sinusoidally, like a sine wave, with a period of about 100 days. Why did they call a press conference to announce that a distant radio source was varying? Because they thought it was an extraterrestrial civilization of immense powers. That is worth calling a press conference for. This was before even the word "quasar" existed. Today we know that CTA-102 is a quasar. We don't know very well what quasars are; and there is more than one mutually exclusive explanation for them in the scientific literature. Nevertheless, few seriously consider that a quasar, like CTA-102, is some galaxy-girdling extraterrestrial civilization, because there are a number of alternative explanations of their properties that are more or less consistent with the physical laws we know without invoking alien life. The extraterrestrial hypothesis is a hypothesis of last resort. Only if everything else fails do you reach for it.

Second example: British scientists in 1967 found a nearby bright radio source that is fluctuating on a much shorter time scale, with a period constant to ten significant figures. What was it? Their first thought was that it was something like a message being sent to us, or an interstellar navigational beacon for spacecraft that ply the spaces between the stars. They even gave it, among themselves at Cambridge University, the wry designation LGM-1—Little Green Men, LGM. However (they were wiser than the Soviets), they did not call a press conference, and it soon became clear that what we had here was what is now called a "pulsar." In fact it was the first pulsar, the Crab Nebula pulsar. Well, what's a pulsar? A pulsar is a star shrunk to the size of a city, held up as no other stars are, not by gas pressure, not by electron degeneracy, but by nuclear forces. It is in a certain sense an atomic nucleus the size of Pasadena. Now that, I maintain, is an idea at least equally as bizarre as an interstellar navigational beacon. The answer to what a pulsar is has to be something mighty strange. It isn't an extraterrestrial civilization, it's something else; but a something else that opens our eyes and our minds and indicates possibilities in nature that we had never guessed at.

Then there is the question of false positives. Frank Drake in his original Ozma experiment, Paul Horowitz in the META (Megachannel Extraterrestrial Assay) program sponsored by the Planetary Society, the Ohio University group and many other groups have all had anomalous signals that make the heart palpitate. They think for a moment that they have picked up a genuine signal. In some cases we have not the foggiest idea what it was; the signals did not repeat. The next night you turn the same telescope to the same spot in the sky with the same modulation and the same frequency and bandpass, everything else the same, and you don't hear a thing. You don't publish that data. It may be a malfunction in the detection system. It may be a military AWACS plane flying by and broadcasting on frequency channels that are supposed to be reserved for radio astronomy. It may be a diathermy machine down the street. There are many possibilities. You don't immediately declare that you have found extraterrestrial intelligence because you find an anomalous signal.

And if it were repeated, would you then announce? You would not. Maybe it's a hoax. Maybe it is something you haven't been smart enough to figure out that is happening to your system. Instead, you would then call scientists at a bunch of other radio telescopes and say that at this particular spot in the sky, at this frequency and bandpass and modulation and all the rest, you seem to be getting something funny. Could they please look at it and see if they get something similar? And only if several independent observers get the same kind of information from the same spot in the sky do you think you have something. Even then you don't know that the something is extraterrestrial intelligence, but at least you could determine that it's not something on Earth. (And that it's also not something in Earth orbit; it's further away than that.) That's the first sequence of events that would be required to be sure that you actually had a signal from an extraterrestrial civilization.

Now notice that there is a certain discipline involved. Skepticism imposes a burden. You can't just go off shouting "little green men," because you are going to look mighty silly, as the Soviets did with CTA-102, when it turns out to be something quite different. A special caution is necessary when the stakes are as high as here. We are not obliged to make up our minds before the evidence is in. It's okay not to be sure.

I'm often asked the question, "Do you think there is extraterrestrial intelligence?" I give the standard arguments—there are a lot of places out there, and use the word *billions,* and so on. And then I say it would be astonishing to me if there weren't extraterrestrial intelligence, but of course there is as yet no compelling evidence for it. And then I'm asked, "Yeah, but what do you really think?" I say, "I just told you what I really think." "Yeah, but what's your gut feeling?" But I try not to think with my gut. Really, it's okay to reserve judgment until the evidence is in.

*   *   *

After my article "The Fine Art of Baloney Detection" came out in *Parade* (Feb. 1, 1987), I got, as you might imagine, a lot of letters. Sixty-five million people read *Parade*. In the article I gave a long list of things that I said were "demonstrated or presumptive baloney"—thirty or forty items. Advocates of all those positions were uniformly offended, so I got lots of letters. I also gave a set of very elementary prescriptions about how to think about baloney —arguments from authority don't work, every step in the chain of evidence has to be valid, and so on. Lots of people wrote back, saying, "You're absolutely right on the generalities; unfortunately that doesn't apply to my particular doctrine." For example, one letter writer said the idea that intelligent life exists outside the earth is an excellent example of baloney. He concluded, "I am as sure of this as of anything in my experience. There is no conscious life anywhere else in the Universe. Mankind thus returns to its rightful position as center of the Universe."

Another writer again agreed with all my generalities, but said that as an inveterate skeptic I have closed my mind to the truth. Most notably I have ignored the evidence for an Earth that is six thousand years old. Well, I haven't ignored it; I considered the purported evidence and *then* rejected it. There is a difference, and this is a difference, we might say, between prejudice and postjudice. Prejudice is making a judgment before you have looked at the facts. Postjudice is making a judgment afterwards. Prejudice is terrible, in the sense that you commit injustices and you make serious mistakes. Postjudice is not terrible. You can't be perfect of course; you may make mistakes also. But it is permissible to make a judgment after you have examined the evidence. In some circles it is even encouraged.

*   *   *

I believe that part of what propels science is the thirst for wonder. It's a very powerful emotion. All children feel it. In a first grade classroom everybody feels it; in a twelfth grade classroom almost nobody feels it, or at least acknowledges it. Something happens between first and twelfth grade, and it's not just puberty. Not only do the schools and the media not teach much skepticism, there is also little encouragement of this stirring sense of wonder. Science and pseudoscience both arouse that feeling. Poor popularizations of science establish an ecological niche for pseudoscience.

If science were explained to the average person in a way that is accessible and exciting, there would be no room for pseudoscience. But there is a kind of Gresham's Law by which in popular culture the bad science drives out the good. And for this I think we have to blame, first, the scientific community ourselves for not doing a better job of popularizing science, and second, the media, which are in this respect almost uniformly dreadful. Every newspaper in America has a daily astrology column. How many have even a weekly astronomy column? And I believe it is also the fault of the educational system. We do not teach how to think. This is a very serious failure that may even, in a world rigged with 60,000 nuclear weapons, compromise the human future.

I maintain there is much more wonder in science than in pseudoscience. And in addition, to whatever measure this term has any meaning, science has the additional virtue, and it is not an inconsiderable one, of being true.

ISAAC ASIMOV

The Perennial Fringe

I doubt that any of us really expects to wipe out pseudoscientific beliefs. How can we when those beliefs warm and comfort human beings?

Do you enjoy the thought of dying, or of having someone you love die? Can you blame those people who convince themselves that there is such a thing as life-everlasting and that they will see all those they love in a state of perpetual bliss?

Do you feel comfortable with the daily uncertainties of life; with never knowing what the next moment will bring? Can you blame people for convincing themselves they can forewarn and forearm themselves against these uncertainties by seeing the future clearly through the configuration of planetary positions, or the fall of cards, or the pattern of tea-leaves, or the events in dreams?

Inspect every piece of pseudoscience and you will find a security blanket, a thumb to suck, a skirt to hold. What have we to offer in exchange? Uncertainty! Insecurity!

For those of us who live in a rational world, there is a certain strength in understanding; a glory and comfort in the effort to understand where the understanding does not as yet exist; a beauty even in the most stubborn unknown when it is at least recognized as an *honorable* foe of the thinking mechanism that goes on in three pounds of human brain, one that will gracefully yield to keen observation and subtle analysis, once the observation is keen enough and the analysis subtle enough.

Yet there is an odd paradox in all this that amuses me in a rather sardonic way.

We, the rationalists, would seem to be wedded to uncertainty. We know that the conclusions we come to, based, as they must be, on rational evidence, can never be more than tentative. The coming of new evidence, or of the recognition of a hidden fallacy in the old evidence, may quite suddenly overthrow a long-held conclusion. Out it must go, however attached to it one may be.

That is because we have *one* certainty, and that rests not with any conclusion, however fundamental it must seem, but in the process whereby such conclusions are reached and, when necessary, changed. It is the scientific process that is certain, the rational view that is sure.

The fringers, however, cling to *conclusions* with bone-crushing strength. They have no evidence worthy of the name to support those conclusions, and no rational system for forming or changing them. The closest thing they have to a process of reaching conclusions is the acceptance of statements they consider authoritative. Therefore, having come to a belief, particularly a security-building belief, they have no other recourse but to retain it, come what may.

When we change a conclusion it is because we have built a better conclusion in its place, and we do so gladly—or possibly with resignation, if we are emotionally attached to the earlier view.

When the fringers are faced with the prospect of abandoning a belief, they see that they have no way of fashioning a successor and, therefore, have nothing but vacuum to replace it with. Consequently, it is all but impossible for them to abandon that belief. If you try to point out that their belief goes against logic and reason, they refuse to listen and are quite likely to demand that you be silenced.

Failing any serviceable process of achieving useful conclusions, they turn to others in their perennial search for authoritative statements that alone can make them (temporarily) comfortable.

I am quite commonly asked a question like this: "Dr. Asimov, you are a scientist. Tell me what you think of the transmigration of souls?" Or of life after death, or of UFOs, or of astrology—anything you wish. What they want is for me to tell them that scientists have worked out a rationale for the belief and now know, and perhaps have always known, that there is some truth to it.

The temptation is great to say that, as a scientist, I am of the belief that what they are asking about is a crock of unmitigated nonsense—but that is just a matter of supplying them with another kind of authoritative statement, and one they won't under any circumstances accept. They will just grow hostile.

Instead, I invariably say, "I'm afraid that I don't know of a single scrap of scientific evidence that supports the notion of transmigration of souls"— or whatever variety of fringe they are trying to sell.

This doesn't make them happy, but unless they can supply me with a piece of credible scientific evidence—which they never can—there is nothing more to do. And who knows, my remark might cause a little germ of doubt to grow in their minds, and there is nothing so dangerous to fringe beliefs as a bit of honest doubt.

Perhaps that is why the more "certain" a fringer is, the more angry he

seems to get at any expression of an opposing view. The most deliriously
certain fringers are, of course, the creationists, who presumably get the word
straight from God by way of the Bible that creationism is correct. You can't
get a more authoritative statement than that, can you?

I get furious letters from creationists occasionally, letters that are filled
with opprobrious adjectives and violent accusations. The temptation is great
to respond with something like this: "Surely my friend, you know that you
are right and I am wrong, because God has told you so. Surely, you also
know that you are going to heaven and I am going to hell, because God
has told you that, too. Since I am going to hell, where I will suffer unimaginable
torments through all of eternity, isn't it silly for you to call me bad names?
How much can your fury add to the infinite punishment that is awaiting
me? Or is it that you are just a little bit uncertain and think that God may
be lying to you and you would feel better to apply a little torment of your
own (just in case he *is* lying) by burning me at the stake, as you could have
in the good old days when creationists controlled society?"

However, I never send such a letter. I merely grin and tear up the one
I got.

But, then, is there nothing to fight? Do we simply shrug and say that
the fringers will always be with us and we might just as well ignore them
and simply go about our business?

No, of course not. There is always the new generation coming up. Every
child, every new brain, is a possible field in which rationality can be made
to grow. We must therefore present the view of reason, not out of a hope
of reconstructing the deserts of ruined minds that have rusted shut, which
is all but impossible—but to educate and train new and fertile minds.

Furthermore, we must fight any attempt on the part of the fringers and
irrationalists to call to their side the force of the state. We cannot be defeated
by reason, and the fringers don't know how to use that weapon anyway;
but we can be defeated (temporarily, at any rate) by the thumbscrew and
the rack, or whatever the modern equivalents are.

That we must fight to the death.

PAUL KURTZ

Reflections on the "Transcendental Temptation"

Little did we expect when the Committee for the Scientific Investigation of Claims of the Paranormal was first established in 1976 that it would grow and flourish as it has. We are especially gratified by the positive reception from the scientific community and by the enthusiastic support of the distinguished Fellows and Scientific Consultants who have enlisted in our cause.

Unfortunately, uncritically held beliefs in paranormal phenomena seem to be endemic to the contemporary social landscape; and thus CSICOP has by now become an essential organization attempting to ferret out facts from fictions. Given the high level of literacy and education, the high incidence of belief in pseudoscience and the paranormal is a surprising phenomenon, especially since it seems to affect all strata of society. It is of interest to note that uncritical belief in the supernatural and the paranormal in one guise or another has been pervasive in human culture. The recent growth of paranormal belief is perhaps simply a repetition of age-old beliefs and attitudes in new forms.

How do we account for the strength of paranormal belief-systems in this modern age? Several causal explanations have been offered. The one most often heard is that paranormal claims have been sensationalized by irresponsible media. Psychic miracles, UFO encounters, and other strange anomalies are dramatized in color and sound; they are made to appear so real to the viewer that any standards for judging truth from fiction are weakened and often ineffective.

Those of us affiliated with CSICOP have said many times that we cannot reject a priori any responsible claims of paranormal phenomena and that many of them deserve patient examination to find out if something genuinely extraordinary might actually be taking place. In one sense, if a scientific explanation of a paranormal claim is found, it immediately becomes a nonparanormal event and part of the natural universe, even though J. B. Rhine and others have insisted that only a "nonphysical" universe could account for such phenomena. It *may be* that psychokinesis, precognition, and remote viewing will someday be verified by patient observation and experimental work in the laboratory. Until they have, we continue to be skeptical. Skepticism is the lifeblood of the scientific enterprise, but it must not be taken as a new orthodoxy and it must give a fair hearing to unorthodox claims.

II

In dealing with the vagaries of paranormal belief, I have found that one salient fact emerges—the persistence of what I call the "transcendental temptation," the tendency for human beings to resort to magical thinking and to ascribe occult, mysterious, hidden, or unknown causes to events they cannot fathom. This is surely not a recent development in human history; it has been present throughout the long evolution of the species. Primitive men and women were overwhelmed, no doubt, by the contingent character of human existence: Often confronted by brutal tragedies they could not comprehend—disease, floods, volcanic eruptions, death—they attributed these to the wrath of unknown deities and demons, and attempted to propitiate these unseen postulated forces and powers by sacrifice and prayer. This is the seedbed of mythic religious systems, but it also has its parallels today in the attribution of paranormal causes to otherwise inexplicable events.

Whether the transcendental temptation is genetic in origin—some have even suggested a sociobiological explanation—or a product of human culture is an important question that I will not attempt to resolve here. The point is that there are striking similarities between the religious mythologies that abound in contemporary society and paranormal belief-systems. Some interesting data that may shed some light on this question are pertinent, however.

III

Ever since the founding of CSICOP, I have offered a course at the State University of New York at Buffalo called "Philosophy, Parapsychology, and the Paranormal." Most of the students who register for the course (from 35 to 50 each semester) begin as believers in the paranormal. In some years, more than 80 percent of those who respond to a questionnaire I distribute on the first day say they believe in ESP, precognition, and the existence of ghosts, psychokinesis, levitation, or other such phenomena. By the end of the course, however, there is a massive reversal of belief: 80 to 90 percent have become skeptics.

I begin the course by asking the students if they have had strange psychic or paranormal experiences or if they know of someone who has, and they relate this anecdotal information to the class. In the first half of the course, I present the case for a paranormal universe, and I attempt to be fair-minded and neutral. As we proceed, I begin to introduce alternative skeptical explanations.

The high point of the course, I believe, is the term project. I ask students to undertake a research experiment, usually in teams of two to four students. They are to report on the progress of their research project in subsequent weeks. I caution them to tighten up the protocol, to have a sufficient number of runs or trials, and to guard against any sensory leakage or fraud. The students have been highly creative in their research projects. They have performed ganzfeld and remote-viewing experiments; they have tested for psychokinesis, using everything from dice to random-number generators; and they have used Zener cards innumerable times to test for clairvoyance, precognition, and telepathy.

Many diverse hypotheses have come in for scrutiny: Do identical twins have some kind of telepathic affinity? What about parents and children, brothers and sisters, husbands and wives? Is there any difference between children, adolescents, and adults in regard to psi phenomena? Some students have analyzed psychic or tarot-card readings by practitioners in the area; others have made studies of astrological predictions; still others have analyzed reports of apparitions or poltergeists, or have examined claims that psychics help detectives, of the lunar effect, or of UFOs.

I was surprised to find that of almost 100 experiments conducted the results have invariably been negative. I should add that we also do Zener card tests in class (for clairvoyance and precognition) and we try to enter into the data bank at least 1,000 trials for each student. Again, the results both for individual students and for the class as a whole have always been negative. Never has there been any significant deviation from chance expectation.

Lest anyone counter that I am a "goat" and hence am wielding a negative influence on the data, let me point out that a large percentage of the student-experimenters begin as "sheep," believing in the phenomena. They are as astonished as I am that no one has ever had positive results. Just as the British parapsychologist Susan Blackmore became a skeptic after a decade of trying to elicit psi phenomena and getting only negative results, so the students in my classes have become highly skeptical. It appears that a process of deconversion sets in: at the end of the course most of the students are skeptical about any evidence for paranormal claims, though open-minded about the possibility of psi phenomena.

One of the reasons I assigned these projects was that I wanted to see for myself whether any positive results could be obtained. These continuous negative findings puzzled me, for I thought that at least on some occasions they might achieve some above-chance results. What are we to conclude from this? Is it that ESP is so weak that it does not come up very often? Or that ESP does not exist at all?

I have long felt that one of the most important papers the *Skeptical*

Inquirer ever published was by B. F. Singer and V. A. Benassi, "Fooling Some of the People All of the Time" (Winter 1980-81, pp. 17-24), in which they reported the ready tendency among students to engage in magical thinking. Their students were shown psychic demonstrations by a magician posing as a "psychic," and even when they were told that he was a conjurer and not really a psychic, many nonetheless continued to believe in the reality of his psychic feats. This suggests that there could be something very deep in human psychology that makes it difficult to overcome the tendency toward magical thinking—the *transcendental temptation.* Singer and Benassi, however, did this in only one class during each course, and the students were not exposed to any kind of sustained criticism. I have found that if there is intensive exposure (of, say, 15 weeks' duration) to the rational view, if alternative critiques are available, and if the students are allowed to test the claims empirically for themselves, this will do more than anything else to guard against the transcendental temptation. In other words, more elaborate skeptical scrutiny of paranormal claims does have an important therapeutic effect; it diminishes the tendency to believe without evidence.

I shall end on this positive note. It indicates that the fair-minded and open presentation of paranormal subjects, together with skeptical scrutiny, will tend to develop rational defenses—at least for many people. This is all the more reason for CSICOP to continue its effort to bring scientific findings and alternative explanations to the attention of the public. And that has been our task: to foster an appreciation of the aims of science and, by continuing to present results of our inquiries, to raise the level of rationality in society. Critical inquiry has proved to be the best antidote for misperception and misconception for a significant number of educated people. The transcendental temptation can thus be moderated.

L. SPRAGUE DE CAMP

The Uses of Credulity

We debunkers have long lamented the invincible willingness of our fellow primates to believe in the absence of evidence, or even contrary to evidence. Suppose we could teach everybody to think logically all the time and never to believe without evidence, thus slaying our pet dragon of pseudoscience once and for all. Would that be a good thing? Not necessarily. Our species is caught in a paradoxical dilemma.

All human societies above the hunter-gatherer stage have ideologies, either religious or secular. All these ideologies contain irrational elements—tenets that must be accepted on faith, such as the Christian's belief in the divinity of Jesus the Nazarene, or the Buddhist's concept of reincarnation, or the Marxist's faith in the malleability and perfectability of men and women. Any of these beliefs might be true, but none can be scientifically demonstrated.

When a characteristic like human credulity becomes so widespread in a species, we must suspect that it plays a part in enabling the species to survive, even though we may not know what that function is. For example, people long thought that the bull mammoth's spirally curved tusks, crossing at the tips, were a useless excrescence, good for neither digging nor fighting. Then it was realized that they were useful as snow shovels to get at food in winter.

The same with ideologies. Human beings on the most primitive level seem actuated by two main drives, less compelling and more inhibitable than true instincts but still effective. One is the drive of self-interest, without which no species could survive.

The other is the altruistic drive—the drive to help and defend others of one's species. Some social drive or instinct is necessary for any pack-hunting species, like lions and wolves; it is lesser or wanting in solitary hunters, like tigers and foxes. Having spent millions of years hunting in packs, people have a natural drive to form hierarchical, cooperative groups. However

---

necessary the drive of self-interest, a certain minimum of altruism is needed to make any group, from a family to a nation, function successfully.

Natural or innate altruism, however, seems confined to one's kith and kin—that is, to the number of persons, usually several score, that make up a hunting band. Among primitives, where a tribe typically calls itself "the real human beings," altruism usually stops at the band or tribal boundary. Those beyond, being thought subhuman, are considered fair game.

With the Agricultural Revolution of about 10,000 years ago and the subsequent rise of civilization, it became necessary to organize people in groups much larger than the hunting band. To persuade people to act altruistically toward persons beyond their own families and friends, ideologies were devised. After many centuries of ineffectual experiments by priesthoods, well-thought-out ideologies began to be devised about the eighth century B.C.E. by Isaiah and Zarathustra, followed within the next couple of centuries by Gautama, Mahavira, Confucius, and Lao-dze. All preached benevolence and altruism toward fellow human beings. Since it is a matter of universal observation that virtue is not always rewarded and that the wicked often flourish like a green bay tree, the prophets combined these commandments with promises of rewards for altruistic behavior in Heaven, or in the next incarnation, or in benefits to one's descendants, as well as with threats of punishment for acts they held wicked. Prophets have been at it ever since.

Most prophets have built their ideologies upon tribal myths and legends, which the priesthoods of early cities compiled and tried to rationalize and render self-consistent. That they were not altogether successful is shown by the first verse of Genesis, literally "In the beginning, the gods created. . . ."

A few prophets have composed secular ideologies, ignoring or denying the gods. The most successful have been Confucianism, Stoicism, and Marxism. Of these, the most effective in the long run has been Confucianism; but none has been conspicuously more successful in getting men to act altruistically toward all mankind than religious ideologies.

A completely rational ideology would leave its adherents free ruthlessly to pursue their own selfish desires without scruple or limit. Many do so now; we call them criminals. If everybody did, we should have a *bellum omnia contra omnes* and life, in Hobbes's phrase, would be "solitary, poore, nasty, brutish, and shorte."

A realistic appraisal of the role of the irrational in ideology was made by the geographer Strabon, a contemporary of Augustus: "The great mass of women and common people cannot be induced by mere force of reason to devote themselves to piety, virtue, and honesty. Superstition must therefore be employed, and even this is insufficient without the aid of the marvelous and the terrible."

We must excuse Strabon's male chauvinism, since as a Classical Greek he could not help it. Niccolò Machiavelli voiced a similar sentiment, albeit more cautiously, since he lived in the days of the Inquisition. In *Discourses on the First Ten Books of Livy,* he said that rulers should foster the current religion and uphold its principles for the sake of the unity and good order of the state, even though they themselves did not believe it.

Since some credulity is needed for a people to embrace an ideology, such credulity, up to a point, may be a survival trait. Ideology is one of the lubricants, like liquor and hypocrisy, that enable men to live together in numbers vastly greater than those the species was evolved to cope with.

In view of humankind's demonstrated credulity and capacity for wishful thinking, the possibility that all people will adopt a coldly and selfishly rational viewpoint seems the least of our present worries. The greater danger is that an ideology will get out of hand and lead to self-destructive mass behavior, as when the Uwet of West Africa nearly exterminated themselves by poison ordeals, the Balengi of the same region killed off most of their tribe by executions for witchcraft, the Christians burned Serveto and Bruno among thousands of others, and the twentieth-century Germans set out to conquer the world on the basis of faith in Aryan superiority.

So we must continue to combat the more destructive irrationalities. The scientific debunker's job may be compared to that of the trash collector. The fact that the garbage truck goes by today does not mean that there will not be another load tomorrow. But if the garbage were not collected at all, the results would be much worse, as some cities found when the sanitation workers went on strike.

PHILLIPS STEVENS, JR.

The Appeal of the Occult: Some Thoughts on History, Religion, and Science

Popular interest in phenomena variously labeled as "occult" has been intensifying recently in Western society to a degree probably unparalleled since the sixteenth century. Great numbers of people today, at all levels of society, are both willing and apparently eager to accept, uncritically, occult explanations and to assign paranormal causes to poorly understood or imperfectly perceived phenomena.

Analyses in the social sciences tend to focus on correlations between social conditions and patterns of belief. Many conclude with reference to "the will to believe"; the notion that people need to believe in something, and that they can reestablish a lost sense of social community by sharing a particular belief (Catherine Albanese's notion of "homesteads of the mind" [Albanese 1981, pp. 163ff.]). In this view, if the thing loses its appeal it must be replaced by something else, to alleviate cognitive dissonance or to satisfy human needs for social affect and for explanation of the natural world. Often, so this argument can run, a magical explanation is easier or more comfortable than admitting there is no explanation or confronting the relatively complex and apparently discomfiting task of examining alternative explanations, as Singer and Benassi (1981) found in their widely cited classroom experiments.

Most of these sociological analyses and explanations have some validity in their basic premises. Although few have been successfully subjected to rigorous examination, many surely contribute to understanding various manifestations of the current "occult explosion." In our search for explanations for the current appeal of the occult, however, we need to look deeper into three areas: the historical background, the structure of religion, and the role of science in human life.

The Occult in History

Most social analysts who investigate the general public's interest in the occult deal with the subject as if the trend represents something sudden or unique to the second half of the twentieth century. Albanese (1981) is one notable exception; another is Galbreath (1983). Paul Kurtz (1986) notes that what he calls the "transcendental temptation" is as old as humanity, but under this rubric he is referring to the development of mainstream religion as well as "fringe" beliefs and activities. The definition problem is important: We will seriously confuse our investigations if we do not make a distinction between mainstream religion, as adhered to by a majority of the population and sanctioned by mainstream institutions, and occult beliefs, as operating on the social fringe without mainstream sanction.

The specific forms current occult beliefs assume may be new, even unique, but in their general nature such trends have long and fairly continuous historical precedent. My perspective on occult beliefs is anthropological; and a great strength of anthropology is its view of culture as a dynamic system, its recognition of culture as process. This combination of systemic and historical perspective gives anthropology special insight among the social sciences. I think we will make better progress toward understanding the current interest in the occult if we place it in historical perspective and recognize what Albanese (1981, p. 164) refers to as "the occult tradition" in Western history. In a historical context, *occult* should mean not only "hidden," and hence mysterious (and magical), but also aberrant, somehow at odds with mainstream tenets. We should also recognize varying social factors in the definition and application of the term.

We tend to regard periods of history as self-defining, bounded units rather than as influenced by the legacy of their own past, and we view them through lenses of our own making. But in our view of history, the methods of anthropology must apply: We must be certain that our terms (e.g., *occult*) are applicable to other historical periods—which are, in fact, other cultures. Throughout the first thousand years of the Christian era, beliefs we would call "occult" were relatively unimportant on a broad social scale, in spite of our label "Dark Ages" to the latter half of this period. During the early centuries of the second millennium, such beliefs moved rapidly from the fringe to a central position in the mainstream. The seventeenth century saw the beginnings of modern science—but also the height of executions for witchcraft.

Seventeenth-century thinking was heir to at least 700 years of the centrality of concern with magic, witchcraft, demonology, and heretical religious movements. Will and Ariel Durant (1961, p. 575) wrote: "Religions are born and may die, but superstition is immortal. . . . Kepler believed in witchcraft, and Newton wrote less on science than on the Apocalypse." Chadwick Hansen (1969, p. 7), discussing the historical context of Salem witchcraft, noted:

The difficulty is that we tend to remember these men only for the ideas we still value, forgetting the other contents of their minds.

We forget that Bacon believed you could cure warts by rubbing them with a rind of bacon and hanging it out of a window that faced south, and that witchcraft may take place "by a tacit operation of malign spirits." We forget that Boyle believed in an astonishing and repulsive variety of medicaments, including stewed earthworms, a worsted stocking that has long been worn next to the flesh, and human urine. . . . It was Boyle who proposed that English miners be interviewed as to whether they "meet with any subterraneous demons; and if they do, in what shape and manner they appear; what they portend, and what they do." And Newton, the greatest scientist of his age, spent more of his time on the study of the occult than he did in the study of physics. He explicated, for example, apocalyptic passages in the Bible, and interpreted the measurements of Solomon's temple, hoping in both cases that a mystic reading of the scriptures would lead him to the inmost secrets of the universe. . . .

Hansen goes on, quoting from Locke's *Essay Concerning Human Understanding* (our capacity for reason does not equip us to "deny or doubt the existence of such spirits") and Hobbes's *Leviathan* (we may doubt the efficacy of their powers, but witches should be punished for their evil ambitions) to show that even principles and applications of logic may vary among historical periods. He summarizes his discussion of the seventeenth-century world-view: "Scientists, philosophers, lawyers, physicians, the learned community in general believed firmly in the existence of an invisible world, and in the capacity of the inhabitants of that world to intrude on this one" (Hansen 1969, p. 8). What was "occult" was the misguided—and potentially dangerous—efforts of people to contact and manipulate such entities or forces.

Historians tend to give the impression that, with the rapid decline of the "witchcraft craze" and the firm emplacement of the Industrial Revolution and the Age of Reason in the eighteenth and nineteenth centuries, occult beliefs in general retreated to an insignificant fringe. In fact, this was not so, as both Albanese (1981, pp. 163-187) and Galbreath (1983) point out. Instead, *different manifestations* of occult interest rose and faded. The eighteenth, nineteenth, and early twentieth centuries saw fairly continuous interest in occult dabblings, but largely of sorts that operated quietly, without the sensationalism of their forebears. They made appeal to intellect, reason, philosophical sophistication, and selected principles of science ("pseudoscience" has old roots!). Examples include European Swedenborgianism, mesmerism, secret societies, and phrenology in the late eighteenth and early nineteenth centuries; American spiritualism from 1848; Theosophy and "psychical research" by the mid-1880s; Asian religions, magic, and various forms of divination by the turn of the century; and the various interests of this century, with which we are more familiar. Many of these, on examination, can be seen to be new forms of old cosmologies and traditions (e.g., Neoplatonism and various medieval dualisms); many have bases in the same premises of material/spiritual interaction addressed by inquisitions and witch hunts of earlier periods.

We should recognize that there is a fairly continuous "tradition" of adherence to phenomena variously labeled as "occult" through history. We should understand history, the historical legacy of any period we investigate, and the contemporary factors that shape our view of history. We should understand how the heresies and fantasies of one age, or one segment of a population, may be the dogma of another.

So there is nothing very significant *per se* in the fact of the current interest in the occult. There has always been an occult fringe, manifested in some ways or others, its adherents seeking satisfaction of whatever particular needs. What is significant is that today, in an era of unparalleled scientific discovery, occult interest seems to be moving from the fringe into the mainstream and seems now to be attracting, possibly, a majority of the people. A Gallup poll in June 1984 showed that 55 percent of teenagers aged 13 to 18 believe that astrology works, up from 40 percent in 1978. In April 1985 (*New York Times,* April 18 and April 25, 1985) a beleaguered Procter & Gamble elected to remove its 100-year-old man-in-the-moon logo from its packaging, in reaction to thousands of phone calls and letters denouncing it as a satanic emblem; and today the public and the news media are alarmingly open to satanic interpretations of various disturbing or bizarre events. Shirley MacLaine's 1987 television movie *Out on a Limb* and her several books have received serious reviews, in which the words *dream* and *fantasy* are scarce. In February 1987, NBC-TV's "Today" show presented a special on "crystal consciousness"; and in March, a two-part series on "channeling" (which Kurtz [1987, p. 3] has identified as simply spiritualism with no frills). *Time*'s December 7, 1987, cover story on the New Age was written quite fairly; but the language in Time-Life Books' recent deluge of promotion for their new *Mysteries of the Unknown* series is unconscionably pro-paranormal.

Galbreath (1983) cautions against relying on superficial statistics, pointing out the difficulty of measuring the depth of respondents' *real commitment* to occult explanations. But clearly there is growing public confusion and an increased willingness at all levels of society to abandon or sidestep principles of reason and jump to extremely fuzzy supernatural or occult explanations. In the preface to his monumental work *Religion and the Decline of Magic* (1971), Keith Thomas states, "Astrology, witchcraft, magical healing, divination, ancient prophecies, ghosts and fairies, are now all rightly disdained by intelligent persons" (p. ix). The evidence today indicates that he is dead wrong. The recognition that such practices and beliefs have long historical precedent, however, is not sufficient to explain what we must now regard as a general crisis of reason.

A Dynamic Model of Religion

In my courses on the anthropology of religion, magic, and witchcraft, I have developed a model of religion. (See figure.) It is a heuristic device only, not meant to be self-explanatory. But with careful discussion it has proved to be quite effective. It is meant to be universally applicable—and to be both descriptive and dynamic. The model depicts religion as operating simultaneously along two spatial dimensions. The "vertical" dimension is the relationship between people and elevated divinities; the "horizontal," the recognition of supernatural agencies operating on the level of human society.

As a framework for description, the model can help to show how, for any religious system, various categories of divinities, powers, and forces are conceived; and the nature of their cosmological roles, their interrelationships, and their relationships with human society. The categories of the supernatural are intentionally broad, allowing for inclusion of many variant forms.

Not all religious systems will contain beings or concepts that fit in all categories. Some societies do not distinguish clearly between ghosts and ancestors; some conceive of several agencies in similar terms; some place differing degrees of importance on different ones. Sorcery, the malevolent human manipulation of natural "forces" through magical means, is universal, but a few societies have no beliefs in witches—people possessed of special powers enabling them to work evil directly, without magic. Relatively few religious systems postulate an active "underworld"; hence my vertical arrows do not extend *below* the level of human society. As a framework for description the model is of most utility to an anthropological analysis of religion in culture.

As a guide to historical process, the model can help us understand how religion responds to social trends, both ideological and technological. Specifically, it can help us to conceptualize responses of both popular cosmology and mainstream religion to the Enlightenment and the Industrial Revolution of the seventeenth and eighteenth centuries.

We are heir to a religious tradition that provided a system of *total cosmology*—explanations for the workings of the natural world and of people's place within it—for its adherents, acknowledging and accommodating both the vertical and the horizontal dimensions. Such a tradition continued, as we have seen, through the seventeenth century. Now, in terms of the model: as science and technology advanced, mainstream religion adapted by collapsing steadily inward toward an exclusively vertical focus, allowing less and less formal accommodation for beliefs in agencies along the horizontal dimension. The observance of religion itself has narrowed and become compartmentalized, conducted in specially designated places generally on only one day of the week and for a very short time on that day. Agencies along the horizontal dimension, which once were regarded as real and immediate, have become relegated by the mainstream to "superstition"—or "occult." The realm of "the occult" has thus broadened considerably since the seventeenth century, and organized religion today offers little counsel about it.

The presumption has been—and it continues, as an ideal—that science and technology would provide satisfactory explanations for those areas of cosmology previously explained by reference to agencies and forces along the horizontal dimension. Indeed, this presumption is the underlying premise of

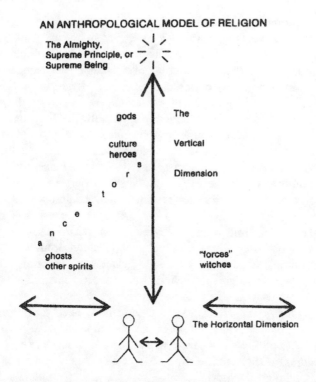

AN ANTHROPOLOGICAL MODEL OF RELIGION

modern mainstream religious institutions and is tacitly fundamental to all others. The presumption—hence the premise—was, and is, of course, false. The implications of this falsity can be seriously disorienting to the tens of millions of people—not only recent immigrants from Third World cultures, although theirs is the clearest case—for whom spiritual agencies, magic, and witchcraft are real, as I have indicated elsewhere (Stevens 1982).

Science and the Public

There seems to be general agreement among scholars and educators today that general science education in the United States has failed miserably; that the failure was accelerated during the "do your own thing" ethos of the late sixties and early seventies; and that, as an academic premise at least, the increasing appeal of the occult directly correlates with the failure of science education. We should consider carefully some aspects of the conduct of science and the public's attitude toward it.

An understanding of the nature and processes of science is not generally accessible to the public; it is itself esoteric knowledge. Jon D. Miller, in the Spring 1983 *Daedalus,* a special issue devoted to scientific literacy, defined "the attentive public" as "individuals interested in a particular policy area, and willing to become and remain knowledgeable about the issue" (Miller 1983, p. 44). According to a survey conducted by the National Science Foundation, the attentive public for science policy in 1979 constituted about 27 million adults, or about 18 percent of the adult population. Of this number, according to a test of the understanding of fundamental concepts and issues administered by the NSF, approximately 70 percent did not meet the minimal criteria for scientific literacy. Stated more directly, the findings showed that a bare 5.4 percent of adults in the United States qualified as scientifically literate.[1] A more recent poll of "technological literacy," also conducted by Miller and sponsored by the NSF, reported in *Science Indicators—1985,* showed large numbers of people expressing "little" understanding of certain basic technological concepts and strong reliance on luck, alternative medicine, UFOs, and some other occult-related indicators. This research has recently been published in full (Miller 1987). Miller and others conclude that this situation is not only nationally disgraceful but potentially dangerous.

Singer and Benassi (1981, p. 54) conclude their important study of occult beliefs by acknowledging the validity of various sociological and psychological explanations, but asserting that a large share of credit for the current popularity of the occult lies with science education. Their main conclusions are: (1) Science is taught as an academic or clinical exercise, to be applied strictly within the parameters of a particular classroom or laboratory project, its specific procedures to be memorized by rote as part of the particular package. Science is not taught as a general "cognitive tool," a way of reaching a deeper understanding of our environment. (2) "Many occult claims could be countered with even an elementary knowledge of scientific facts, but even an elementary knowledge may be largely lacking." (3) Scientists themselves, by overstressing the limitations of science and underplaying its achievements, have contributed to a widespread impression that science is largely subjective, vacillating, and able to assess only clinically measurable facts.

To these observations I will add some others:

• Science is characterized by exceedingly narrow specializations. Scientists come to know more and more about less and less.

• The avenues of communication within science, and between scientists and the public—except when scientists step outside their own areas of expertise, as we shall see below—are extremely restricted. Scientists tend to talk only to one another, generally within their own disciplines, and in esoteric language. Very little of what they communicate trickles down in terms comprehensible to the lay public. Gerald Holton, in a recent issue of *Daedalus* on Art and Science, notes: "The thought processes and operations of both [science and technology] have moved behind a dark curtain. There they have taken on a new form of autonomy—isolated from the active participation or real intellectual contact of all but the highly trained. Contrary to eighteenth-century expectations, the scientists are losing what should be their most discerning audience, their wisest and most humane critics" (Holton 1986, p. 92).

• There is usually a substantial lag, often years, between the news of a scientific breakthrough and its application—if any—to people's lives.

• The two foregoing factors contribute to the public's sense of alienation from science, and mistrust of it. William J. Broad, in a *New York Times* review of Richard Rhodes's comprehensive *The Making of the Atomic Bomb* (a book I would recommend as part of a remedial package for "the attentive public for science policy") points out the next logical danger:

> All too often the moral drawn from the atomic saga and its legacy of arms development is, simply put, that science can lead to evil, and that since its temptations cannot be resisted, its powers should be sharply constricted. (Broad 1987, p. 39)

Or take fluoridation of public drinking water in the 1950s, or genetic engineering today.

• Singer and Benassi have observed that science has become packaged as an academic or clinical exercise and that the process of empirical reasoning and the search for alternative explanations for apparently mysterious phenomena are not carried into the everyday world as tests of empirical reality. I would observe, further, that this is a problem not only at the student level; some established scientists violate their own principles when they step into other fields.

• The public has not the understanding to differentiate among "scientists," and an advanced degree or other distinction can become acceptable qualification for making pronouncements beyond one's area of expertise, e.g., Linus Pauling and vitamin C, astronaut James Irwin and the search for Noah's Ark, and books on "creation science" written by holders of doctorates in civil engineering. Now distinguished popular publishers, most recently Time-Life Books, can be similarly faulted.

These are observations on the conduct of science, not criticisms deserving of remedies, although some of them should certainly be used as cautionary. A major scientific enterprise, such as the manned space flight program or AIDS research, is necessarily a highly complex system, its parts at once rigidly compartmentalized and vitally interdependent. Its broad pyramidal base is virtually hidden from view, even probably from the purview of those at the peak of the pyramid where the breakthrough takes place. And, of course, the peak is illusory; it is not really there, and the pyramidal structure of the whole enterprise cannot be defined until the breakthrough does occur. This is the nature of science, and general descriptions of it and observations about the resultant public attitude toward it ought themselves to be part of a general science curriculum.

But the basic principles of the scientific method, modified very little since Francis Bacon, are not at all abstruse. At the level of a specific scientific project, Holton is quite right in noting that "the connection between phenomena and theory, the theory itself, and the way it is constructed, confirmed, and elaborated are, and have to be, fully controlled by the scientific community." But his concluding phrase can be misleading: "and understanding them comes only with long immersion" (Holton 1986, p. 93). The educated public need not be immersed in a specific project in order to understand the basic structure of science itself. The fundamental principles of the scientific method are simple, straightforward, and easily taught. Science education must start, and continue for some time, at the very basics: the vocabulary of science ("evolution is just a theory" is demonstrative of ignorance of the meaning of "theory") and the principles of logic and reason. As a university instructor for 17 years, I agree with Jon Miller (1983, p. 46) that the place to start is elementary school.

Isn't it paradoxical that our age is witness at once to the most profound scientific discoveries of all time *and* such a burgeoning of interest in the occult? Considered at one level, it would seem so. But when we look a bit deeper, at the general structure and content of knowledge, as we did for the seventeenth century, we can see that it is not. There is a grave crisis in science education. The basic principles of the scientific method are not being taught in a manner that can enable otherwise well-educated people to apply them to problems in their daily experiences. The public are the avid consumers of the products of science, but have little awareness of its processes.

The expectation that science and technology would replace supernatural explanations on the horizontal dimension of people's relations with one another and with the natural world has not been fulfilled. But organized mainstream religion has come to focus almost exclusively on a vertical dimension, and "occult" beliefs are increasingly accepted by the science-deprived public.

Notes

Earlier versions of this paper were presented to a meeting of the Northeastern Anthropological Association at Lake Placid, New York, April 1985, and to a meeting of the Western New York Skeptics, Buffalo, New York, June 1986.

1. Kendrick Frazier, reporting on Miller's findings in the *Skeptical Inquirer* (Summer 1986, p. 296), is more generous, giving the figure as 7 percent. But according to Miller's figures, the "attentive public for science policy" constituted 18 percent of the adult population; 70 percent of them, or 12.6 percent of the population, did not meet the minimal criteria; 5.4 percent, therefore, did.

References

Albanese, Catherine L. 1981. *America: Religions and Religion.* Belmont, Calif.: Wadsworth.
Broad, William J. 1987. The men who made the sun rise. Review of Richard Rhodes, *The Making of the Atomic Bomb* (New York: Simon & Schuster). *New York Times Book Review,* February 8, pp. 1, 39.
Durant, Will and Ariel. 1961. *The Age of Reason Begins.* Vol. 7 of *The Story of Civilization.* New York: Simon & Schuster.
Galbreath, Robert. 1983. Explaining modern occultism. In *The Occult in America: New Historical Perspectives,* ed. by Howard Kerr and Charles L. Crow. Urbana: University of Illinois

Press.

Hansen, Chadwick. 1969. *Witchcraft at Salem.* New York: Braziller.

Holton, Gerald. 1986. The advancement of science, and its burdens. *Daedalus,* Summer, pp. 75-104.

Kurtz, Paul. 1986. CSICOP after ten years: Reflections on the "transcendental temptation." *Skeptical Inquirer,* 10:229-232, Spring.

———. 1987. Is there intelligent life on Earth? *Skeptical Inquirer,* 12:2-5, Fall.

Miller, Jon D. 1983. Scientific literacy: A conceptual and empirical review. *Daedalus,* Spring, pp. 29-48.

———. 1987. The scientifically illiterate. *American Demographics,* 9:26-31, June.

Singer, Barry, and Victor A. Benassi. 1981. Occult beliefs. *American Scientist,* 69:49-55, January-February.

Stevens, Phillips, Jr. 1982. Some implications of urban witchcraft beliefs. *New York Folklore,* 8(3-4):29-45, Winter.

Thomas, Keith. 1971. *Religion and the Decline of Magic.* New York: Scribner's.

Part 2: Encouraging Critical Thinking

A Field Guide to Critical Thinking

There are many reasons for the popularity of paranormal beliefs in the United States today, including: (1) the irresponsibility of the mass media, who exploit the public taste for nonsense, (2) the irrationality of the American world-view, which supports such unsupportable claims as life after death and the efficacy of the polygraph, and (3) the ineffectiveness of public education, which generally fails to teach students the essential skills of critical thinking. As a college professor, I am especially concerned with this third problem. Most of the freshman and sophmore students in my classes simply do not know how to draw reasonable conclusions from the evidence. At most, they've been taught in high school *what* to think; few of them know *how* to think.

In an attempt to remedy this problem at my college, I've developed an elective course called "Anthropology and the Paranormal." The course examines the complete range of paranormal beliefs in contemporary American culture, from precognition and psychokinesis to channeling and cryptozoology and everything between and beyond, including astrology, UFOs, and creationism. I teach the students very little about anthropological theories and even less about anthropological terminology. Instead, I try to communicate the essence of the anthropological perspective, by teaching them, indirectly, what the scientific method is all about. I do so by teaching them how to evaluate evidence. I give them six simple rules to follow when considering any claim, and then show them how to apply those six rules to the examination of any paranormal claim.

The six rules of evidential reasoning are my own distillation and sim-plification of the scientific method. To make it easier for students to remem-ber these half-dozen guidelines, I've coined an acronym for them: Ignoring the vowels, the letters in the word "FiLCHeRS" stand for the rules of Falsi-fiability, Logic, Comprehensiveness, Honesty, Replicability, and Sufficiency. Apply these six rules to the evidence offered for any claim, I tell my students, and no one will ever be able to sneak up on you and steal your belief. You'll be filch-proof.

Falsifiability

It must be possible to conceive of evidence that would prove the claim false.

It may sound paradoxical, but in order for any claim to be true, it must be falsifiable. The rule of falsifiability is a guarantee that *if* the claim is *false*, the evidence will prove it false; and if the claim is *true*, the evidence will not disprove it (in which case the claim can be tentatively accepted as true until such time as evidence is brought forth that does disprove it). The rule of falsifiability, in short, says that the evidence must matter, and as such it is the first and most important and most fundamental rule of evidential reasoning.

The rule of falsifiability is essential for this reason: If *nothing* conceivable could *ever* disprove the claim, then the evidence that does exist would not matter; it would be pointless to even examine the evidence, because the conclusion is already known—the claim is invulnerable to any possible evidence. This would not mean, however, that the claim is true; instead it would mean that the claim is meaningless. This is so because it is impossible—logically impossible—for *any* claim to be true *no matter what*. For every true claim, you can always *conceive* of evidence that would make the claim untrue—in other words, again, every true claim is falsifiable.

For example, the true claim that the life span of human beings is less than 200 years is falsifiable; it would be falsified if a single human being were to live to be 200 years old. Similarly, the true claim that water freezes at 32° F is falsifiable; it would be falsified if water were to freeze at, say, 34° F. Each of these claims is firmly established as scientific "fact," and we do not expect either claim ever to be falsified; however, the point is that either *could* be. Any claim that could *not* be falsified would be devoid of any propositional content; that is, it would not be making a factual assertion—it would instead be making an emotive statement, a declaration of the way the claimant feels about the world. Nonfalsifiable claims do communicate information, but what they describe is the claimant's value orientation. They communicate nothing whatsoever of a factual nature, and hence are neither true nor false. Nonfalsifiable statements are propositionally vacuous.

There are two principal ways in which the rule of falsifiability can be violated—two ways, in other words, of making nonfalsifiable claims. The first variety of nonfalsifiable statements is the *undeclared claim:* a statement that is so broad or vague that it lacks any propositional content. The undeclared claim is basically unintelligible and consequently meaningless. Consider, for example, the claim that crystal therapists can use pieces of quartz to restore balance and harmony to a person's spiritual energy? What does it mean to have unbalanced spiritual energy? How is the condition recognized and diagnosed? What evidence would prove that someone's unbalanced spiritual energy had been—or had *not* been—balanced by the application of crystal therapy? Most New Age wonders, in fact, consist of similarly undeclared claims that dissolve completely when exposed to the solvent of rationality.

The undeclared claim has the advantage that virtually any evidence that could be adduced could be interpreted as congruent with the claim, and for that reason it is especially popular among paranormalists who claim precognitive powers. Jeane Dixon, for example, predicted that 1987 would be a year "filled with changes" for Caroline Kennedy. Dixon also predicted that Jack Kemp would "face major disagreements with the rest of his party" in 1987 and that "world-wide drug terror" would be "unleashed by narcotics czars" in the same year. She further revealed that Dan Rather "may [or may not] be hospitalized" in 1988, and that Whitney Houston's "greatest problem" in 1986 would be "balancing her personal life against her career." The undeclared claim boils down to a statement that can be translated as "Whatever will be, will be."

The second variety of nonfalsifiable statements, which is even more popular among paranormalists, involves the use of the *multiple out,* that is, an inexhaustible series of excuses intended to explain away the evidence that would seem to falsify the claim. Creationists, for example, claim that the universe is no more than 10,000 years old. They do so despite the fact that we can observe stars that are billions of light-years from the earth, which means that the light must have left those stars billions of years ago, and which proves that the universe must be billions of years old. How then do the creationists respond to this falsification of their claim? By suggesting that God must have created the light already on the way from those distant stars at the moment of creation 10,000 years ago. No conceivable piece of evidence, of course, could disprove that claim.

Additional examples of multiple outs abound in the realm of the paranormal. UFO proponents, faced with a lack of reliable physical or photographic evidence to buttress their claims, point to a secret "government conspiracy" that is allegedly preventing the release of evidence that would support their case. Psychic healers say they can heal you if you have enough faith in their psychic powers. Psychokinetics say they can bend spoons with their minds if they are not exposed to negative vibrations from skeptical observers. Tarot readers can predict your fate if you're sincere in your desire for knowledge. The multiple out means, in effect, "Heads I win, tails you lose."

Logic

Any argument offered as evidence in support of any claim must be sound.

An argument is said to be "valid" if its conclusion follows unavoidably from its premises; it is "sound" if it is valid and if all the premises are true. The rule of logic thus governs the validity of inference. Although philosophers have codified and named the various forms of valid arguments, it is not necessary to master a course in formal logic in order to apply the rules of inference consistently and correctly. An invalid argument can be recognized by the simple method of counterexample: If you can conceive of a single imaginable instance whereby the conclusion would not necessarily follow from the premises *even if* the premises were true, then the argument is invalid. Consider the following syllogism, for example: All dogs have fleas; Xavier has fleas; therefore Xavier is a dog. That argument is invalid, because a single flea-ridden feline named Xavier would provide an effective counterexample. If an argument is invalid, then it is, by definition, unsound. Not all valid arguments are sound, however. Consider this example: All dogs have fleas; Xavier is a dog; therefore Xavier has fleas. That argument is unsound, even though it is valid, because the first premise is false: All dogs do *not* have fleas.

To determine whether a valid argument is sound is frequently problematic; knowing whether a given premise is true or false often demands additional knowledge about the claim that may require empirical investigation. If the argument passes these two tests, however—if it is both valid and sound—then the conclusion can be embraced with certainty.

The rule of logic is frequently violated by pseudoscientists. Erich von Däniken, who singlehandedly popularized the ancient-astronaut mythology in the 1970s, wrote many books in which he offered invalid and unsound arguments with benumbing regularity (see Omohundro 1976). In *Chariots of the Gods?* he was not above making arguments that were *both* logically invalid and factually inaccurate—in other words, arguments that were doubly unsound. For example, von Däniken argues that the map of the world made by the sixteenth-century Turkish admiral Piri Re'is is so "astoundingly accurate" that it could only have been made from satellite photographs. Not only is the argument invalid (any number of imaginable techniques other than satellite photography could result in an "astoundingly accurate" map), but the premise is simply wrong—the Piri Re'is map, in fact, contains many gross inaccuracies (see Story 1981).

Comprehensiveness

The evidence offered in support of any claim must be exhaustive—that is, all of the available evidence must be considered.

For obvious reasons, it is never reasonable to consider only the evidence that supports a theory and to discard the evidence that contradicts it. This rule is straightforward and self-apparent, and it requires little explication or justification. Nevertheless, it is a rule that is frequently broken by proponents of paranormal claims and by those who adhere to paranormal beliefs.

For example, the proponents of biorhythm theory are fond of pointing to airplane crashes that occurred on days when the pilot, copilot, and/or navigator were experiencing critically low points in their intellectual, emotional, and/or physical cycles. The evidence considered by the biorhythm apologists, however, does not include the even larger number of airplane crashes that occurred when the crews were experiencing high or neutral points in their biorhythm cycles (Hines 1988:160). Similarly, when people believe that Jeane Dixon has precognitive ability because she predicted the 1988 election of George Bush (which she did, two months before the election, when every social scientist, media maven, and private citizen in the country was making the same prognostication), they typically ignore the thousands of forecasts that Dixon has made that have failed to come true (such as her predictions that John F. Kennedy would not win the presidency in 1960, that World War III would begin in 1958, and that Fidel Castro would die in 1969). If you are willing to be selective in the evidence you consider, you could reasonably conclude that the earth is flat.

Honesty

The evidence offered in support of any claim must be evaluated without self-deception.

The rule of honesty is a corollary to the rule of comprehensiveness. When you have examined all of the evidence, it is essential that you be honest with yourself about the results of that examination. If the weight of the evidence contradicts the claim, then you are required to abandon belief in that claim. The obverse, of course, would hold as well.

The rule of honesty, like the rule of comprehensiveness, is frequently violated by both proponents and adherents of paranormal beliefs. Parapsychologists violate this rule when they conclude, after numerous subsequent experiments have failed to replicate initially positive psi results, that psi must be an elusive phenomenon. (Applying Occam's Razor, the more honest conclusion would be that the original positive result must have been a coincidence.) Believers in the paranormal violate this rule when they conclude, after observing a "psychic" surreptitiously bend a spoon with his hands, that he only cheats *sometimes*.

In practice, the rule of honesty usually boils down to an injunction against breaking the rule of falsifiability by taking a multiple out. There is more to it than that, however: The rule of honesty means that you must accept the obligation to come to a rational conclusion once you have examined all the evidence. If the overwhelming weight of all the evidence falsifies your belief, then you must conclude that the belief is false, and you must face the implications of that conclusion forthrightly. In the face of overwhelmingly negative evidence, neutrality and agnosticism are no better than credulity and faith. Denial, avoidance, rationalization, and all the other familiar mechanisms of self-deception would constitute violations of the rule of honesty.

In my view, this rule alone would all but invalidate the entire discipline of parapsychology. After more than a century of systematic, scholarly research, the psi hypothesis remains wholly unsubstantiated and unsupportable; parapsychologists have failed, as Ray Hyman (1985:7) observes, to produce "any consistent evidence for paranormality that can withstand acceptable scientific scrutiny." From all indications, the number of parapsychologists who observe the rule of honesty pales in comparison with the number who delude themselves. Veteran psychic investigator Eric Dingwall (1985:162) summed up his extensive experience in parapsychological research with this observation: "After sixty years' experience and personal acquaintance with most of the leading parapsychologists of that period I do not think I could name a half dozen whom I could call objective students who honestly wished to discover the truth."

Replicability

If the evidence for any claim is based upon an experimental result, or if the evidence offered in support of any claim could logically be explained as coincidental, then it is necessary for the evidence to be repeated in subsequent experiments or trials.

The rule of replicability provides a safeguard against the possibility of error, fraud, or coincidence. A single experimental result is never adequate in and of itself, whether the experiment concerns the production of nuclear fusion or the existence of telepathic ability. Any experiment, no matter how carefully designed and executed, is always subject to the possibility of implicit bias or undetected error. The rule of replicability, which requires independent observers to follow the same procedures and to achieve the same results, is an effective way of correcting bias or error, even if the bias or error remains permanently unrecognized. If the experimental results are the product of deliberate fraud, the rule of replicability will ensure that the experiment will eventually be performed by honest researchers.

If the phenomenon in question could conceivably be the product of coincidence, then the phenomenon must be replicated before the hypothesis of coincidence can be rejected. If coincidence is in fact the explanation for the phenomenon, then the phenomenon will not be duplicated in subsequent trials, and the hypothesis of coincidence will be confirmed; but if coincidence is not the explanation, then the phenomenon may be duplicated, and an explanation other than coincidence will have to be sought. If I correctly predict the next roll of the dice, you should demand that I duplicate the feat before granting that my prediction was anything but a coincidence.

The rule of replicability is regularly violated by parapsychologists, who are especially fond of misinterpreting coincidences. The famous "psychic sleuth" Gerard Croiset, for example, allegedly solved numerous baffling crimes and located hundreds of missing persons in a career that spanned five decades, from the 1940s until his death in 1980. The truth is that the overwhelming majority of Croiset's predictions were either vague and nonfalsifiable or simply wrong. Given the fact that Croiset made thousands of predictions during his lifetime, it is hardly surprising that he enjoyed one or two chance "hits." The late Dutch parapsychologist Wilhelm Tenhaeff, however, seized upon those "very few prize cases" to argue that Croiset possessed demonstrated psi powers (Hoebens 1986a:130). That was a clear violation of the rule of replicability, and could not have been taken as evidence of Croiset's psi abilities even if the "few prize cases" had been true. (In fact, however, much of Tenhaeff's data was fraudulent—see Hoebens 1986b.)

Sufficiency

The evidence offered in support of any claim must be adequate to establish the truth of that claim, with these stipulations: (1) the burden of proof for any claim rests on the claimant, (2) extraordinary claims demand extraordinary evidence, and (3) evidence based upon authority and/or testimony is always inadequate for any paranormal claim.

The burden of proof always rests with the claimant for the simple reason that the absence of disconfirming evidence is not the same as the presence of confirming evidence. This rule is frequently violated by proponents of paranormal claims, who argue that, because their claims have not been disproved, they have therefore been proved. (UFO buffs, for example, argue that because skeptics have not explained every UFO sighting, some UFO sightings must be extraterrestrial spacecraft.) Consider the implications of that kind of reasoning: If I claim that Adolf Hitler is alive and well and living in Argentina, how could you *disprove* my claim? Since the claim is logically possible, the best you could do (in the absence of unambiguous forensic evidence) is to show that the claim is highly improbable—but that would not disprove it. The fact that you cannot prove that Hitler is not living in Argentina, however, does not mean that I have proved that he is. It only means that I have proved that he could be—but that would mean very little; logical possibility is not the same as established reality. If the absence of disconfirming evidence were sufficient proof of a claim, then we could "prove" anything that we could imagine. Belief must be based not simply on the absence of disconfirming evidence but on the presence of confirming evidence. It is the claimant's obligation to furnish that confirming evidence.

Extraordinary claims demand extraordinary evidence for the obvious reason of balance. If I claim that it rained for ten minutes on my way to work last Tuesday, you would be justified in accepting that claim as true on the basis of my report. But if I claim that I was abducted by extraterrestrial aliens who whisked me to the far side of the moon and performed bizarre medical experiments on me, you would be justified in demanding more substantial evidence. The ordinary evidence of my testimony, while sufficient for ordinary claims, is not sufficient for extraordinary ones.

In fact, testimony is *always* inadequate for *any* paranormal claim, whether it is offered by an authority or a layperson, for the simple reason that a human being can lie or make a mistake. No amount of expertise in any field is a guarantee against human fallibility, and expertise does not preclude the motivation to lie; therefore a person's credentials, knowledge, and experience cannot, in themselves, be taken as sufficient evidence to establish the truth of a claim. Moreover, a person's sincerity lends nothing to the credibility of his or her testimony. Even if people are telling what they sincerely believe to be the truth, it is always possible that they could be mistaken. Perception is a selective act, dependent upon belief, context, expectation, emotional and biochemical states, and a host of other variables. Memory is notoriously problematic, prone to a range of distortions, deletions, substitutions, and amplifications. Therefore the testimony that people offer of what they remember seeing or hearing should always be regarded as only provisionally and approximately accurate; when people are speaking about the paranormal, their testimony should never be regarded as reliable evidence in and of itself. The possibility and even the likelihood of error are far too extensive (see Connor 1986).

Conclusion

The first three rules of FiLCHeRS—falsifiability, logic, and comprehensiveness—are all *logically* necessary rules of evidential reasoning. If we are to have confidence in the veracity of any claim, whether normal or paranormal, the claim must be propositionally meaningful, and the evidence offered in support of the claim must be rational and exhaustive.

The last three rules of FiLCHeRS—honesty, replicability, and sufficiency—are all *pragmatically* necessary rules of evidential reasoning. Because human beings are often motivated to rationalize and to lie to themselves, because they are sometimes motivated to lie to others, because they can make mistakes, and because perception and memory are problematic, we must demand that the evidence for any factual claim be evaluated without self-deception, that it be carefully screened for error, fraud, and appropriateness, and that it be substantial and unequivocal.

What I tell my students, then, is that you can and should use FiLCHeRS to evaluate the evidence offered for any claim. If the claim fails any one of these six tests, then it should be rejected; but if it passes all six tests, then you are justified in placing considerable confidence in it.

Passing all six tests, of course, does not guarantee that the claim is true (just because you have examined all the evidence available today is no guarantee that there will not be new and disconfirming evidence available tomorrow), but it does guarantee that you have good reasons for believing the claim. It guarantees that you have sold your belief for a fair price, and that it has not been filched from you.

Being a responsible adult means accepting the fact that almost all knowledge is tentative, and accepting it cheerfully. You may be required to change your belief tomorrow, if the evidence warrants, and you should be willing and able to do so. That, in essence, is what skepticism means: to believe if and only if the evidence warrants.

References

Connor, John W. 1984. Misperception, folk belief, and the occult: A cognitive guide to understanding. *Skeptical Inquirer*, 8:344-354, Summer.

Dingwall, E. J. 1985. The need for responsibility in parapsychology: My sixty years in psychical research. In *A Skeptic's Handbook of Parapsychology*, 161-174, ed. by Paul Kurtz. Buffalo, N.Y.: Prometheus Books.

Hines, Terence. 1988. *Pseudoscience and the Paranormal*. Buffalo, N.Y.: Prometheus Books.

Hoebens, Piet Hein. 1981. Gerard Croiset: Investigation of the Mozart of "psychic sleuths." *Skeptical Inquirer*, 6(1):17-28, Fall.

————. 1981-82. Croiset and Professor Tenhaeff: Discrepancies in claims of clairvoyance. *Skeptical Inquirer*, (2):21-40, Winter.

Hyman, Ray. 1985. A critical historical overview of parapsychology. In *A Skeptic's Handbook of Parapsychology*, 3-96, ed. by Paul Kurtz. Buffalo, N.Y.: Prometheus Books.

Omohundro, John T. 1976. Von Däniken's chariots: Primer in the art of cooked science. *Skeptical Inquirer*, 1(1):58-68, Fall.

Story, Ronald D. 1977. Von Däniken's golden gods, *Skeptical Inquirer*, 2(1):22-35, Fall/Winter.

RAY HYMAN

Assessing Arguments and Evidence

Many fellow skeptics have asserted, in my presence, that if everyone were properly educated in critical thinking no one would be taken in by inadequate arguments for paranormal claims. Although these skeptics do not provide details on just what training in critical thinking ought to include, I get the impression that they have in mind what is currently taught in courses in elementary logic and scientific reasoning. Such courses teach the students rules and procedures that can be applied to the premises and conclusions of arguments to see if the conclusions are warranted by the premises.

Psychologists have been accumulating evidence that suggests that such courses unfortunately fail to protect the students from falling for the same illogical and unscientific arguments that convince those who have not had such training. One problem may be that these courses focus on what to do *after* the argument has been carefully specified. Psychological research has been discovering that the major problem most of us face is how to recognize and specify the premises and the type of argument that must be dealt with. Knowing how to deal with argument of type X is not of much help if we do not recognize that the argument before us is of type X.

The current psychological research on the importance of the ability to adequately recognize what sort of argument one is dealing with is consistent with the earlier research on problem-solving, creativity, and reasoning. The early investigators of critical thinking focused on the thought processes that occurred after the thinker was presented with a carefully specified problem. Again and again the results were disappointing in that individual differences in such thinking processes did not seem to be critical in succeeding or failing to solve the problem. More important was how the thinker represented or formulated the problem in the first place. Once the problem was formulated properly, most thinkers could arrive at an adequate solution.

Properly formulating the problem is a skill that is difficult to reduce to specific rules of the sort used to teach logical and scientific procedures for testing claims. It has to be developed with practice and requires experience with the domain in question. In my attempts over the past 20 years to teach students how to cope with paranormal and borderline claims, I have learned that it does little good to try to teach them formal procedures for logically evaluating arguments or for testing specific hypotheses. I have had much more success in drilling them on systematically asking the right sorts of questions. In K. D. Moore's *A Field Guide to Inductive Arguments* (Kendall/Hunt Publishing, 1986), she wisely writes (p. 2):

> In order to evaluate an argument it is necessary to understand it. And in order to understand an argument, it is often necessary to work quite hard to find out what is being argued for, and why, and how. Many arguments are confusingly presented, with major parts omitted, unclear, or tangled together. As a result, a careful reader must begin by sorting through an argument, looking for its most important parts.

Moore provides the following list of questions to help the student analyze an argument: (1) What is the issue? (2) What is the speaker arguing for? (3) What reasons is the speaker offering? (4) Are the premises true? (5) How well do the reasons support the conclusion?

I suspect that most people, regardless of prior training in logic or critical thinking, would greatly improve their ability to deal adequately with paranormal claims if they subjected the arguments to such questions before attempting to draw any conclusions. The reader may find it worthwhile to compare Moore's set of questions with the following ones I developed for the students in my course on pseudopsychologies:

1. What pseudopsychology (or pseudopsychologies) is under discussion?

2. What are the explicit and/or the implicit claims being made by the proponents of the pseudopsychologies being discussed?

3. What sorts of evidence and arguments are used by proponents to justify the claims?

4. How well do the evidence and the arguments justify the claims?

5. What sorts of evidence and arguments would be required to justify the claims for this pseudopsychology?

6. What alternative reasons can be hypothesized to account for beliefs in the pseudopsychology even if it is invalid?

My first five questions achieve the same ends as Moore's. My sixth goes beyond evaluating the argument and tries to get the student to think in terms of those psychological factors that can create belief. My experience indicates that the students require practice and some feedback before they acquire the knack of using these questions effectively. I have them read three books. For each book they produce a short paper by writing answers to the six questions. The books I currently use for this purpose are Vogt and Hyman's *Water Witching U.S.A.*, Blackmore's *Beyond the Body,* and Marks and Kammann's *The Psychology of the Psychic.* By the time they have completed evaluating the pseudopsychologies discussed in these three books, almost all the students are able to successfully complete a term paper in which they evaluate a pseudo-psychology of their own choosing.

Both inductive and deductive logic provide guidelines for evaluating how well a conclusion follows from the premise. But logic does not help us decide whether the premise is true or false. An argument might be logically impeccable, but rely on factually incorrect evidence. So it is important to gauge the trustworthiness of the "facts" that the claimant puts forth to buttress his or her claim. One of the most unfortunate aspects of many skeptical attempts to provide a normal "explanation" for a paranormal claim is that the "explanation" is based on the events *as reported* rather than as they *actually happened.* It is futile to try to evaluate an argument for the paranormal if the "facts" are not of the highest quality.

How can we judge the quality of the evidence? This is not an easy matter. Here again, no simple set of universal rules exists. In addition, the judging of the reliability of the data can require highly technical skills in statistics, instrumentation, and scientific methodology. However, some simple guidelines can often quickly identify many arguments that are not worth evaluating. I once tried to supply a simple set of guidelines for this purpose ("Scientists and Psychics," in *Science and the Paranormal,* ed. by G. O. Abell and B. Singer, 1981, pp. 137-141). The following questions were the key to my guidelines:

1. How reliable is the source?

2. How recent is the research upon which the claim is based?

3. Has the original investigator been able to successfully replicate the findings?

4. Have the phenomena been replicated by an independent investigator in another laboratory?

5. Does the original report conform to the standards required for observing human performance?

I maintain that if such questions are addressed to all the known cases in which scientists endorsed the paranormal claims of psychics, not one of the instances over the past 130 years would survive as worthy of further evaluation. Even the best reasoning procedure will produce nonsense if the material to which it is applied is untrustworthy. And, outside of the four major parapsychological journals, it is almost impossible to encounter paranormal claims based on evidence that can be trusted to be as reported. The moral is that, in most situations, it is a waste of time to try to evaluate the underlying argument.

Part 3: Evaluating the Anomalous Experience

BARRY L. BEYERSTEIN

The Brain and Consciousness: Implications for Psi Phenomena

Men ought to know that from the brain and from the brain only arise our pleasures, joys, laughter, and jests as well as our sorrows, pains, griefs and tears. . . . It is the same thing which makes us mad or delirious, inspires us with dread and fear, whether by night or by day, brings us sleeplessness, inopportune mistakes, aimless anxieties, absent-mindedness and acts that are contrary to habit. . . .

Hippocrates (c. 460–c. 377 B.C.), *The Sacred Disease*

Anomalous subjective experiences contribute strongly to paranormal beliefs. They may be powerful, unprovoked emotions or apparently spontaneous percepts that others cannot verify. For some these experiences are accompanied by a feeling that consciousness is estranged from the body or that an alien force is "usurping the seat of the will." These interludes are variously construed as divine or diabolical, enlightening or foreboding, a mere curiosity or a calling to a sacred mission. Those acknowledging guidance from such "revelations" range from Plato, St. Paul, Muhammad, Joan of Arc, Columbus, Mozart, and Newton, on the one hand, to Atilla the Hun, Hitler, Stalin, Idi Amin, and Charles Manson, on the other. Many ancient supernatural beliefs probably have their origins in revelations of this sort. They still nurture many mystical beliefs today (Greeley and McCready 1975).

We now know that both normal and diseased brains will generate, from time to time, spontaneous sensations and emotions that seem to originate externally, even in other minds. These compelling experiences continue to be cited as evidence for the paranormal, despite the cogent objections of Michael Scriven (1961), a philosopher otherwise favorably disposed toward psi phenomena.

Psychophysiologists are challenged by paranormal "explanations" for these occurrences because, if correct, the implications for the neurosciences' view of the mind-brain relationship are profound. If, as many occultists assert, mind can exist free of the body, directly influence other minds or matter at a distance, and receive information by other than the conventional senses, several fundamental tenets of neuroscience are sadly incomplete, if not totally erroneous. While I doubt that studies of allegedly paranormal anomalies of consciousness will overturn the foundations of neuroscience, they could add to our conventional understanding of perception, memory, and emotion. Such studies can also eventually help people understand the true causes of "extraordinary" experiences that seem so real to them.

It is proper to demand stronger evidence for newly tendered "facts" if accepting them entails abandoning a substantial amount of better-established data. Neuroscience cannot rule out psi phenomena, but it is difficult, logically, to embrace both.

The Underpinnings of Neuroscience—Psychoneural Identity (PNI)

In 1949, Donald O. Hebb enunciated the creed to which an overwhelming majority of neuroscientists would still subscribe:

> Modern psychology takes completely for granted that behavior and neural function are perfectly correlated, that one is completely caused by the other. There is no separate soul or lifeforce to stick a finger into the brain now and then and make neural cells do what they would not otherwise. Actually, of course, this is a working assumption only. . . . It is quite conceivable that someday the assumption will have to be rejected. But it is important also to see that we have not reached that day yet: the working assumption is a necessary one and there is no real evidence opposed to it. Our failure to solve a problem so far does not make it insoluble. One cannot logically be a determinist in physics and biology, and a mystic in psychology. [Hebb 1949, p. xiii]

While views of determinism have modified since Hebb wrote, his conviction that consciousness is inseparable from the functioning of individual brains remains the cornerstone of physiological psychology. A discussion of the philosophical issues underlying Hebb's working assumption—psychoneural identity (PNI) theory—is beyond the scope of this paper. Several good treatments are available (Bunge 1980; Campbell 1970; Churchland 1984; Uttal 1978).

Though PNI cannot be proved empirically (cf. Malcolm 1971), psychophysiology offers an impressive array of data supporting its claim that thinking, perceiving, remembering, desiring, and feeling are brain functions. Research supporting mind-brain identity is summarized in numerous texts (e.g., Oakley and Plotkin 1979; Rosenzweig and Leiman 1982; Uttal 1978).

Briefly, that evidence falls into the following categories:

Phylogenetic: There is an evolutionary relationship between brain complexity and species' cognitive attributes (Russell 1979).

Developmental: Abilities emerge with brain maturation; failure of the brain to mature arrests mental development (Parmalee and Sigman 1983).

Clinical: Brain damage from accidental, toxic, or infectious sources, or from deprivation of nutrition or stimulation during brain development, results in predictable and largely irreversible losses of mental function (Kolb and Whishaw 1985; Sacks 1987).

Experimental: Mental operations correlate with electrical, biochemical, biomagnetic, and anatomical changes in the brain. When the human brain is stimulated electrically or chemically during neurosurgery, movements, percepts, memories, and appetites are produced that are like those arising from ordinary activation of the same cells (Valenstein 1973).

Experiential: Numerous natural and synthetic substances interact chemically with brain cells. Were these neural modifiers unable to affect consciousness pleasurably and predictably, the recreational value of nicotine, alcohol, caffeine, LSD, cocaine, and marijuana would roughly equal that of blowing soap bubbles.

Despite their abundance, diversity, and mutual reinforcement, the foregoing data cannot, by themselves, entail the truth of PNI. Nevertheless, the theory's parsimony and research productivity, the range of phenomena it accounts for, and the lack of credible counter-evidence are persuasive to virtually all neuroscientists (Uttal 1978). Brain researchers are apt to view rejections of PNI much as paleontologists do suggestions of "Creation Scientists" that the fossil record was merely "salted" in the strata by a suspicious deity to test believers' faith—possible, but credulity is strained.

The evidence for PNI is such that many parapsychologists admit the only hope for dualistic alternatives (i.e., that brain and consciousness are separable and not subject to the same natural laws) lies in documenting telepathy, clairvoyance, or psychokinesis. Given the centrality and implications of PNI, most psychophysiologists, not surprisingly, doubt the existence of disembodied minds and other psi phenomena.

Split Brains, Brain Damage, and Consciousness

Compelling support for PNI is found when brain tracts connecting the left and right hemispheres are severed to alleviate seizures. If information is presented uniquely to one hemisphere in these "split-brain" patients, the other hemisphere is unaware of it and unable to comprehend the informed side's reactions (LeDoux et al. 1979). Two mental systems, each with independent memories, percepts, and desires, coexist in one body and are able to initiate (with no sense of conflict) mutually contradictory actions with opposite hands (Dimond 1979).

If consciousness is not tied to brain function, it is difficult to understand how interrupting nerve tracts could compartmentalize it. If a "free-floating" mind exists, why can't it maintain unity of consciousness by providing an information conduit between the disconnected hemispheres? Parapsychologists claim that a mind can span continents to communicate with other minds; why is it patently unable to jump a few millimeters of uncoupled neural tissue?

Similarly, after brain damage, why is an allegedly separate mind unable to compensate for lost faculties when brain cells die? Having observed the devastation of brain injuries, it seems to me a cruel joke to suggest that only the input-output channels of a still intact mind have been damaged. The fortunate few who recover from reversible brain syndromes certainly recall no such serene redoubt (Gardner 1974; Linge 1980).

Alpha and Omega

Psychobiologists are also suspicious of claims that mature consciousness exists before birth and beyond death. Alleged scientific evidence for an afterlife (e.g., Moody 1975; Osis and Haraldsson 1977) is flawed, logically and empirically (Alcock 1979; Puccetti 1979; Siegel 1981). In addition to inconsistencies and methodological defects, most survivalist claims suffer from an outmoded conception of death.

By modern neurological criteria, patients who supposedly "returned from the other side" were never dead, only resuscitated from cardiopulmonary arrest (CPA)—temporary interruption of heartbeat and respiration. Because brain cells do not cease functioning immediately following CPA, mental activity can continue (albeit degraded by oxygen/glucose deprivation and other neurochemical changes) for several minutes after the last pulse and breath are detected.

Dying is a multistage process, reversible until critical cells in the brain stem or neocortex succumb (Walker 1981). Thus a patient with a silent neocortex can still breathe and show a pulse but be clinically dead; one suffering from *temporary* CPA lacks two so-called vital signs but is not brain dead. Memories from the period before resuscitation do not entail an afterlife because, fortunately, these patients never succumbed to brain death. Irreversible breakdown of communication among critical brain cells is now the criterion of legal death in most modern societies. They have conceded a corollary of PNI—that "human life" presupposes a brain capable of sustaining the essential attributes of consciousness.

In other pseudosciences, complex cognitive and motivational capacities are ascribed to the unborn—parental conversations are allegedly comprehended by fetuses who can then suffer persistent psychological scars. L. Ron Hubbard's (1968) morbid Scientological musings about life in the womb began as science fiction and are now marketed, appropriately, as religion, but it is especially worrisome when supposedly knowledgeable professionals make similar claims in the face of established neurological caveats. Their data generally are gleaned from "recollections" of adults with psychological complaints severe enough to require therapy. Psychiatrist Thomas Verny (1981) fashions his theories of psychopathology from what patients tell him are fetal memories. Arthur Janov (1970), founder of the suspect "Primal Scream" movement, asserts that neuroses stem from memories of birth trauma, and "rebirther" Leonard Orr offers the cure: reliving one's nativity while hyperventilating in a warm bathtub (for good critiques, see Rosen 1977). Similarly, Stanislav Grof (1985) explains sexual perversions involving excrement as consequences of contact with maternal feces while exiting the birth canal (see Richard Morrock's critique, *Skeptical Inquirer*, Spring 1986).

Memories from the womb are extremely doubtful, given the immaturity of the fetal brain. The auditory system attains rudimentary functioning by the last trimester of pregnancy, and by shortly after birth infants can be trained to make different movements in response to various speech sounds (Aslin et al. 1983). However, extrapolating from these simple abilities to the conjecture that fetuses understand adult utterances, and years later resent them, offends common sense and considerable research in child development.

Language competence emerges as certain indices of brain development reach about 65 percent of mature values (Lenneberg 1969). Newborns, let alone fetuses, are far short of this (Hirsch and Jacobson 1975; Parmalee and Sigman 1983)—the brain increases fourfold in size and weight from birth to maturity. This, coupled with research on sensory and cognitive abilities of neonates, makes the mentalities presumed by Hubbard, Verny, Janov, Orr, and Grof highly dubious.

Verny goes even further, however, asserting that "everything a woman thinks, feels, says, and hopes influences her unborn child" (quoted in Cannon 1981). This mystical bond between maternal and fetal consciousness is incompatible with PNI because there is no neural link between their brains.

While severe maternal stress during pregnancy can adversely affect offspring by altering intrauterine chemistry, it is hard to imagine how specific thoughts and feelings of the mother could reach and be recognized by the fetal brain. Verny's speculations amount to claims of telepathy between the mother and an unbelievably precocious fetal mind. They are reminiscent of old superstitions that pregnant mothers frightened by elephants have deformed babies and that those who steal bear thieves.

While psychophysiologists are merely amused by Verny's conjectures, it is unfortunate that his psychiatric credentials engender widespread trust. I have met several mothers of children with developmental disorders whose burdens he has needlessly compounded with guilt—they believed their ambivalent thoughts during difficult pregnancies must have caused their children's plight. Verny's latest enterprise is marketing soothing musical recordings for mother and unborn child, implying future benefits for her progeny.

Competent studies of childhood memory do not inspire confidence in alleged pre- or perinatal recollections (White and Pillemer 1979). There are alternative explanations for why people believe they recall life in the womb or previous incarnations (Alcock 1981; Loftus 1980; Zusne and Jones 1982). "Demand characteristics" in psychotherapy could easily extract fantasies masquerading as veridical memories (Orne 1969; Hilgard and Loftus 1979).

There is evidence that memories are stored as structural modifications in neural circuits (Squire 1986). Improbable as it is that these mechanisms would be fully functional prenatally, it is logically impossible for experiences so stored to survive disintegration of the brain. Prevalent beliefs that knowledge can be tapped from previous incarnations or from a "universal mind" (the repository of all past wisdom and creativity) not only are implausible but also unfairly demean the stunning achievements of individual human brains.

Points of Departure

Many people believe their "psychic selves" periodically leave their bodies to retrieve distant information. If true, this would challenge PNI gravely, but critics of the literature on "out-of-body experiences" (OBEs) find the evidence unconvincing (Blackmore 1982; Neher 1980). OBE descriptions are consistent with known neural and psychological phenomena that evoke vivid hallucinations and temporarily impair reality testing. Neher (1980) even offers relaxation and imagery exercises for those wishing to experience an OBE for themselves.

In the last century, the neurologist Hughlings Jackson reported that aberrancies in the temporal lobes of the brain can produce floating, disembodied sensations, including viewing one's body from a distance (MacLean 1970). OBEs have since been produced by electrical stimulation of the temporal lobes during neurosurgery. They are also associated with a variety of drugs, epileptic seizures, hypoglycemic and migraine episodes, and neurochemical changes near death. Occasionally, OBEs occur spontaneously in normal, awake individuals, probably due to random activation of temporal lobe systems. OBEs seem less mysterious when we consider that the brain generates similar imagery during dreams and even in visual memories, where we routinely view ourselves from positions we never actually occupied. It is primarily the clarity or "realness" of the OBE (related to frontal and temporal lobe activity) that distinguishes it from related forms of imagery, including those of "daydreams," which can themselves be quite vivid (Singer 1975; Kolb and Whishaw 1985, ch. 10).

OBEs can also be triggered by miscues when the brain's arousal mechanisms shift from drowsiness to sleep, sleep to waking, nondream to dream sleep, and so on. In such a multicomponent system (Cohen 1979), occasional desynchronizations are to be expected—resulting here in dreamlike activity during quasi-wakefulness. Sleep-onset (hypnogogic) and sleep-offset (hypnopompic) images are often bizarre, but seemingly real, mixes of genuine percepts and hallucinations (Stoyva 1973).

The Physiology of Hallucination

Except during dreams, OBEs, and so on, it is usually easy to distinguish authentic percepts from self-produced images. Occasionally it can be difficult though, because brain systems that generate images from memory share neural circuitry with those that decipher sensory input from the environment (Finke 1986). Many factors can temporarily disable higher brain mechanisms that confirm the reality status of percepts.

Hallucinations result when the sensory cortex is activated without input to peripheral receptors. This can arise from electrical or drug stimulation of the brain, hypnotic suggestion, high fever, narcolepsy, migraine, epilepsy, schizophrenia, and sensory overload or prolonged isolation (Horowitz 1975; Johnson 1978; Siegel and West 1975). Hallucinations can ensue when internal imagery overwhelms external sensory input in shared neural pathways, or when indistinct perceptual fractions are embellished in accordance with expectations and belief (Horowitz 1975). They are also possible in situations that affect our normal alternation between external vigilance and attention to imagery (used in recollection, problem-solving, day-dreaming, and so on). Strong conflict, emotional threat, fear, or desire can lend an intensely real quality to imagistic thinking. Meditation, by reducing sensory input while suppressing verbal modes of consciousness, can have similar results.

Schatzman (1980) found objective support for the notion that hallucinations are processed in the visual areas of the brain. A patient who experienced vivid hallucinations was presented with a visual stimulus. The electrical response of her visual cortex while she viewed it normally was compared to that when she hallucinated something that obscured it from view. In the latter condition, the trace of the stimulus in the recording essentially disappeared as her visual cortex began to process her hallucinated image. Simply asking her not to attend to the stimulus had no comparable effect on this "visual evoked response."

Many occult beliefs stem from the misconception that everything seen or heard must necessarily exist outside ourselves. Fatigue, stress, monotony, or fervent desire can obscure the "tags" that designate internal and external origin as messages pass through the brain—blurring thereby the demarcation between reality and fancy.

Perception—Normal and Extrasensory

A vast literature sustains the PNI corollary that perception is a brain process (Uttal 1973). For the conventional senses (vision, hearing, taste, smell, touch) we know much about how different energies are transduced by receptors into neural codes and how brain systems distribute and analyze their content (Coren et al. 1984).

Damage to specific analyzers in the brain obliterates perception of the qualities they encode. If minds can abandon bodies and retain full awareness on the voyage, why should a mere hardware defect in the brain leave neurological patients insentient? On the other hand, if only peripheral receptors are damaged, crude prostheses are possible by stimulating the sensory cortex with patterned electrical impulses (Dobelle et al. 1974). That this evokes simple visual patterns supports PNI, but the crudity of the percepts produced[1] by even the most advanced prosthetic stimulators underscores the enormous task putative telepathic "energies" would have to accomplish in order for ESP to be compatible with PNI. A "message" bypassing conventional neurosensory routes to consciousness would still have to impose precisely patterned activity across millions of brain cells.

A theorist trying to marry ESP and PNI would need to suggest plausible mechanisms in order to respond to the following questions: (a) How is the "message" generated by the "sender's" brain in telepathy and by inanimate objects in clairvoyance? (b) What kind of energy is involved that could carry the message, without loss, over immense distances and through intervening objects?[2] (c) What is the propagating medium for the signal; what prevents "crosstalk" among simultaneous messages and what addresses them to recipients? (d) Once at the recipient, what directs the message to the appropriate sense modality—e.g., to vision rather than smell—let alone to produce a meaningful percept? (e) What conceivable form of energy would have the informational capacity to impose the necessary spatio-temporal patterns on the astronomical number of neurons involved in even a simple percept? How would it duplicate the subtle movements of neurochemicals across the cell membranes that constitute the neural code?[3]

These demands of PNI are rarely addressed by ESP enthusiasts. In fact, avoiding them is one of the attractions of dualism—if mind is nonphysical, these restraints need not apply. Tart (1977), to his credit, faces some of these issues, but his proposed solutions are essentially the ancient principles of Sympathetic and Contact Magic restated in high-tech jargon. He argues that "channels," "decoders," etc., for ESP must exist in the brain because ESP is an established ability, but he does not suggest where and what they might be.

Godbey (1975) is correct that proof of telepathy or clairvoyance would be insufficient, by itself, to refute PNI. The brain could conceivably be put in a physical state of "knowing something" by some as yet undiscovered material force. However, as I have argued, this would entail a form of energy quite unlike those known to physicists, operating on neural mechanisms in ways that seem equally bizarre to psychobiologists. While both may eventually be confirmed, at present they are required only to "explain" phenomena for which there are more credible naturalistic interpretations (Alcock 1981; Blackmore 1982; Marks and Kammann 1980; Neher 1980; Zusne and Jones 1982).

The 10-percent Solution

In arguing that current theories of brain function cast suspicion on ESP, psychokinesis, reincarnation, and so on, I am frequently challenged with the most popular of all neuro-mythologies—the notion that we ordinarily use only 10 percent of our brains. "Enlightened ones" supposedly tap the remainder for levitation, spoon-bending, precognition, telepathy, and other fantastica inconceivable to those subsisting on the drudgelike 10 percent.

Origins of the 10-percent myth are obscure, but the concept was widely disseminated in courses like Dale Carnegie's and canonized in public utterances by no less a personage than Albert Einstein. I believe the error arose from misinterpretations of research in the 1930s showing that, with evolutionary advancement, a progressively smaller proportion of the brain is tied to strictly sensory or motor duties. For methodological reasons, the enlarged nonsensory, nonmotor areas were referred to as the "silent cortex," though they are anything but silent. They are responsible for our most human characteristics, including language and abstract thought. Areas of maximal activity shift in the brain as we engage in different tasks, and there can be some reorganization of functional regions after brain damage; but there are normally no dormant regions awaiting new assignments.

This "cerebral spare tire" concept continues to nourish the clientele of "pop psychologists" and their many recycling self-improvement schemes. As a metaphor for the fact that few of us fully exploit our talents, who could deny it? As a refuge for occultists seeking a neural basis of the miraculous, it leaves much to be desired.

Conclusion

Extraordinary claims demand extraordinary proof. There are many examples of outsiders who eventually overthrew entrenched scientific orthodoxies, but they prevailed with irrefutable data. More often, egregious findings that contradict well-established research turn out to be artifacts. I have argued that accepting psychic powers, reincarnation, "cosmic consciousness," and the like, would entail fundamental revisions of the foundations of neuroscience. Before abandoning materialist theories of mind that have paid handsome dividends, we should insist on better evidence for psi phenomena than presently exists, especially when neurology and psychology themselves offer more plausible alternatives.

Notes

1. Stimulating the visual cortex produces dots of light that can be connected to simulate objects; stimulation of the temporal cortex produces more lifelike hallucinations but their content

is not controllable.

2. Brain-generated electromagnetic fields drop to infinitesimal strength within millimeters of the scalp. Electromagnetic fields pass through many materials, but they obey the inverse square law and are blocked by appropriate shielding, neither of which is true, proponents claim, of ESP "energies," whatever they may be.

3. In normal perception, this is accomplished by known environmental mechanisms interacting with the anatomy/physiology of the sensory pathways. Different energies match selective receptors whose output travels via separate tracts to specialized cortical areas for each modality —all but the last of which are allegedly bypassed in ESP.

References

Alcock, J. E. 1979. Psychology and near-death experiences. *Skeptical Inquirer,* 3(3):25-41.

———. 1981. *Parapsychology: Science or Magic?* Oxford: Pergamon.

Aslin, R., D. Pisoni, and P. Jusczyc. 1983. Auditory development and speech perception in infancy. In *Handbook of Child Psychology,* vol. 2, *Infancy and Developmental Psychobiology,* ed. by P. Mussen, 573-688. New York: Wiley.

Blackmore, S. J. 1982. *Beyond the Body.* London: Grenada.

Bunge, M. 1980. *The Mind-Body Problem.* Oxford: Pergamon.

Campbell, K. 1970. *Body and Mind.* London: Macmillan.

Cannon, M. 1981. Tapping memories of life in the womb: A psychiatrist claims fetuses possess feelings. *Maclean's Magazine,* September 28, pp. 46-47.

Churchland, P. M. 1984. *Matter and Consciousness.* Cambridge, Mass.: MIT Press.

Cohen, D. B. 1979. *Sleep and Dreaming: Origins, Nature and Functions.* Oxford: Pergamon.

Coren, S., C. Porac, and L. Ward. 1984. *Sensation and Perception,* 2nd ed. New York: Academic Press.

Dimond, S. J. 1979. Symmetry and asymmetry in the vertebrate brain. In *Brain, Behaviour, and Evolution,* ed. by D. Oakley and H. Plotkin, 189-218. London: Methuen.

Dobelle, W., M. Mladejovsky, and J. Girvin. 1974. Artificial vision for the blind: Electrical stimulation of the visual cortex offers hope for a functional prosthesis. *Science,* 183:440-444.

Finke, R. A. 1986. Mental imagery and the visual system. *Scientific American,* 254(3): 88-95.

Gardner, H. 1974. *The Shattered Mind.* New York: Vintage Books.

Godbey, J. W. 1975. Central-state materialism and parapsychology. *Analysis,* 36: 22-25.

Greeley, A., and McCready. 1975. Are we a nation of mystics? *New York Times Magazine,* January 16.

Grof, S. 1985. *Beyond the Brain: Birth, Death and Transcendence in Psychotherapy.* Albany, N.Y.: State University of New York Press.

Hebb, D. O. 1949. *Organization of Behavior: A Neuropsychological Theory.* New York: Wiley.

Hilgard, E., and E. Loftus. 1979. Effective interrogation of the eyewitness. *Int. J. of Clinical & Experimental Hypnosis,* 27(4): 342-357.

Hirsch, H., and M. Jacobson. 1975. The perfectable brain: Principles of neuronal development. In *Handbook of Psychobiology,* ed. by M. Gazzaniga and C. Blakemore, 107-137. New York: Academic Press.

Horowitz, M. 1975. Hallucinations: An information-processing approach. In *Hallucinations: Behavior, Experience, and Theory,* ed. by R. K. Siegel and L. J. West, 163-194. New York: Wiley.

Hubbard, L. R. 1968. *Scientology: A History of Man.* Los Angeles: American Saint Hill.

Janov, A. 1970. *The Primal Scream.* New York: Delta.

Johnson, F. H. 1978. *The Anatomy of Hallucinations.* Chicago: Nelson-Hall.

Kolb, B., and I. Whishaw. 1985. *Fundamentals of Human Neuropsychology,* 2nd ed. New York: W. H. Freeman.

LeDoux, J., D. Wilson, and M. Gazzaniga. 1979. Beyond commisurotomy: Clues to consciousness. In *Handbook of Behavioral Neurobiology,* vol. 2, ed. by M. Gazzaniga, 543-554.

New York: Plenum.

Lenneberg, E. 1969. On explaining language. *Science,* 164: 635-643.

Linge, F. 1980. What does it feel like to be brain damaged? *Canada's Mental Health,* September.

Loftus, E. 1980. *Eyewitness Testimony.* Cambridge: Harvard University Press.

MacLean, P. 1970. The limbic brain in relation to the psychoses. In *Physiological Correlates of Emotion,* ed. by P. Black, 129-146. New York: Academic Press.

Malcolm, N. 1971. *Problems of Mind.* New York: Harper & Row.

Marks, D., and R. Kammann. 1980. *The Psychology of the Psychic.* Buffalo, N.Y.: Prometheus Books.

Moody, R. 1975. *Life After Life.* Atlanta: Mockingbird Books.

Neher, A. 1980. *The Psychology of Transcendence.* Englewood Cliffs, N.J.: Prentice-Hall.

Oakley, D., and H. Plotkin. 1979. *Brain, Behaviour, and Evolution.* London: Methuen.

Orne, M. 1969. Demand characteristics and the concept of quasi-controls. In *Artifact in Behavioral Research,* ed. by R. Rosenthal and R. Rosnow, 143-179. New York: Academic Press.

Osis, K., and E. Haraldsson. 1977. *At the Hour of Death.* New York: Avon.

Parmelee, A., and M. Sigman. 1983. Perinatal brain development and behavior. In *Handbook of Child Psychology,* vol. 2, *Infancy and Developmental Psychobiology,* ed. by P. Mussen, 95-155. New York: Wiley.

Puccetti, R. 1979. The experience of dying. *The Humanist,* July-August, pp. 62-65.

Rosen, R. D. 1977. *Psychobabble: Fast Talk and Quick Cure in the Era of Feeling.* New York: Atheneum.

Rosenzweig, M., and A. Leiman. 1982. *Physiological Psychology.* Lexington, Mass.: D. C. Heath.

Russell, I. S. 1979. Brain size and intelligence: A comparative perspective. In *Brain, Behaviour, and Evolution,* ed. by D. Oakley and H. Plotkin, 126-153. London: Methuen.

Sacks, O. 1987. *The Man Who Mistook His Wife for a Hat and Other Clinical Tales.* New York: Harper & Row.

Schatzman, M. 1980. Evocations of unreality. *New Scientist,* September 25, pp. 935-937.

Scriven, M. 1961. New frontiers of the brain. *Journal of Parapsychology,* 25: 305-318.

Siegel, R. K. 1981. Life after death. In *Science and the Paranormal,* ed. by G. Abell and B. Singer, 159-184. New York: Scribner.

Siegel, R. K., and L. J. West. 1975. *Hallucinations: Behavior, Experience and Theory.* New York: Wiley.

Singer, J. L. 1975. Navigating the stream of consciousness: Research in daydreaming and related experience. *Amer. Psychologist,* July, pp. 727-738.

Squire, L. R. 1986. Mechanisms of memory. *Science,* 232:1,612-1,619.

Stoyva, J. 1973. Biofeedback techniques. In *The Psychophysiology of Thinking,* ed. by F. McGuigan and R. Schoonover, 399-414. New York: Academic Press.

Tart, C. T. 1977. *Psi: Scientific Studies of the Psychic Realm.* New York: E. P. Dutton.

Uttal, W. 1973. *The Psychobiology of Sensory Coding.* New York: Harper & Row.

———. 1978. *The Psychobiology of Mind.* Hillsdale, N.J.: L. Erlbaum.

Valenstein, E. 1973. *Brain Control.* New York: Wiley-Interscience.

Verny, T. 1981. *The Secret Life of the Unborn Child.* New York: Dell.

Walker, A. E. 1981. *Cerebral Death,* 2nd ed. Baltimore, Md.: Urban & Schwarzenberg Medical Publ.

White, S. H., and B. P. Pillemer. 1979. Childhood amnesia and the development of a socially accessible memory system. In *Functional Disorders of Memory,* ed. by J. F. Kihlstrom and F. J. Evans. Hillsdale, N.J.: L. Erlbaum.

Zusne, L., and W. Jones. 1982. *Anomalistic Psychology.* Hillsdale, N.J.: L. Erlbaum.

ROBERT A. BAKER

The Aliens Among Us:
Hypnotic Regression Revisited

For the average person walking down the aisle of a modern bookstore or passing through the checkout lane at the nearest supermarket, it would be easy to conclude that aliens from outer space not only are here but also have joined the Baptist church, have put their kids in school, and belong to the Rotary Club. This conclusion is demanded by the recent rash of nonfiction books about UFO contacts, encounters of the third kind, and human abductions by little gray men from outer space or some other parallel universe. Typical of these tomes are *Communion,* by Whitley Strieber; *Intruders,* by Budd Hopkins; and *Light Years: An Investigation into the Extraterrestrial Experience of Eduard Meier,* by Gary Kinder. According to these and other UFO pundits, abductions by "little gray aliens" are so prevalent they will soon become commonplace and generally accepted as a fact of life by a now skeptical public and press.

My friends and colleagues and I, however, are beginning to believe that we have Alien B.O. or something worse, because none of us has been contacted, interviewed, briefed, threatened, kidnapped, or physically examined by any of the little folk. We, sadly enough, have not even had our car stalled by one of their spaceships. It stalls on occasion, but the problem lies in Detroit rather than with the aliens. Could all this alien activity going on around us be overlooked by responsible authorities?

To impress the general reader, all three authors have taken great pains to give as much credibility and authenticity as possible to their claims. Strieber not only took a lie-detector test but also had a psychiatrist write a statement attesting to his sanity.[1] Kinder had professional photographers examine a number of Meier's photographs and also had an IBM metallurgist endorse the unusual quality of a metal fragment from the purported spaceship. To Kinder's credit, however, he admits that he is skeptical about some of Meier's claims—particularly that of journeying back in time and talking to Jesus Christ. As for Hopkins, he not only consulted a number of psychologists and psychiatrists (he even found an abductee among them) but also had medical specialists corroborate the correctness of the medical techniques used to examine the human subjects. Just why aliens should copy human medical approaches is an unanswered question.

One would have thought that Philip Klass's (1981) devastating attack on abductee claims coupled with Robert Sheaffer's (1981) brilliant and calmly reasoned work, *The UFO Verdict*, and Douglas Curran's (1985) *In Advance of the Landing: Folk Concepts of Outer Space*, along with William R. Corliss's (1983) *Handbook of Unusual Natural Phenomena*, would have given the true believers pause and would have dampened somewhat their extravagant claims. But, like a rubber ball, they keep bouncing back.

Sheaffer and Corliss offer credible and scientific explanations of 99 percent or more of the strange lights in the sky, whereas Curran's extensive catalog of aberrant human believers suggests that the true aliens in our midst are not from outer space or a parallel dimension but are our fellow *Homo sapiens* from the edge of town. If you wish to see some excellent photos of aliens study the pictures and read the biographies in Curran's book.[2] Truly, the aliens and the alienated are already among us and have been for a long while, differing from the majority of other Americans only in the extreme nature of their beliefs and convictions. Klass's continuing excellent work on UFO demystification highlights the significance of hypnotic regression in the abductee belief system. For hypnotic regression and the personality pattern Wilson and Barber (1983) call "fantasy-prone," as well as the behavior of individuals undergoing hypnogogic and hypnopompic experiences, furnish, we believe, complete and credible explanations to most—if not all—accounts of UFO contacts and abductions past and present.

Most people seem unaware of the fact that there is an already well established branch of psychology, anomalistic psychology, that deals specifically with the kind of experiences had by Strieber, Meier, and the other UFO abductees. This psychology provides naturalistic and satisfying explanations for the entire range of such behaviors. Let us examine these explanations a little more closely and in a little more detail.

Hypnosis and Hypnotic Regression

In France in the 1770s, when Mesmerism was in its heyday, the king appointed two commissions to investigate Mesmer's activities. The commissions included such eminent men as Benjamin Franklin, Lavoisier, and Jean-Sylvain Bailly, the French astronomer. After months of study the report of the commissioners concluded that it was *imagination*, not magnetism, that accounted for the swooning, trancelike rigidity of Mesmer's subjects. Surprisingly enough, this conclusion is still closer to the truth about hypnosis than most of the modern definitions found in today's textbooks.

So-called authorities still disagree about "hypnosis." But whether it is or is not a "state," there is common and widespread agreement among all the major disputants that "hypnosis" is a situation in which people set aside critical judgment (without abandoning it entirely) and engage in make-believe and fantasy, that is, they use their imagination (Sarbin and Andersen 1967; Barber 1969; Gill and Brenman 1959; Hilgard 1977). As stated earlier, there are great individual differences in the ability to fantasize, and in recent years many authorities have made it a *requirement* for any successful "hypnotic" performance. Josephine Hilgard (1979) refers to hypnosis as "imaginative involvement," Sarbin and Coe (1972) term it "believed-in imaginings," and Sutcliffe (1961) has gone so far as to characterize the hypnotizable individual as someone who is "deluded in a descriptive, nonpejorative sense" and he sees the hypnotic situation as an arena in which people who are skilled at make-believe and fantasy are provided with the opportunity and the means to do what they enjoy doing and what they are able to do especially well. Even more recently Perry, Laurence, Nadon, and Labelle (1986) concluded that "abilities such as imagery/imagination, absorption, disassociation, and selective attention underlie high hypnotic responsivity in yet undetermined combinations." The same authors, in another context dealing with past-lives regression, also concluded that "it should be expected that any material provided in age regression (which is at the basis of reports of reincarnation) may be fact or fantasy, and it is most likely an admixture of both." The authors further report that such regression material is colored by issues of confabulation, memory creation, inadvertent cueing, and the regressee's current psychological needs. (See also Nicholas Spanos's article in this volume.)

Confabulation

Because of its universality, it is quite surprising that the phenomenon of confabulation is not better known. Confabulation, or the tendency of ordinary, sane individuals to confuse fact with fiction and to report fantasized events as actual occurrences, has surfaced in just about every situation in which a person has attempted to remember very specific details from the past. A classical and amusing example occurs in the movie *Gigi*, in the scene where Maurice Chevalier and Hermione Gingold compare memories of their courtship in the song "I Remember It Well." We remember things not the way they really were but the way we would like them to have been.

The work of Elizabeth Loftus and others over the past decade has demonstrated that the human memory works not like a tape recorder but more like the village storyteller—i.e., it is both creative and recreative. We can and we do easily forget. We blur, shape, erase, and change details of the events in our past. Many people walk around daily with heads full of "fake memories." Moreover, the unreliability of eyewitness testimony is not only legendary but well documented. When all of this is further complicated and compounded by the impact of suggestions provided by the hypnotist plus the social-demand characteristics of the typical hypnotic situation, little wonder that the resulting recall on the part of the regressee bears no resemblance to the truth. *In fact, the regressee often does not know what the truth is.*

Confabulation shows up without fail in nearly every context in which hypnosis is employed, including the forensic area. Thus it is not surprising that most states have no legal precedents on the use of hypnotic testimony. Furthermore, many state courts have begun to limit testimony from hypnotized witnesses or to follow the guidelines laid down by the American Medical Association in 1985 to assure that witnesses' memories are not contaminated by the hypnosis itself. For not only do we translate beliefs into memories when we are wide awake, but in the case of hypnotized witnesses with few specific memories the hypnotist may unwittingly suggest memories and create a witness with a number of crucial and vivid recollections of events that never happened, i.e., pseudo-memories. It may turn out that the recent Supreme Court decision allowing the individual states limited use of hypnotically aided testimony may not be in the best interests of those who seek the truth. Even in their decision the judges recognized that hypnosis may often produce incorrect recollections and unreliable testimony.

There have also been a number of clinical and experimental demonstrations of the creation of pseudo-memories that have subsequently come to be believed as veridical. Hilgard (1981) implanted a false memory of an experience connected with a bank robbery that never occurred. His subject found the experience so vivid that he was able to select from a series of photographs a picture of the man he thought had committed the robbery. At another time, Hilgard deliberately assigned two concurrent—though spatially different—life experiences to the same person and regressed him at separate times to *that date*. The individual subsequently gave very accurate accounts of both experiences, so that anyone believing in reincarnation who reviewed the two accounts would conclude the man *really had* lived the two assigned lives.

In a number of other experiments designed to measure eyewitness reliability, Loftus (1979) found that details supplied by others invariably contaminated the memory of the eyewitness. People's hair changed color, stop signs became yield signs, yellow convertibles turned to red sedans, the left side of the street became the right-hand side, and so on. The results of these studies led her to conclude, "It may well be that the legal notion of an independent recollection is a psychological impossibility." As for hypnosis, she says: "There's no way even the most sophisticated hypnotist can tell the difference between a memory that is real and one that's created. If a person is hypnotized and highly suggestible and false information is implanted in his mind, it may get embedded even more strongly. One psychologist tried to use a polygraph to distinguish between real and phony memory, but it didn't work. Once someone has constructed a memory, he comes to believe it himself."

Cueing: Inadvertent and Advertent

Without a doubt, inadvertent cueing also plays a major role in UFO-abduction fantasies. The hypnotist unintentionally gives away to the person being regressed exactly what response is wanted. This was most clearly shown in an experimental study of hypnotic age regression by R. M. True in 1949. He found that 92 percent of his subjects, regressed to the day of their tenth birthday, could accurately recall the day of the week on which it fell. He also found the same thing for 84 percent of his subjects for their fourth birthday. Other investigators, however, were unable to duplicate True's findings. When True was questioned by Martin Orne about his experiment, he discovered that the editors of *Science,* where his report had appeared, altered his procedure section without his prior consent. True, Orne discovered, had inadvertently cued his subjects by following the unusual technique of asking them, "Is it Monday? Is it Tuesday? Is it Wednesday?" etc., and he monitored their responses by using a perpetual desk calendar in full view of all his subjects. Further evidence of the prevalence and importance of such cueing came from a study by O'Connell, Shor, and Orne (1970). They found that in an existing group of four-year-olds not a single one knew what day of the week it was. The reincarnation literature is also replete with examples of such inadvertent cueing. Ian Wilson (1981), for example, has shown that hypnotically elicited reports of being reincarnated vary as a direct function of the hypnotist's belief about reincarnation. Finally, Laurence, Nadon, Nogrady, and Perry (1986) have shown that pseudo-memories were elicited also by inadvertent cueing in the use of hypnosis by the police.

As for advertent, or *deliberate,* cueing, one of my own studies offers a clear example. Sixty undergraduates divided into three groups of twenty each were hypnotized and age-regressed to previous lifetimes. Before each hypnosis session, however, suggestions very favorable to and supportive of past-life and reincarnation beliefs were given to one group; neutral and noncommittal statements about past lives were given to the second group; and skeptical and derogatory statements about past lives were given to the third group. The results clearly showed the effects of these cues and suggestions. Subjects in the first group showed the most past-life regressions and the most past-life productions; subjects in the third group showed the least (Baker 1982).

Regression subjects take cues as to how they are to respond from the person doing the regressions and asking the questions. If the hypnotist is a believer in UFO abductions the odds are heavily in favor of him eliciting UFO-abductee stories from his volunteers.

Fantasy-Prone Personalities and Psychological Needs

"Assuming that all you have said thus far *is* true," the skeptical observer might ask, "why would hundreds of ordinary, mild-mannered, unassuming citizens suddenly go off the deep end and turn up with cases of amnesia and then, when under hypnosis, all report nearly identical experiences?" First, the abductees are not as numerous as we are led to believe; and, second, even though Strieber and Hopkins go to great lengths to emphasize the diversity of the people who report these events, they are much more alike than these taxonomists declare. In an afterword to Hopkins's *Missing Time*, a psychologist named Aphrodite Clamar raises exactly this question and then adds, "All of these people seem quite ordinary in the psychological sense—*although they have not been subjected to the kind of psychological testing that might provide a deeper understanding of their personalities*" (italics added). And herein lies the problem. If these abductees were given this sort of intensive diagnostic testing it is highly likely that many similarities would emerge—particularly an unusual personality pattern that Wilson and Barber (1983) have categorized as "fantasy-prone." In an important but much neglected article, they report in some detail their discovery of a group of excellent hypnotic subjects with unusual fantasy abilities. In their words:

> Although this study provided a broader understanding of the kind of life experiences that may underlie the ability to be an excellent hypnotic subject, it has also led to a serendipitous finding that has wide implication for all of psychology—it has shown that there exists a small group of individuals (possibly 4% of the population) who fantasize a large part of the time, who typically "see," "hear," "smell," and "touch" and fully experience what they fantasize; and who can be labeled *fantasy-prone personalities*.

Wilson and Barber also stress that such individuals experience a reduction in orientation to time, place, and person that is characteristic of hypnosis or trance during their daily lives whenever they are deeply involved in a fantasy. They also have experiences during their daily ongoing lives that resemble the classical hypnotic phenomena. In other words, the behavior we would normally call "hypnotic" is exhibited by these fantasy-prone types (FPs) all the time. In Wilson and Barber's words: "When we give them 'hypnotic suggestions,' such as suggestions for visual and auditory hallucinations, negative hallucinations, age regression, limb rigidity, anesthesia, and sensory hallucinations, we are asking them to do for us the kind of thing they can do independently of us in their daily lives."

The reason we do not run into these types more often is that they have learned long ago to be highly secretive and private about their fantasy lives. Whenever the FPs do encounter a hypnosis situation it provides them with a social situation in which they are encouraged to do, and are rewarded for doing, what they usually do only in secrecy and in private. Wilson and Barber also emphasize that regression and the reliving of previous experiences is something that virtually all the FPs do naturally in their daily lives. When they recall the past, they relive it to a surprisingly vivid extent, and they all have vivid memories of their experiences extending back to their early years.

Fantasy-prone individuals also show up as mediums, psychics, and religious visionaries. They are also the ones who have many realistic "out of body" experiences and prototypic "near-death" experiences.

In spite of the fact that many such extreme types show FP characteristics, the overwhelming majority of FPs fall within the broad range of normal functioning. It is totally inappropriate to apply a psychiatric diagnosis to them. In Wilson and Barber's words: "It needs to be strongly emphasized that our subjects with a propensity for hallucinations are as well adjusted as our comparison group or the average person. It appears that the life experiences and skill developments that underlie the ability of hallucinatory fantasy are more or less independent of the kinds of life experience that leads to pathology." In general, FPs are "normal" people who function as well as others and who are as well adjusted, competent, and satisified or dissatisfied as everyone else.

Anyone familiar with the the fantasy-prone personality who reads *Communion* will suffer an immediate shock of recognition. Strieber is a classic example of the genre: he is easily hypnotized; he is amnesiac; he has vivid memories of his early life, body immobility and rigidity, a very religious background, a very active fantasy life; he is a writer of occult and highly imaginative novels; he has unusually strong sensory experiences—particularly smells and sounds—and vivid dreams. More interesting still is the comment made by Strieber's wife during her questioning under hypnosis by Budd Hopkins (p. 197). In referring to some of Strieber's visions she says: "Whitley saw a lot of things that I didn't see at that time." "Did you look for it?" "Oh, no. Because I knew it wasn't real." "How did you know it wasn't real? Whitley's a fairly down-to-earth guy—" "No, he isn't." . . . "It didn't surprise you hearing Whitley, that he sees things like that [a bright crystal in the sky]?" "No." It seems if anyone really knows us well it's our wives. But even more remarkable are the correspondences between Strieber's alien encounters and the typical hypnopompic hallucinations to be discussed later.

It is perfectly clear, therefore, why most of the UFO abductees, when given cursory examinations by psychiatrists and psychologists, would turn out to be ordinary, normal citizens as sane as themselves. It is also evident why the elaborate fantasies woven in fine cloth from the now universally familiar UFO-abduction fable—a fable known to every man, woman, and child newspaper reader or moviegoer in the nation—would have so much in common, so much consistency in the telling. Any one of us, if asked to pretend that he had been kidnapped by aliens from outer space or another dimension, would make up a story that would vary little, either in its details or in the supposed motives of the abductors, from the stories told by any and all of the kidnap victims reported by Hopkins. As for the close encounters of the third kind and conversations with the little gray aliens described in *Communion* and *Intruders,* again, our imaginative tales would be remarkably similar in plot, dialogue, description, and characterization. The means of transportation would be saucer-shaped; the aliens would be small, humanoid, two-eyed, and gray, white, or green. The purpose of their visits would be: (1) to save our planet; (2) to find a better home for themselves; (3) to end nuclear war and the threat we pose to the peaceful life in the rest of the galaxy; (4) to bring us knowledge and enlightenment; and (5) to increase their knowledge and understanding of other forms of intelligent life. In fact, the fantasy-prone abductees' stories would be much more credible if some of them, at least, reported the aliens as eight-foot-tall, red-striped octapeds riding bicycles and intent upon eating us for dessert.

Finally, what would or could motivate even the FPs to concoct such outlandish and absurd tales, tales that without fail draw much unwelcome attention and notoriety? What sort of psychological motives and needs would underlie such fabrications? Perhaps the best answer to this question is the one provided by the author-photographer Douglas Curran. Traveling from British Columbia down the West Coast and circumscribing the United States along a counterclockwise route, Curran spent more than two years questioning ordinary people about outer space. Curran writes:

> On my travels across the continent I never had to wait too long for someone to tell me about his or her UFO experience, whether I was chatting with a farmer in Kansas, Ruth Norman at the Unarius Foundation, or a cafe owner in Florida. What continually struck me in talking with these people was how positive and ultimately life-giving a force was their belief in outer space. Their belief reaffirmed the essential fact of human existence: the need for order and hope. It is this that establishes them—and me—in the continuity of human experience. It brought to me a greater understanding of Oscar Wilde's observation. "We are all lying in the gutter—but some of us are looking at the stars."

Jung (1969), in his study of flying saucers, first published in 1957, argues that the saucer represents an archetype of order, wholeness, deliverance, and salvation—a symbol manifested in other cultures as a sun wheel or magic circle. Further in his essay, Jung compares the spacemen aboard the flying saucers to the angelic messengers of earlier times who brought messages of hope and salvation—the theme emphasized in Strieber's *Communion*. Curran also observes that the spiritual message conveyed by the aliens is, recognizably, our own. None of the aliens Curran's contactees talked about advocated any moral or metaphysical belief that was not firmly rooted in the Judeo-Christian tradition. As Curran says, "Every single flying-saucer group I encountered in my travels incorporated Jesus Christ into the hierarchy of its belief system." No wonder Eduard Meier had to travel back in time and visit the Savior. Many theorists have long recognized that whenever world events prove to be psychologically destabilizing, men turn to religion as their only hope. Jung, again, in his 1957 essay, wrote: "In the threatening situation of the world today, when people are beginning to see that everything is at stake, the projection-creating fantasy soars beyond the realm of earthly organization and powers into the heavens, into interstellar space, where the rulers of human

The Power of Suggestion on Memory

In my own work on hypnosis and memory, the power of suggestion on the evocation of false memories was clearly and dramatically evident. Sixty volunteers observed a complex visual display made up of photographs of a number of common objects, e.g. a television set, a clock, a typewriter, a book, and so on, and eight nonsense syllables. They were instructed to memorize the nonsense syllables in the center of the display and were given two minutes to accomplish it. Nothing was said about the common objects. Following a 40-minute delay the students were questioned about the nonsense syllables and the other objects on display. They were also asked to state their confidence in the accuracy of their answers. Some were questioned under hypnosis and others while they were wide awake.

As a secondary part of the study the extent of the student's suggestibility was also studied. This was done by asking them to report on the common objects (as well as their primary task of memorizing the nonsense syllables) and asking specific questions about objects that *were not on the display*. Since their attention was not directed at the objects *specifically*, they were of course unsure about what they saw and didn't see. Therefore, when they were asked the questions "What color was the sports car?" and "Where on the display was it located?" they immediately assumed there must have been a sports car present or I wouldn't be asking the question. Similarly with a suggested lawnmower and calendar. Although 35 subjects reported the color of the suggested automobile in the hypnoidal condition, 34 reported the color while awake. Similarly, although 26 subjects reported the suggested lawnmower's color and position in the hypnoidal state, 27 reported its color and position while awake. For the nonexistent calendar, 24 reported the month and date while hypnotized, and 23 did so while awake.

As for suggestibility *per se* under all conditions, 50 out of 60 volunteers reported seeing something that wasn't there with a confidence level of 2 (a little unsure) or greater, while 45 out of 60 reported seeing something that wasn't there with a confidence level of 3 (sure) or greater, whereas 25 out of 60 reported seeing something that was not there with a confidence level of 4 (very sure) or greater. Finally, 8 out of the 60 reported something not there with a confidence level of 5 (absolute certainty). Interestingly enough, 5 of the 8 reported they were certain of the object's existence even though they were wide awake; and, when they were allowed to see the display again, they were shocked to discover their error (Baker, Haynes, and Patrick 1983).—*R.A.B.*

fate, the gods, once had their abode in the planets."

The beauty and power of Curran's portraits of hundreds of true UFO believers lies in his sympathetic understanding of their fears and frailties. As psychologists are well aware, our religions are not so much systems of objective truths about the universe as they are collections of subjective statements about humanity's hopes and fears. The true believers interviewed by Curran are all around us. Over the years I have encountered several. One particularly memorable and poignant case was that of a federal prisoner who said he could leave his body at will and sincerely believed it. Every weekend he would go home to visit his family while (physically) his body stayed behind in his cell. Then there was the female psychic from the planet Xenon who could turn electric lights on and off at will, especially traffic signals. Proof of her powers? If she drove up to a red light she would concentrate on it intently for 30 to 40 seconds and then, invariably, it would turn green!

Hypnogogic and Hypnopompic Hallucinations

Another common yet little-publicized and rarely discussed phenomenon is that of hypnogogic (when *falling asleep*) and hypnopompic (when *waking up*) hallucinations. These phenomena, often referred to as "waking dreams," find the individual suddenly awake, but paralyzed, unable to move, and most often encountering a "ghost." The typical report goes somewhat as follows, "I went to bed and went to sleep and then sometime near morning something woke me up. I opened my eyes and found myself wide awake but unable to move. There, standing at the foot of my bed was my mother, wearing her favorite dress—the one we buried her in. She stood there looking at me and smiling and then she said: 'Don't worry about me, Doris, I'm at peace at last. I just want you and the children to be happy.' " Well, what happened next? "Nothing, she slowly faded away." What did you do then? "Nothing, I just closed my eyes and went back to sleep."

There are always a number of characteristic clues that indicate a hypnogogic or hypnopompic hallucination. First, it always occurs before or after falling asleep. Second, one is paralyzed or has difficulty in moving; or, contrarily, one may float out of one's body and have an out-of-body experience. Third, the hallucination is unusually bizarre; i.e., one sees ghosts, aliens, monsters, and such. Fourth, after the hallucination is over the hallucinator typically goes back to sleep. And, fifth, the hallucinator is unalterably convinced of the "reality" of the entire experience.

In Strieber's *Communion* (pp. 172-175) is a classic, textbook description of a hypnopompic hallucination, complete with the awakening from a sound sleep, the strong sense of reality and of being awake, the paralysis (due to the fact that the body's neural circuits keep our muscles relaxed and help preserve our sleep), and the encounter with strange beings. Following the encounter, instead of jumping out of bed and going in search of the strangers he has seen, Strieber typically goes back to sleep. He even reports that the burglar alarm was still working—proof again that the intruders were mental rather than physical. Strieber also reports an occasion when he awakes and believes that the roof of his house is on fire and that the aliens are threatening his family. Yet his only response to this was to go peacefully back to sleep. Again, clear evidence of a hypnopompic dream. Strieber, of course, is convinced of the reality of these experiences. This too is expected. If he was not convinced of their reality, then the experience would not be hypnopompic or hallucinatory.

The point cannot be more strongly made that ordinary, perfectly sane and rational people have these hallucinatory experiences and that such individuals are in no way mentally disturbed or psychotic. But neither are such experiences to be taken as incontrovertible proof of some sort of objective or consensual reality. They may be subjectively real, but objectively they are nothing more than dreams or delusions. They are called "hallucinatory" because of their heightened subjective reality. Leaving no rational explanation unspurned, Strieber is nevertheless forthright enough to suggest at one point the possibility that his experiences indeed could be hypnopompic. Moreover, in a summary chapter he speculates, correctly, that the alien visitors could be "from within us" and/or "a side effect of a natural phenomenon . . . a certain hallucinatory wire in the mind causing many different people to have experiences so similar as to seem to be the result of encounters with the same physical phenomena" (p. 224).

Interestingly enough, these hypnopompic and hypnogogic hallucinations do show individual differences in content and character as well as a lot of similarity: ghosts, monsters, fairies, friends, lovers, neighbors, and even little gray men and golden-haired ladies from the Pleiades are frequently encountered. Do such hallucinations appear more frequently to highly imaginative and fantasy-prone people than to other personality types? There is some evidence that they do (McKellar 1957; Tart 1969; Reed 1972; Wilson and Barber 1983), and there can certainly be no doubt that Strieber is a highly imaginative personality type.

"Missing" Time?

As for the lacunae or so-called "missing time" experienced by all the UFO abductees, this too is a quite ordinary, common, and universal experience. Jerome Singer (1975) in his *Inner World of Daydreaming* comments:

> Are there ever any truly "blank periods" when we are awake? It certainly seems to be the case that under certain conditions of fatigue or great drowsiness or extreme concentration upon some physical act we may become aware that we cannot account for an interval of time and have no memory of what happened for seconds and sometimes minutes.

Graham Reed (1972) has also dealt with the "time-gap" experience at great length. Typically, motorists will report after a long drive that at some point in the journey they wake up to realize they have no awareness of a preceding period of time. With some justification, people still will describe this as a "gap in time," a "lost half-hour," or a "piece out of my life." Reed writes:

> A little reflection will suggest, however, that our experience of time and its passage is determined by *events*, either external or internal. What the time-gapper is reporting is not that a slice of time has vanished, but that he has failed to register a series of events which would normally have functioned as his time-markers. If he is questioned closely he will admit that his "time-gap" experience did not involve his realization at, say, noon that he had somehow "lost" half an hour. Rather, the experience consists of "waking up" at, say, Florence and realizing that he remembers nothing since Bologna. . . . To understand the experience, however, it is best considered in terms of the absence of *events*. If the time-gapper had taken that particular day off, and spent the morning sitting in his garden undisturbed, he might have remembered just as little of the half-hour in question. He might still describe it in terms of lost time, but he would not find the experience unusual or disturbing. For he would point out that he could not remember what took place between eleven-thirty and twelve simply because nothing of note occurred.

In fact, there is nothing recounted in any of the three works under discussion that cannot be easily explained in terms of normal, though somewhat unusual, psychological behavior we now term *anomalous*. Different and unusual? Yes. Paranormal or otherworldly, requiring the presence of extraterrestrials? No. Diehard proponents may find these explanations unsatisfying, but the open-minded reader will find elaboration and illumination in the textbooks and other works in anomalistic psychology. Strongly recommended are Reed (1972), Marks and Kammann (1980), Corliss (1982), Zusne and Jones (1982), Radner and Radner (1982), Randi (1982), Gardner (1981), Alcock, (1981), Taylor (1980), and Frazier (1981).

If one looks at the psychodynamics underlying the confabulation of Hopkins's contactees and abductees it is easy to see how even an ordinary, non-FP individual can become one of his case histories. How does Hopkins, for instance, locate such individuals in the first place? Typically, it is done through a selection process; i.e., those individuals who are willing to talk about UFOs—the believers—are selected for further questioning. Those who scoff are summarily dismissed. Once selected for study and permission to volunteer for hypnosis is obtained, a response-anticipation process sets in (Kirsch 1985), and the volunteer is now set up to supply answers to anything that might be asked. Then, during the hypnosis sessions, something similar to the Hawthorne Effect occurs: The volunteer says to himself, "This kindly and famous writer and this important and prestigious doctor are interested in poor little old unimportant me!" And the more the volunteer is observed and interrogated, the greater is the volunteer's motivation to come up with a cracking "good story" that is important and significant and pleasing to these important people. Moreover, as we have long known, it is the perception of reality not the reality itself that is truly significant in determining behavior. If the writer and the doctor-hypnotist are on hand to encourage the volunteer and to suggest to him that his fantasy really happened, who is he to question their interpretation of his experience? Once they tell the contactee how important his fantasy is, he now—if he ever doubted before—begins to believe it himself and to elaborate and embellish it every time it is repeated.

Consequences and Summary

Many readers might feel compelled to ask: "Well, what is so bad about people having fantasies anyway? What harm do they do? You certainly cannot deny they are entertaining. And, as far as the psychiatrists' clients are concerned, whether the fantasies are true or false is of little matter—it's the clients' perceptions of reality that matter and it is this that you have to treat." True, if the client believes it is so, then you have to deal with that belief. The only problem with this lies in its potential for harm. On the national scene today too many lives have been negatively affected and even ruined by well-meaning but tragically misdirected reformers who believe the fantasies of children, the alienated, and the fantasy-prone personality types and have charged innocent people with rape, child molestation, assault, and other sorts of abusive crimes. Nearly every experienced clinician has encountered such claims and then much later has discovered to his chagrin that none of these fantasized events ever happened. Law-enforcement officials are also quite familiar with the products of response expectancies and overactive imaginations in the form of FPs who confess to murders that never happened or to murders that did happen but with which they have no connection. Another problem with the UFO abductee literature is that it is false, misleading, rabble-rousing, sensationalistic, and opportunistically money-grubbing. It takes advantage of people's hopes and fears and diverts them from the literature of science. Our journeys to the stars will be made on spaceships created by determined, hardworking scientists and engineers applying the principles of science, not aboard flying saucers piloted by little gray aliens from some other dimension.

Need we be concerned about an invasion of little gray kidnappers? Amused, yes. Concerned, no.

Should we take Strieber, Hopkins, Kinder, et al. seriously? Not really. They are a long, long way from furnishing reliable and replicable data and their rather shaky hypotheses are miles from anything resembling proof.

Should we insist that such semi-hysterical and poorly informed journalistic efforts not be published? Only if we all are a bunch of wet blankets and party-poopers. After all, it has been dull lately and these pseudoscientific thrillers have added a welcome note of excitement. And without these works there would be no puzzles to solve. As the old disclaimer says, "It's fun to be fooled, but it's more fun to know!"

Is the human mind a weird and wonderful place and human behavior a billion-ring circus of astounding events? Unquestionably, yes!

One cannot help but be struck by the thought that, in their way, the UFOnaut creations are of some redeeming value. They, besides their value as entertainment, do provide the useful—albeit unintended—service of directing our attention to the extremities of human belief and the perplexing and perennial problem we have in detecting deception. In spite of all our vaunted scientific accomplishments, we have today no absolutely certain, accurate, or reliable means for getting at the truth—for simply determining whether or not someone is lying. Not only are the polygraph and the voice-stress analyzer notoriously unreliable and inaccurate; but the professional interrogators, body-language experts, and psychological testers are also the first to admit their lack of predictive skill. If these abductee claims do no more than stimulate greater efforts toward the development of better "truth detectors," then they will have made an important contribution.

When one man has a private conversation with an angel in the corner, we consider it hallucinatory; when twenty people simultaneously see and talk with this angel, we then have good reason to suspect it may not be hallucinatory. When one man *never* sees an angel in the corner until and unless he is hypnotized and regressed, even then such reports are not considered hallucinatory. They are merely confabulations. Nor do we classify him as psychologically disturbed or even as lying. He most likely is as normal and mentally healthy as any one of us. If he has been properly primed with powerful suggestions, he may sincerely believe in the truth of his confabulations.

When all things are considered, we shouldn't be too upset with the creators of and believers in what Martin Gardner (1987) calls "the new science-fiction religion." Tolerance *is* the mark of a civilized mind. We can nevertheless demand that the bookstores and supermarkets classify all such material properly. All UFO, UFO-abductee, past-life, and hypnotic-regression accounts should be taken from the nonfiction counters and moved to the science-fiction shelves.

Notes

1. People familiar with the unreliability of the polygraph will not be impressed. As for Strieber's sanity, there can be no doubt of this. As *Omni* magazine reported, he received a million-dollar advance from his publisher.

2. The dictionary defines an alien as "one who is strange, wholly different in nature, incongruous. . . ."

References

Alcock, James E. 1981. *Parapsychology: Science or Magic?* New York: Pergamon.

AMA Council on Scientific Affairs. 1985. Scientific status of refreshing recollection by use of hypnosis. *Journal of the AMA,* 253 (13), April 5.

Baker, Robert A. 1982. The effect of suggestion on past-lives regression. *American Journal of Clinical Hypnosis,* 25(1):71-76.

Baker, Robert A., B. Haynes, and B. Patrick. Hypnosis, memory, and incidental memory. *American Journal of Clinical Hypnosis,* 25(4):253-262.

Barber, Theodore X. 1969. *Hypnosis: A Scientific Approach.* New York: D. Van Nostrand.

Corliss, William R. 1982. *The Unfathomed Mind: A Handbook of Unusual Mental Phenomena.* New York: Sourcebook.

————. 1983. *Handbook of Unusual Natural Phenomena.* New York: Arlington House.

Curran, Douglas. 1985. *In Advance of the Landing: Folk Concepts of Outer Space.* New York: Abbeville Press.

Frazier, Kendrick, ed. 1981. *Paranormal Borderlands of Science.* Buffalo, N.Y.: Prometheus Books.

Gardner, Martin. 1981. *Science: Good, Bad and Bogus.* Buffalo, N.Y.: Prometheus Books.

————. 1987 Science-fantasy religious cults. *Free Inquiry,* 7(3):31-35, Summer.

Gill, M. M., and M. Brenman. 1959. *Hypnosis and Related States.* New York: International Universities Press.

Hilgard, Ernest R. 1977. *Divided Consciousness: Multiple Controls in Human Thought and Action.* New York: Wiley.

————. 1981. Hypnosis gives rise to fantasy and is not a truth serum. *Skeptical Inquirer,* 5(3), Spring.

Hilgard, Josephine R. 1979. *Personality and Hypnosis: A Study of Imaginative Involvement,* 2nd ed. Chicago, Ill.: University of Chicago Press.

Jung, Carl. 1969. *Flying Saucers: A Modern Myth of Things Seen in the Sky.* Signet Books.

Kirsch, Irving. 1985. Response expectancy as a determinant of experience and behavior. *American Psychologist,* 40(11):1189-1202.

Klass, Philip J. 1981. Hypnosis and UFO abductions. *Skeptical Inquirer,* 5(3), Spring.

Laurence, Jean-Roch, Robert Nadon, Heather Nogrady, and Campbell Perry. 1986. Duality, dissociation, and memory creation in highly hypnotizable subjects. *International Journal of Clinical and Experimental Hypnosis,* 34(4):295-310.

Loftus, Elizabeth. 1979. *Eyewitness Testimony.* Cambridge, Mass.: Harvard University Press.

Marks, David, and Richard Kammann. 1980. *The Psychology of the Psychic.* Buffalo, N.Y.: Prometheus Books.

McKellar, Peter. 1957. *Imagination and Thinking.* London: Cohen & West.

O'Connell, D. N., R. E. Shor, and M. T. Orne. 1970. Hypnotic age regression: An empirical and methodological analysis. *Journal of Abnormal Psychology Monograph,* 76(3), Part 2:1-32.

Perry, Campbell, Jean-Roch Laurence, Robert Nadon, and Louise Labelle. 1986. Past lives regression. In *Hypnosis: Questions and Answers,* ed. by Bernie Zilbergeld, M. G. Edelstein, and D. L. Araoz. New York: Norton.

Radner, Daisie, and Michael Radner. 1982. *Science and Unreason.* Belmont, Calif.: Wadsworth.

Randi, James. 1982. *Flim-Flam!* Buffalo, N.Y.: Prometheus Books.

Reed, Graham. 1972. *The Psychology of Anomalous Experience.* Boston: Houghton Mifflin. (Revised paperback edition published by Prometheus Books, Buffalo, N.Y., 1988.)

Sarbin, T. R., and W. C. Coe. 1972. *Hypnosis: A Social Psychological Analysis of Influence Communication.* New York: Holt, Rinehart & Winston.

Sarbin, T. R., and M. L. Andersen. 1967. Role-theoretical analysis of hypnotic behavior. In *Handbook of Clinical and Experimental Hypnosis,* ed. by Jesse E. Gordon. New York: Macmillan.

Sheaffer, Robert. 1981. *The UFO Verdict.* Buffalo, N.Y.: Prometheus Books.

Singer, Jerome. 1975. *The Inner World of Daydreaming.* New York: Harper & Row.

Spanos, N. P., and T. X. Barber. 1974. Toward a convergence in hypnotic research. *American Psychologist,* 29(3):500-511.

Sutcliffe, J. P. 1961. "Credulous" and "skeptical" views of hypnotic phenomena: Experiments on esthesia, hallucinations, and delusion. *Journal of Abnormal and Social Psychology,* 62(2):189-200.

Tart, Charles, ed. 1969. *Altered States of Consciousness: A Book of Readings.* New York: Wiley.

Taylor, John. 1980. *Science and the Supernatural: An Investigation of Paranormal Phenomena.* New York: Dutton.

True, R. M. 1949. Experimental control in hypnotic age regression states. *Science,* 110, pp. 583-584.

Wilson, Ian. 1981. *Mind Out of Time.* London: Gollancz.

Wilson, Sheryl C., and T. X. Barber. 1983. The fantasy-prone personality: Implications for understanding imagery, hypnosis, and parapsychological phenomena. In *Imagery: Current Theory, Research and Application,* ed. by A. A. Sheikh. New York: Wiley.

Zusne, Leonard, and Warren H. Jones. 1982. *Anomalistic Psychology: A Study of Extraordinary Phenomena of Behavior and Experience.* Hillsdale, N.Y.: Erlbaum.

BILL ELLIS

The Varieties of Alien Experience

Scholars who study the folklore of past and present cultures may have some useful perspectives to offer about the many recent reports of abduction by aliens. At the conclusion of his review of Whitley Strieber's *Communion,* Ernest H. Taves (1987) suggests that either Strieber is mentally ill or he is consciously perpetrating a hoax, "playing a joke on his readers." Taves then invokes Occam's Razor to opt for the latter and tacitly pare away any other alternative. (Taves may be correct, but he is not logical. Occam's Razor is a method for analyzing alternatives and choosing the simplest explanation that will explain the evidence. It cannot be used to eliminate alternatives without analysis.) Robert A. Baker (1987 and in this volume) suggests a far simpler explanation: that Strieber is a fantasy-prone personality sincerely describing what he believes he remembers. From the perspective of psychology, Baker suggests several recognized and well-understood mechanisms for such "memories," including hypnogogic hallucination, confabulation, and inadvertent cueing by the hypnotist (the "Clever Hans" phenomenon in yet another form).

Folklorists familiar with accounts of supranormal experiences can, however, suggest two additional mechanisms at work: one may explain Strieber's experience the other certainly accounts for his actions since his "abduction."

1. On October 4 and December 26, 1985, Strieber may have actually experienced an event, common to many other cultures and individuals, in which he felt paralyzed and then believed he was levitated and subjected to indignities by nonhuman agents.

2. Whether Strieber experienced this event or not, he did undergo, during the period from January to March 1986, an experience identical to that of religious conversion.

Collections of legends and folktales, both European and otherwise, contain a variety of "real life" accounts that contain close parallels to elements of modern abduction stories. Most of these are anonymous, migratory tales that have no weight as evidence. Some, however, contain alleged firsthand experiences.

Strieber himself notes these parallels and cites them as support for the "reality" of his abduction. The more carefully recorded cases, however, make it clear that these earlier abductions, like Strieber's, were subjective in nature.

Anne Jeffries (ca. 1626-1698), an illiterate country girl from Cornwall, was one such celebrated abductee. In 1645 she apparently suffered a convulsion and was found, semi-conscious, lying on the floor. As she recovered, she began to recall in detail how she was accosted by a group of six little men. Paralyzed, she felt them swarm over her, kissing her, until she felt a sharp pricking sensation. Blinded, she found herself flying through the air to a palace filled with people. There, one of the men (now her size) seduced her, and suddenly an angry crowd burst in on them and she was again blinded and levitated. She then found herself lying on the floor surrounded by her friends.

Significantly, the accounts note that the experience left her ill for some time, and only after she regained her health did she "recall" this experience. Still, like Strieber, Jeffries claimed that this encounter was followed by further contacts with the "fairies," and she was taken seriously enough by the local authorities in 1646 to be arrested for witchcraft and imprisoned (Briggs 1971, pp. 176-177; Briggs 1976, pp. 239-242).

A more recent incident, with some connections to Strieber's alleged experience, was reported by theologian Henry James, Sr., in May 1844. While relaxing one afternoon in his chair, James suddenly felt the presence of some invisible, ineffably evil being squatting in the room with him. Rationally, he recognized that his emotion "was a perfect insane and abject terror, without ostensible cause"; still he found himself completely paralyzed while (as in Strieber's October 4 experience) his mind was flooded with images of "doubt, anxiety, and despair" (Edel 1953, p. 30). The senior James eventually found release in the fringe religion of Swedenborgianism, while his sons dealt with the impact of this experience in their own ways. William James provided one of the first rational anatomies of paranormal encounters in *The Varieties of Religious Experience;* Henry James, Jr., dealt with the lingering threat of such events in his fiction, ranging from *The Turn of the Screw* to "The Jolly Corner," both of which contain suggestive parallels to *Communion*.

Some light has been thrown on such experiences by folklorist David Hufford. In Newfoundland, he found, the term "Old Hag" referred to a fairly common phenomenon in which a person who is (as Strieber was on October 4 and December 26) relaxed but apparently awake suddenly finds himself paralyzed and in the presence of some nonhuman entity. Often the sensation is accompanied by terrifying hallucinations—of shuffling sounds, of humanoid figures with prominent eyes, even (rarely) of strange, musty smells. Often the figure even sits on the victim's chest, causing a choking sensation.

Like Baker, Hufford at first assumed that the consistencies present in victims' accounts of the Old Hag could be explained by previous exposure to oral traditions—the "cultural source hypothesis." It is interesting, though, that when he moved his base of research to the United States, Hufford (1982, p. 245) found the experience just as common here as in Newfoundland—

affecting perhaps more than 15 percent of the population. Despite the absence of a folk tradition naming and explaining it, many of the specific details of the Old Hag hallucination recurred in victims' experiences, leaving them profoundly confused and reluctant to talk for fear of ridicule. Hence the phenomenon remains largely unstudied by psychologists and practically unknown (as a general phenomenon) to the general public.

Surveying the psychological and psychiatric literature relating to the experience, Hufford found no evidence that the Old Hag was linked to neurological or psychotic illnesses. During this event, evidently, the brain functions as if asleep, producing the characteristic paralysis and apnea; hence it is similar to the hypnogogic hallucination (Baker 1987-88, and in this volume). The peculiar stability of the hallucinations' content across cultural boundaries and in the absence of traditions concerning it, however, remains unexplained. Hufford suggests that the most likely explanation is that the Old Hag might be the side-effect of a documented but poorly understood derangement of the sleep pattern, akin to narcolepsy.

Strieber's October 4 visitation, along with the similar experiences of paralysis and physical manipulation reported by many abductees, might be explained by some form of abnormal sleep pattern, producing a distinctive set of hallucinatory events. Future studies need to focus carefully on the phenomenology of such events, which may reveal genuine correspondences among "abduction" events. In this regard, the main value of Strieber's book to folklorists is that much of it was committed to writing soon after the experiences themselves. The interpretations accepted after the fact by the victim (or imposed on him by others) are of less value, as they tend to force the details of the experience into a culturally acceptable mold.

On the other hand, it is not necessary to assume that a neurological experience could provoke detailed memories of abduction. Strieber may have confabulated either or both experiences; we cannot tell for sure. Nevertheless, there is no question that early in 1986 Strieber underwent a quasi-religious conversion, assisted (probably innocently) by Budd Hopkins and his analyst, Robert Klein.

The classic process of indoctrination is described by William Sargant (1961): When the human nervous system is stimulated beyond its normal capacity ("transmarginally") for long periods of time—either deliberately by agents wanting to indoctrinate a person or unintentionally by the person undergoing a lengthy period of psychological stress—it eventually begins to operate in paradoxical ways. Typically, the individual begins to over-respond to weak stimuli; ultimately, his or her previous thought and behavior patterns begin to change and a state of hysteria results, during which the individual is highly susceptible to new concepts and philosophies. It is unrealistic, Sargant warns, to expect a person to resist the process of conversion, once it has begun. Even recognition that one is being indoctrinated, he notes, may not delay breakdown.

This process has been institutionalized in the religious rites of many cultures

(Turner 1969), and the pattern frequently occurs in the narratives of "born-again" Christians (Clements 1982). Strieber's account of events, evidently based on a journal kept before and during his hypnotic sessions, is structurally identical to such narratives. In January we find him in psychological disarray, alienated from his wife, unable to read or write, and suffering from a variety of physical symptoms. At this point, Strieber tells us, images began to float into his mind. In a state of extreme suggestibility, then, Strieber began reading and talking to friends about UFOs, a process climaxing with his discovery of an account of an abduction experience that contained some minor correspondences with the images he was "recalling."

This point of contact evidently led to a psychological crisis a few days later:

> . . . I was sitting at my desk when things just seemed to cave in on me. Wave after wave of sorrow passed over me. I looked at the window with hunger. I wanted to jump. I wanted to die. I just could not bear this memory, and I could not get rid of it. (Strieber 1987, p. 40)

At this precise moment, Strieber contacted Hopkins, who gave him assurances that his memories were indeed similar to those of others. Strieber wept in relief and "went from wanting to hide it all to wanting to understand it." Then Hopkins introduced the idea of looking for a previous encounter, and Strieber—for the first time—began to look at the October 4 events as possibly paranormal. Given this task, Strieber left this interview "a happy man" (Strieber 1987, p. 41).

Sargant's research leaves little doubt that Strieber was, when he contacted Hopkins, transmarginally excited. Loss of sleep combined with obsessive, uncontrollable thought patterns "overloaded" his brain and left him susceptible to the slightest idea that would give his anxieties a licit avenue. Further, the extreme significance put on small details—the slim correspondences that suddenly seem concrete proof of the visitations' reality—exhibits the paradoxical phase of this process. In Strieber's words, "There did seem to be a lot of confusion . . . and perhaps even an emotional response on my part greatly out of proportion to what seemed a minor disturbance" (Strieber 1987, p. 51).

It is not surprising that the hint provided by Hopkins led to the intense moment during Strieber's first hypnosis session in which he suddenly "remembers" the little man by his bed and responds with 20 seconds of prolonged screams. This reaction, common to many other hypnotized "abductees," represents the moment of abreaction, in which the convert's pent-up emotions are released in a controlled way through emotionally reliving the event that the indoctrinator (in this case, Hopkins) has suggested actually caused the anxieties.

This process has actually been used therapeutically since World War II to treat stubborn cases of battle shock and trauma. Significantly, Sargant reports, it was found that it was not necessary to make the patient recall real-life incidents. Rather, "it would often be enough to create in him a state of excitement analogous to that which had caused his neurotic condition and keep it up until he collapsed; he would then start to improve. Thus *imagination would have to be used in inventing artificial situations, or distorting actual events. . . ."* (p. 51; emphasis added).

Recognizing this pattern in *Communion* explains why Strieber acts less like a playful hoaxer than a quasi-religious convert. Indeed, judging from psychological tests, the conversion experience largely restored his mental health, dispelling his self-destructive tendencies and restoring his writing abilities. Further, Strieber was left with the status of a "chosen one" and a mission whose quasi-religious nature is explicit in the book's title.

From a folklorist's perspective, the two alternatives are not mutually exclusive; indeed, a confusing neurological attack may require a conversion experience to dispel the anxiety produced. Henry James, Sr., we note, took the first steps toward regaining his mental health when he learned from a certain Mrs. Chichester that the encounter he had had with evil was known among Swedenborgians as "vastation" (Edel 1953, p. 32). And, Hufford (1982, p. 161) reports, one surgeon unnerved by an Old Hag experience was literally reduced to tears when he found it described in psychological literature as "idiopathic SP." Strieber may, then, have fallen into the abductees' camp exactly for the reasons he describes: to find convenient cultural language for a psychological event that otherwise would have to be labeled "fraud" or "madness." If Strieber sincerely believes that he is not consciously fabricating his experience, and if he is not mentally ill, then the hard-line rationalist position, as stated by Taves, gives him no alternative but to proceed on the assumption that the aliens are real. We actually leave him no other psychologically sound option.

The pity is, though, that concepts like "vastation" or "Old Hag" derive from cultural systems with complex psychological checks and balances. To accept the Newfoundland conception of Old Hag, for instance, one must also accept the reality of witchcraft. But the tradition also comes prepared with countercharms—sleeping with a sharp knife, for instance—known to be effective against repeat attacks (Hufford 1982, pp. 3-4; Hyatt 1965, pp. 270, 273). Such a practice, like any fetish, would materially reduce the anxieties of the victim (though perhaps not those of his bedmate). The concept "alien abduction," by contrast, leaves the victim unprotected against future visits, which no open knives, strings of garlic, burglar alarms, or concentrated skeptical thought patterns can repel. So accepting the concept may immediately reduce anxieties, but at the cost of inviting recurrent attacks.

The progress of Strieber's "visitations" after his conversion shows him gaining some degree of psychological control over his visitors—making the visionary face move as he pleases or, in the last scene, actually inviting them to return so that he can show his lack of fear. But it is unclear whether his missionary role will communicate the same control to other troubled souls who may have experienced—or who may find relief in "remembering"—similar events. If the Triad Group that Strieber has formed to collect and analyze abduction accounts actually turns over to qualified professionals a corpus of similar experiences, some good may come from Strieber's missionary work. Competent psychologists may be able to examine the phenomenology of the events described and determine more exactly what mechanisms lie behind them. This in turn may suggest more specific and appropriate psychological treatment for the victims. Time will tell.

In the meantime, the rationalist community needs to be cautious not to commit itself too quickly to a presumption of fraud. Even Baker, as sympathetic as he is to abductees, still suggests that books based on their accounts should be labeled "science fiction," a move that has the effect of calling their stories conscious fictions. Perhaps rationalists (and bookstores) ought to abandon the simple dualism of classifying narratives into "fiction" and "nonfiction" and follow folklorists in their more complex scheme: "tale" (conscious fantasy), "history" (unquestionable fact), and—in the middle—"legend" (alleged but disputable fact). The need of Strieber and other abductees to hedge their accounts with proofs of their veracity is itself proof of the debatable status of their narratives, just as oral accounts of ghosts, manlike apes, and other anomalous phenomena are spiked with details, corroborations, and even disclaimers, to the point of losing the forward motion of the story (Bennett 1988).

Sargant (1961, p. 233) notes that a sense of humor is one of the surest blocks to conversion, and mirth is doubtless our first line of defense against works like *Communion* that convey a patently missionary message. But humor, like an oversharpened razor carelessly used, may turn on its user. We need to admit that sane, intelligent people may sincerely perceive, or come to believe, that they have been attacked or abducted by paranormal agents. In the case of persons who (like Strieber and, before him, the Hills) seem to be objectively disturbed by memories of abduction, the proper response is not amusement but concern—not over the risk of UFO invaders, but over the treatment of such victims.

We should insist that they receive appropriate professional evaluation and treatment. Otherwise such victims will continue, as we all must, to adjust to life at a high level of uncertainty. At present this means they will seek out those who will listen to their experiences without assuming they are either lying or mentally ill. Unfortunately this leads them to the UFOlogists, whose sympathy inevitably must be less for the suffering individuals than for the value their testimony may have for supporting the extraterrestrial hypothesis.

Whether this price is a fair one for maintaining our own fiction that "intelligent" people do not experience apparently paranormal events, I leave the skeptical community to decide.

Postscript

By the time this essay appeared in 1988, Whitley Strieber had already broken off relations with Budd Hopkins and most other abductee researchers. In a move that surprised and alienated many UFOlogists, he came to terms with Philip J. Klass. Klass agreed to acknowledge publicly that Strieber was not intentionally falsifying his experiences; Strieber warned fellow abductees to avoid untrained investigators like Hopkins and, instead, seek professional help from therapists who "are aware that many normal, well integrated people confront the visitors" (Strieber 1988). Both skeptics and believers have shifted from absolute claims of truth or falsehood to more specific, testable hypotheses about the nature of abductions, but their etiology remains for the moment unresolved. Though similar in many ways to earlier supernatural attack reports, modern experiences do show unique elements not present in the historical materials. Still, abductions are probably produced by a transient dislocation of normal sleep/dreaming cycles, as is the "Old Hag" experience.

My call for "appropriate professional evaluation and treatment," however, must be tempered with caution, in the light of the astonishing ease with which American and British therapists have assumed that multiple-personality syndrome is caused by satanic-ritual child-abuse (see Mulhern 1991). The process by which "survivors" of such abuse have "recalled" unverifiable experiences with the help of trained professionals resembles the way in which abductions have been "recovered." Organizations armed with such "memories" have been able to inflict considerable social and mental harm on patients and innocent parties named by them. By contrast, the continuing efforts of Strieber and Hopkins to form self-help networks for abductees seems wrong-headed but relatively benign. Further objective research on the phenomenology of these disorienting events seems necessary before the issue can be resolved.

References

Baker, Robert A. 1987-88. The aliens among us: Hypnotic regression revisited. *Skeptical Inquirer,* 12 (Winter 1987-88):148-161.

Bennett, Gillian. 1988. Legend: Performance and truth. In *Monsters with Iron Teeth: Perspectives on Contemporary Legend III,* ed. by Gillian Bennett, Paul Smith, and J. D. A. Widdowson. Sheffield: Sheffield Academic Press.

Briggs, Katharine. 1971. *A Dictionary of British Folk-Tales.* Part B. *Folk Legends.* vol. 1. Bloomington: Indiana University Press.

———. 1976. *An Encyclopedia of Fairies.* New York: Pantheon.

Clements, William. 1982. "I once was lost": Oral narratives of born-again Christians. *International Folklore Review,* 2:105-111.

Edel, Leon. 1953. *Henry James: 1843-1870.* New York: Lippincott.

Hufford, David J. 1982. *The Terror That Comes in the Night.* Philadelphia: University of Pennsylvania Press.

Hyatt, Harry Middleton. 1965. *Folk-Lore from Adams County, Illinois.* Hannibal, Mo.: Alma Egan Hyatt Foundation.

Mulhern, Sherrill. 1991. Satanism and psychotherapy: A rumor in search of an inquisition.

In *The Satanism Scare,* ed. by James T. Richardson, Joel Best, and David Bromley. New York: Aldine de Gruyter.

Sargant, William. 1961. *Battle for the Mind,* rev. ed. Baltimore: Penguin. Originally published in 1957.

Strieber, Whitley. 1987. *Communion.* New York: Beech Tree Books/Morrow.

———. 1988. *Transformation: The Breakthrough.* New York: Beech Tree Books/Morrow.

Taves, Ernest H. 1987. Communion with the imagination. *Skeptical Inquirer,* 12 (Fall 1987): 90-96.

Turner, Victor. 1969. *The Ritual Process: Structure and Anti-Structure.* Ithaca, N.Y.: Cornell University Press.

NICHOLAS P. SPANOS

Past-Life Hypnotic Regression:
A Critical View

Some people who have been administered hypnotic-induction procedures followed by suggestions to regress back past their birth times report that they experienced past lives. For instance, a 22-year-old Caucasian woman, while recently "regressed" in our laboratory, claimed that the year was 1940 and that "he" (her past-life identity involved a change of sex) was a Japanese fighter pilot. How are reports of this type to be explained? The parsimonious answer is that they are suggestion-induced fantasy creations of imaginative subjects. If the subjects hold prior beliefs about the validity of reincarnation and/or if they are given encouragement to do so by the hypnotist, they may come to interpret their fantasies as evidence for the existence of actual past-life personalities.

For some (e.g., Wambach 1979), the parsimonious answer will not do. Instead, hypnotically engendered past-life reports are taken as evidence for the validity of reincarnation. Certainly this is the interpretation most commonly conveyed in popular books and articles on the topic. A few mental-health professionals also accept the reincarnation interpretation and even offer past-life therapy to alleviate problems in a client's present life that purportedly stem from unresolved difficulties in some previous incarnation (e.g., Wambach 1979).

Although "hypnosis" has gained a good deal of contemporary scientific legitimation, it continues to be uncritically conceptualized by many as involving profound alterations in consciousness (i.e., the "hypnotic trance state") that produce fundamental changes in perceptual and cognitive functioning. For instance, hypnotic procedures are sometimes seen as enabling subjects to transcend normal volitional capacities (e.g., to eliminate pain, to retrieve "repressed" memories) or as causing subjects to lose voluntary control over mental and behavioral functions (e.g., hypnotically amnesic subjects are supposedly *unable* rather than unwilling to remember). If hypnosis can do all of these remarkable things, then perhaps regression to past lives isn't so far-fetched after all. Thus my first concern is to examine what the available experimental data really tell us about the nature of hypnotic phenomena.

Is Hypnosis an Altered State of Consciousness?

After more than a century of research, there is no agreement concerning the fundamental characteristics of the supposed "hypnotic trance state" and there are no physiological or psychological indicators that reliably differentiate between people who are supposedly "hypnotized" and those who are not (Fellows 1986). Despite widespread belief to the contrary, hypnotic procedures do *not* greatly augment responsiveness to suggestions. Nonhypnotic control subjects who have been encouraged to do their best respond just as well as hypnotic subjects to suggestions for pain reduction, amnesia, age-regression, hallucination, limb rigidity, and so on (Spanos 1986a). Hypnotic procedures are no more effective than nonhypnotic relaxation procedures at lowering blood pressure and muscle tension or effecting other behavioral, physiological, or verbal-report indicators of relaxation (Edmonstron 1980). Hypnotic procedures are no more effective than various nonhypnotic procedures at enhancing imagery vividness or at facilitating therapeutic change for such problems as chronic pain, phobic response, cigarette smoking, and so on (Spanos 1986a; Spanos and Barber 1976). In short, the available scientific evidence fails to support the notion that hypnotic procedures bring about unique or highly unusual states of consciousness or that these procedures facilitate responsiveness to suggestions to any greater extent than do nonhypnotic procedures that enhance positive motivation and expectation.

It is important to understand that hypnotic suggestions do not directly instruct subjects to do anything. Instead, suggestions are phrased in the passive voice and imply that something is happening to the subject (e.g., "Your arm is rising," instead of "Raise your arm"). This passive phrasing communicates to subjects the idea that they are supposed to act *as if* the effects suggested are happening automatically. In other words, hypnotic suggestions are tacit requests to become involved in make-believe or *as if* situations. A subject is tacitly instructed to behave as if he is unable to remember, as if his arm is rising, as if he is five years old, and so on. Good hypnotic subjects (a) understand the implications of these tacit requests, and (b) use their imaginative abilities and their acting skills to become absorbed in the make-believe scenarios contained in suggestions. Thus, by actively using their imaginative abilities, good hypnotic subjects can create and convey the impression that they are unable to remember, unable to lift their "heavy" arms, and so on (Spanos 1986b). The method actor who throws himself into the role of Richard III causes himself to experience the thoughts and emotions that are relevant to his character. Good hypnotic subjects throw themselves into generating the experiences and enactments that are relevant to their roles as hypnotized and as responsive to suggestion (Sarbin and Coe 1972).

Hypnotic Age Regression. Age-regression suggestions inform a subject that he is growing younger and younger and returning to an earlier time in his life. Thus a responsive hypnotic subject who is "regressed" to age five states that he is five years old, prints in block letters, and so on. Despite such performances, a good deal of research now indicates that these subjects do *not* in any real sense take on the cognitive, perceptual, or emotional characteristics of actual children (Barber, Spanos, and Chaves 1974). Instead of behaving like real children, age-regressed subjects behave the way they *believe* children behave. To the extent that their expectations about how children behave are inaccurate, their age-regression performances are off the mark. For example, adults commonly overestimate the performance of young children on cognitive and intellectual tasks. Hypnotically age-regressed subjects who are given such tasks usually outperform real children whose ages match those to which the subjects have been regressed (e.g., Silverman and Retzlaff 1986).

In short, age-regression suggestions are invitations to become involved in the make-believe game of being a child once again. People who accept this invitation do not, in any literal sense, revert psychologically to childhood. Instead, they use whatever they know about real children, whatever they remember from their own childhood, and whatever they can glean from the experimental test situation to create and become temporarily absorbed in the fantasy situation of being a child. To the extent that their information about childhood is incorrect, their regressed behavior deviates from the behavior of real children (Barber et al. 1974).

Hidden Selves. Just as subjects can be given suggestions for age regression, amnesia, or pain reduction, they can also be led to develop the idea that they possess "hidden selves" that they didn't earlier know about. For example, in a number of studies (cf. Hilgard 1979) good hypnotic subjects were informed that they possessed "hidden selves" that they were normally unaware of, but who the experimenter could talk to by giving the appropriate signals. When they received these signals, many of these subjects behaved as if they possessed secondary selves that had experiences that differed from those of their "normal selves." When the signals were withdrawn, these subjects often behaved as if they were unable to remember their "hidden self" experiences.

Some investigators (e.g., Hilgard) interpret such findings to mean that good hypnotic subjects really do carry around unconscious hidden selves with certain intrinsic and unsuggested characteristics. However, a good deal of evidence indicates instead that so-called hidden selves are neither intrinsic to hypnotic procedures nor unsuggested. Quite the contrary, hidden-self performances, like other suggested responses, appear to reflect attempts by motivated and imaginative subjects to create the experiences and role behaviors called for by the instructions they are given. By varying such instructions subjects can be easily led to develop "hidden selves" with whatever characteristics the experimenters wish. Thus, depending upon the instructions they are given, good hypnotic subjects will enact "hidden selves" that report very high levels of pain, very low levels of pain, or both high and low levels of pain in succession. Subjects can also be led to act as if they possess hidden selves that can remember concrete words but not abstract words; or the opposite, hidden selves that see stimuli accurately, see stimuli in reverse, or don't see stimuli at all, and so on (e.g., Spanos 1986a; Spanos, Flynn, and Gwynn 1988).

In short, a subject who behaves as though he possesses a "hidden self," like one who behaves as if he has regressed to age five, is acting out a fantasy. The fantasy performance is usually initiated by the suggestions of the hypnotist, it is imaginatively elaborated upon and sustained by the subject, and (frequently) it earns validating feedback from the experimenter/hypnotist who interacts with the subject as if he or she really did possess a hidden self with particular characteristics.

Past-Life Hypnotic Regression

The few experimental studies that have examined past-life regression have yielded findings that are consistent with the picture of hypnotic responding described above. For example, we recently completed two experiments on this topic. In the first, 110 subjects were tested for responsiveness to hypnotic suggestions (i.e., hypnotizability). In separate sessions, all of these subjects were individually administered a hypnotic procedure and suggestions to regress to times before their births and then to describe where and who they were. During their individual sessions, 35 subjects enacted past lives. Each subject told the experimenter that he or she was a different person and was living in a different time. Most went on to provide numerous details about where they lived, their past-life occupations, their families, interests, and so on. Subjects who reported past lives scored higher on hypnotizability than those who did not, and were more likely than those who did not to believe that they had experienced some earlier portents of past lives (e.g., déjà vu experiences, dreams).

Among the 35 subjects who reported past lives, there were wide individual differences both in the vividness of the experiences and in the credibility that subjects assigned to them (i.e., the extent to which they believed them to be real past lives as opposed to fantasies). The vividness of past-life experiences was predicted by the subjects' propensity to be imaginative. Thus the frequency with which subjects reported vivid daydreaming and the frequency with which they reported becoming absorbed in everyday imaginative activities (e.g., reading novels) correlated positively with the vividness of their past-life experiences. The best predictor of how much credibility subjects assigned to their past-life experiences was a composite index of their attitudes and beliefs about reincarnation. People who believed in reincarnation, who thought the idea plausible, and who expected to experience past lives assigned higher credibility to their past-life experiences than did those who scored low on this index.

The past-life reporters in our first experiment almost always indicated that their past-life personalities were the same sex and race as themselves and usually reported that the past-life personalities lived in Westernized societies. In our second experiment, all subjects were given general information about reincarnation. However, those in one group were further informed that it was not uncommon for people to have been of different sexes or races in past lives and to have lived in exotic cultures. Control-group subjects were given no specific information concerning the characteristics they might expect in their past-life personalities. Among subjects who gave past-life reports, those given the specific information were significantly more likely than controls to incorporate one or more of the suggested characteristics into their past-life descriptions.

Wambach (1979) contended that the historical information obtained from hypnotically regressed past-life responders was almost always accurate. To test this idea in both of our experiments we asked subjects questions that were likely to have historically checkable answers (e.g., Was the responder's community/country at peace or war?). Contrary to Wambach (1979), subjects who gave information specific enough to be checked were much more often incorrect than correct, and the errors were often the type that actual persons from the relevant historical epochs would have been unlikely to make. For example, the "Japanese fighter pilot" described at the beginning of this article was unable to name the emperor of Japan and stated incorrectly that Japan was at peace in 1940. A different subject stated that the year was A.D. 50 and that he was Julius Caesar, emperor of Rome. However, Caesar was never crowned emperor, and died in 44 B.C. Moreover, the custom of dating events in terms of B.C. or A.D. did not develop until centuries after A.D. 50.

Kampman and Hirvonoja (1976) also obtained support for the fantasy-construction hypothesis. After obtaining past-life reports from hypnotic subjects these investigators encouraged subjects to connect various elements of their past-life descriptions with events in their current lives. In this way they often uncovered the sources of information used by subjects to construct their fantasies. We obtained similar findings. For instance, during a post-hypnotic interview, the subject who reported having been Julius Caesar indicated that he was taking a history course and found the section on ancient Rome particularly interesting. Other subjects reported post-hypnotically that, during the previous summer, they had visited the countries where their past-life personalities resided, or suddenly remembered that their past-life wives resembled and had the same names as old girlfriends from their current lives, and so on.

In summary, the available data strongly indicate that past-life reports obtained from hypnotically regressed subjects are the fantasy constructions of imaginative subjects who are willing to become absorbed in the make-believe situation implied by the regression suggestions. Not surprisingly, subjects who responded well to other hypnotic suggestions (high hypnotizables) were also relatively likely to respond to regression suggestions. Moreover, those with the most practice at vivid daydreaming and everyday fantasizing were the ones who created the most vivid past-life fantasies. As do subjects who are asked to regress to childhood, past-life reporters construct their fantasies by interweaving information given in the suggestions with information gleaned from their own life experiences and from what they have read and heard that was relevant to their performances. Moreover, just as age-regressed subjects incorporate misinformation into their enactments of being children, so past-life reporters incorporated historical misinformation into their past-life enactments.

People continually interpret their current experiences in term of established conceptual categories. Consequently, whether people interpreted their past-life experiences as real or imaginary depended upon whether they possessed a belief system that accommodated the notion of real past-lives. Those who believed in reincarnation possessed such belief systems, and therefore were relatively likely to interpret their past-life experiences as veridical rather than imaginary.

Since the classic case of Bridey Murphy (Bernstein 1956), the notion of regression to past lives has been legitimized by common and strongly held misconceptions about the nature of hypnotic responding. A more empirically based conceptualization of such responding that emphasizes its goal-directed nature, its *as if* qualities, and its embeddedness in a nexus of social communications allows past-life enactments to be seen for what they are—interesting and imaginative contextually guided fantasy enactments.

References

Barber, T. X., N. P. Spanos, and J. F. Chaves. 1974. *Hypnosis, Imagination and Human Potentialities.* New York: Pergamon.

Bernstein, M. 1956. *The Search for Bridey Murphy.* New York: Pocket Books.

Edmonstron. W. E., Jr. 1980. *Hypnosis and Relaxation.* New York: Wiley.

Fellows, B. J. 1986. The concept of trance. In *What Is Hypnosis?* ed. by P. L. N. Naish. Philadelphia: Open University Press.

Hilgard, E. R. 1979. Divided consciousness in hypnosis: The implications of the hidden observer. In *Hypnosis: Developments in Research and New Perspectives,* ed. by E. Fromm and R. E. Shor. New York: Aldine.

Kampman, R., and R. Hirvonoja. 1976. Dynamic relation of the secondary personality induced by hypnosis to the present personality. In *Hypnosis at Its Bicentennial,* ed. by F. H. Frankel and H. S. Zamansky. New York: Plenum.

Sarbin, T. R., and W. C. Coe. 1972. *Hypnotic Behavior: The Social Psychology of Influence Communication.* New York: Holt, Rinehart & Winston.

Silverman, P. S., and P. D. Retzlaff. 1986. Cognitive stage regression through hypnosis: Are earlier cognitive stages retrievable? *International Journal of Clinical and Experimental Hypnosis,* 34:192-204.

Spanos, N. P. 1986a. Hypnotic behavior: A social psychological interpretation of amnesia, analgesia and "trance logic." *Behavioural and Brain Sciences,* 9:449-502.

———. 1986b. Hypnosis, nonvolitional responding and multiple personality: A social psychological perspective. In *Progress in Personality Research,* 14, ed. by B. Maher and W. Maher. New York: Academic Press.

Spanos, N. P., and T. X. Barber. 1976. Behavior modification and hypnosis. In *Progress in Behavior Modification,* 3, ed. by M. Hersen, R. M. Eisler, and P. Miller. New York:

Academic Press.

Spanos, N. P., D. Flynn, and M. I. Gwynn. 1988. Contextual demands, negative hallucinations, and hidden observer responding: Three hidden observers observed. *British Journal
of Experimental and Clinical Hypnosis,* 5:5-10

Wambach, H. 1979. *Life Before Life.* New York: Bantam.

SARAH G. THOMASON

Past Tongues Remembered?

Suppose you want to convince people that you've discovered a genuine case of reincarnation. If you can prove that your subject can speak the language of an earlier incarnation, that would obviously be strong evidence in favor of the reincarnation claim—provided, of course, that the language is not the subject's present native language and that you can also show that the subject has had no chance to learn the "past life's" language in his or her current lifetime. The reasoning would go like this: Speaking a language is a skill that requires extensive long-term exposure to the language. If a person has that skill, but lacks such exposure in his/her current lifetime, then the skill must have been acquired paranormally—for instance, in a previous lifetime whose memory lingers on.[1]

There are several published case studies in which reincarnation (or the related phenomenon of temporary possession of a subject by another personality) is proposed as the source of a subject's ability to speak a foreign language. The most impressive of these case studies are in two books written by Ian Stevenson (1974; 1984), who is Carlson Professor of Psychiatry at the University of Virginia Medical School. Stevenson has studied two native English-speaking subjects who, under hypnosis, manifest foreign personalities and seem to speak—very haltingly—foreign languages, specifically Swedish and German, respectively. To establish his subjects' linguistic competence in these languages, Stevenson arranged sessions in which native speakers of Swedish and German interviewed the subjects, questioning them about their past lives; in the second case, Stevenson himself participated in the interviews, since he knows some German.

The result of these interviews is what Stevenson calls "responsive xenoglossy"—speaking a language one hasn't learned in one's current lifetime, and speaking it in a responsive way in conversation. He considers *responsive* xenoglossy to be crucial for making the case for the paranormal phenomenon, as opposed to what he calls "recitative xenoglossy"—the mere ability to recite some words in a foreign language one hasn't learned. The reason, he says, is that "one can only acquire the ability to use a language responsively by using it, not by overhearing it spoken" (1984, p. 160). That is, you need practice to acquire the skill of conversing in a foreign language; and, if a subject falsely denies having had such practice, then it should be possible to uncover the fraud by careful investigation.

It must be emphasized that Stevenson is energetic in his search for fraud and also for unconscious recourse by his subjects to former but forgotten experience with the languages in question. For his "German" personality, Gretchen, for instance, he investigated the subject's opportunities for learning German normally. He visited the town she grew up in, interviewed her relatives and old acquaintances, and established (among other things) that the schools she went to did not offer German classes when she attended them. In this respect Stevenson's attention to proper methodology is exemplary, and there is no hint, here or elsewhere, that he is trying to fool anyone; perhaps the strongest evidence of his sincerity is his inclusion in his books of partial transcripts of the sessions in which the subjects were producing their xenoglossic Swedish and German, so that an independent investigator could actually check some of his data.[2] There is also no hint that his subjects are consciously trying to fool anyone.

However, in spite of Stevenson's efforts to provide genuine evidence in support of his paranormal claims, his linguistic evidence is completely unconvincing to a professional linguist. There are two main problems with it. First, his notion of "responsive xenoglossy" is fatally flawed as a methodological criterion for determining a person's ability to speak a language. And, second, most of the explanations he suggests for the obvious inadequacies of his subjects' Swedish and German put his paranormal proposal squarely into the realm of pseudoscience: Ultimately, Stevenson's explanations for the linguistic deficiencies render his hypothesis untestable by emptying it of content. I'll discuss each of these problems in turn, using illustrations from his "German" case. Then, after showing why Stevenson's method doesn't work, I'll outline a method that *would* work as a test of a person's ability to speak any given language.

First, consider the idea that you can't converse in a language without knowing it, and without having practiced speaking it regularly over a considerable period of time. This certainly seems like a reasonable idea, and of course it's also a valid idea if you're thinking of ordinary, normal conversation; anyone who has (say) studied French for a couple of years in high school, and then visited France, has probably noticed that it's hard to carry on a conversation in French with this minimal background. It may even be hard to ask directions to the nearest cathedral, and harder still to understand the answer if it's much more complicated than a pointing gesture. So one must agree with Stevenson that, if his subjects, without having learned Swedish and German, can in fact converse normally in these languages, then a paranormal explanation would seem necessary.

But, as Stevenson himself admits, what his subjects produce is far from normal conversation. He argues that their linguistic behavior is close enough to normal conversation to require a paranormal explanation (barring fraud): I—and I believe this is true of any other linguist who studies the data carefully—would argue that his subjects show no sign of any extensive exposure to Swedish or German, in any lifetime.[3] The issue, of course, revolves around the difference between the lingustic skills manifested by Stevenson's subjects and the linguistic skills manifested by a normal (as opposed to a paranormal) speaker of a language.

Consider what it means to know a language. First, any native speaker of any language has a vocabulary of thousands of words—certainly upwards of 10,000, probably many more. This is true regardless of schooling. Second, a speaker knows grammatical rules—not necessarily, and not only, the rules taught by a grade-school grammar teacher, but rules that enable the speaker to produce and interpret connected utterances that will be readily understood by other speakers of the same language. For instance, any speaker of English knows that a sentence like "Willy doesn't eat horseradish" is a perfectly good English sentence, while a sentence like "Not Willy horseradish eat" is not good English—it doesn't follow the rules of English grammar for any dialect of English. Children born into an English-speaking community, or any other community, have most of their native language's extremely complex grammatical rules under control, and a sizable fraction of its vocabulary too, by the age of four or five.

But compare this normal situation to what Stevenson's subjects are doing. His "German" personality, Gretchen, produced about 120 words in sessions with the hypnotist (her husband, who spoke no German). She produced only a few more words independently in her later sessions with German speakers. A number of these German words were either just like the corresponding English words, e.g., *braun,* which is identical in meaning and very close in pronunciation to English "brown"; or they were similar to the English word, for example, *blü,* the word she used for "blue"—which she pronounced with the non-English German vowel [ü] rather than with the English vowel sound, but *not* with the appropriate German sounds for this word (which in German rhymes with English "cow").

Since Gretchen usually answers with just a word or two rather than in full sentences, her minimal vocabulary does not include the numerous grammatically necessary but semantically empty words like helping verbs; and her answers to many questions indicate that she doesn't understand such words, either. In fact, all she seems to know, either for speaking or for understanding, is a handful of words.

Well, then, how does Gretchen manage to converse? The answer is that she doesn't, in any normal sense of the word *converse.* In the partial transcripts Stevenson provides, Gretchen's spontaneous contributions are almost entirely confined to identical, repeated comments about the danger she's in because people are listening. (Her fears apparently have to do with religious persecution connected with Martin Luther, and Stevenson's own analysis shows them to be completely unrealistic and anachronistic.) Otherwise, she speaks only in short answers to other people's questions. Often her responses are simply repetitions of what the interviewer just said.

Of Gretchen's 172 other responses, 42 are answers to yes/no questions (some asked in German, some asked in English). By "yes/no questions" I mean questions that require only "yes" or "no" as an answer. But yes/no questions don't count for much as a test of language knowledge, because all she has to do is say *ja* for "yes" or *nein* for "no," and she has a 50-50 chance of being right. In any case, since the questions are about her own past life, and no one else present knows anything about it, there is no way to tell whether or not her answers are factually accurate. Furthermore, she can answer any yes/no question even if she doesn't understand the content of the question at all—because the intonation pattern of yes/no questions in German is similar to the intonation of yes/no questions in English and *different* from the intonation of statements and other kinds of questions: Usually (though not always) there is a rise in pitch at the end of a yes/no question, but not in other kinds of sentences. You can check this by saying out loud the questions "Are you hungry?" and "What do you want to eat?" and comparing their intonation patterns. The German pattern is the same. So, for instance, when an interviewer asks Gretchen whether she has a doll, the question is: *Saq mir was von deinen Puppen. . . . Hast du eine?* (Tell me something about your dolls. . . . Do you have one?) Gretchen can recognize the yes/no question by its pitch rise and can safely answer *nein,* though there is nothing in the following discussion about dolls to show that Gretchen has any idea what the interviewer is talking about.

So I think we have to throw out all of Gretchen's answers to yes/no questions as evidence for anything, except for a few answers other than "yes" or "no" that she gave to such questions. This leaves the other questions, those that require a content answer: 102 of these were asked in German, and 28 were asked in English. Gretchen herself speaks only "German," such as it is; but she does much better in answering English content questions than in answering German ones. When the questions were asked in English, she gave 22 appropriate answers, as against 2 inappropriate answers and 4 dubious ones. By contrast, and in sharp contrast to Stevenson's own analysis, I count only 28 appropriate answers to content questions asked in German, as against 45 clearly inappropriate ones and 29 copout answers, such as "I don't understand" and "I don't know." This isn't a very good score of appropriate answers, even before you eliminate some that are repetitions.

Now some of the answers that I consider inappropriate Stevenson considers appropriate, especially because he counts answers as appropriate when they are, in his terms, "appropriate associations to a preceding question, but not direct answers." Here's a typical example. The topic of discussion is food, and specifically what Gretchen eats at different times of day. The German-speaking interviewer asks, *Was gibt es nach dem Schlafen?* (What is there after sleeping? i.e., What do you eat for breakfast?) Gretchen answers, *Schlafen* . . . *Bettzimmer* (Sleep . . . bed-room). Clearly, Gretchen has not understood the question, which contains only entirely ordinary German words and constructions; instead, she has understood only the word *schlafen* (sleep), and she answers as if the question had been about *where* she sleeps—a wrong guess—and with the wrong word: her word *Bettzimmer* is made up of German *Bett* (bed) plus *Zimmer* (room), a literal translation of the English word *bedroom;* but the German word for bedroom is *Schlafzimmer,* literally "sleep room."

This is typical of Gretchen's linguistic performance. She does know a few German words—a tiny fraction of what a teen-age native speaker would know. (She is supposed to be about 14 years old.) She occasionally produces grammatically correct phrases, but in general she neither produces nor understands the simplest German grammatical constructions. When she doesn't understand a content question, which is often, she guesses; sometimes she guesses right—the topics of discussion in these interviews are very limited, so some right guesses aren't surprising—but more often she guesses wrong or says, "I don't understand."

The question is, do we need a paranormal explanation for her knowledge of some German words and phrases? Surely not; Stevenson's research into her background turns up a few opportunities for this amount of very limited exposure to German—World War II movies, a look at a German book— and that's all she shows any evidence of. What evidence there is, furthermore, shows definitely that at least some of her experience with German is with the written language, because some of her pronunciations can only have come from an English speaker's reading of written German, not from a German speaker's pronunciation or reading. And the Gretchen personality can't be responsible for the subject's slight familiarity with written German anyway, because Gretchen says she can't read or write. The point is that Gretchen's level of "responsive xenoglossy" is so very low that Stevenson's argument about the necessity of practice to produce such a skill collapses. At best, she speaks German about as well as someone might who studied the language in high school for a year about 20 years ago.

On the other hand, Stevenson certainly needs some explanation for Gretchen's inadequacies as a German speaker, even with his generous count of appropriate responses to questions. He makes several suggestions to account for her lack of knowledge of her native language. One is that the Gretchen phenomenon represents only a partial manifestation of the foreign personality in the subject, and the part that manifests itself doesn't include much knowledge of the language. I have nothing to say about this, except that it does not seem to be a concept that lends itself to scientific testing.

Another of Stevenson's proposals is that Gretchen may have learned German inadequately because, although her father was supposed to have been a local official who "would presumably . . . have been at least a moderately well-educated man and a speaker of excellent German," Gretchen herself (according to Stevenson's conjecture) "was an illegitimate and neglected child who spent most of her time in the kitchen with a servant"; and since the servant was probably an uneducated person, Gretchen might therefore have come out with poor German-speaking skills (1984, p. 46).

Here Stevenson betrays his profound ignorance about language. Level of education has nothing at all to do with fluency. Even if Gretchen's father spoke educated Standard German and Gretchen herself spoke a substandard German dialect—conjectures which, incidentally, Stevenson makes on the basis of fragmentary and often inconsistent statements of Gretchen's—then their respective German dialects would have differed only in a small number of linguistic features; in most features they would have been identical, and in any case the two people would have been completely equivalent in their abilities to put sentences together coherently. So, though Stevenson could perhaps explain differences between Gretchen's speech and Standard German with such a hypothesis (at least, he could do so if there were any evidence to support his conjectures), he can't in this way explain Gretchen's near-total lack of grammar and her minimal vocabulary.

Stevenson's best attempt at an explanation is also the one he likes best. Perhaps, he says, "the grammatical and other imperfections [in Gretchen's speech] . . . may have arisen from the great difficulties involved in mediumistic communication" (1984, p. 69). Specifically, the earlier incarnation or the possessing personality has to talk through the medium of a native speaker of English, and this presents all the problems one finds (he says) with second-language learning by an adult: The English-speaking medium can't process the foreign language properly because of the subject's own long-ingrained English speech habits, so things come out wrong—just as your pronunciation and grammar would come out wrong, if, with only a year or so of casual study, you tried to speak German. However, the cases he describes, if they were to be accepted as genuine paranormal phenomena, would not resemble second-language learning by an adult; instead, they would be more akin to cases in which an adult tries to speak a language that she/he learned in early childhood but has not spoken for thirty years or more. In both types of cases, pronunciation might well be affected by the subject's English; but, as mentioned earlier, several of Gretchen's pronunciation errors clearly arose from an English speaker's misreading of ordinary German spelling, not from the influence of the English sound system *per se*. In a long-unused native language, grammar could also be affected by the language normally used by the speaker in later life, but many basic grammatical constructions of the speaker's first language would remain intact.

More important, in all kinds of language learning and language loss a speaker's *passive* knowledge—the ability to understand the spoken language—is considerably greater than his/her *active,* or speaking, knowledge of the language being learned or forgotten. Significantly, Gretchen's German does not fit this well-established pattern at all. She clearly understands German just as little as she speaks it: There is no discernible difference between her active knowledge and her passive knowledge. In both speaking and under-standing, her knowledge of the language is limited to words, and not even very many of those. So this proposal of Stevenson's also fails to account for Gretchen's linguistic deficiencies, though it could possibly account for some—just some—of her problems with the actual production of German utterances.

What all this means is that Stevenson's notion of "responsive xenoglossy" is not a good test of a subject's linguistic knowledge, because there is too much room for successful guesswork in a question-and-answer interview. The method also fails for other reasons, such as the investigator's bias in inter-preting the results of the interview—namely, in counting appropriate vs. inappropriate responses. (In other words, you get the experimenter effect in judgments about the responses; and, in this respect, my skeptical judgments—as opposed to my strictly linguistic ones on points where there is clear evidence in the transcript—may be just as suspect as Stevenson's believer judgments.) So if you want a good test of a hypothesis about a subject's knowledge of a language, you need to find a method that makes guessing unhelpful and excludes the experimenter effect.

Here it is. It's very simple. First, take a word list of basic vocabulary (there are standard lists that linguists use in their field-work on previously undescribed languages)—100 or 200 words of the sort that any language is likely to have, e.g., words for "mother," "father," "moon," "water," "walk," "sleep," and so forth. Hypnotize the subject so that she manifests the putative earlier incarnation, and have her translate the word list into the language of that incarnation. Also, get translations of paradigms—e.g., "I walk," "you walk," "they walk"; "I walked," "you walked," etc.; "I will walk," "you will walk," etc.; "I'm walking," etc.; and translations of simple sentences—e.g., "My dog eats bread," "Your dog doesn't eat bread," "Does my dog eat bread?" and so forth. Then wait a month or more. Hypnotize the subject again and have her translate the same items again—without, of course, giving the subject the opportunity beforehand to review what she said in the first session. (In fact, the subject should not be told what will go on in the second session.) If the subject knows the language in question, the translations should be real German, or real Swedish, or whatever the language is. In addition, though they might well show some variation (for instance, because many languages have different words that might translate the same English word), the translations should be mostly identical for the two sessions.

Second, test the subject for comprehension of the language—and remem-ber that, if you want to assume some forgetting during intervening lifetimes, comprehension should be preserved better than production. Read the subject a short story in the language; make sure that the text contains only simple grammatical constructions. Then ask the subject questions about the story—they can be either yes/no questions or content questions, since you can tell what the answers should be, but content questions are preferable.

If the subject fails these tests, she does not know the language. This will not, of course, prove that the case is *not* one of reincarnation or temporary possession; but at least no one will be able to use the linguistic data as evidence in support of a claim that it is such a case.

I should add that I have used the first steps of this method to check the proposals of another hypnotist who believed that his subjects were speaking languages of previous lives under hypnosis. (The later steps turned out to be unnecessary because the first steps yielded a conclusive result; see Thomason 1984 for a description of these cases.) The hypnotist interviewed the "foreign personalities" using a word list I had sent him, and sent me tape recordings of the interviews for analysis. As in Stevenson's case studies, all the participants—hypnotist and subjects alike—seemed to be innocent of any intent to deceive. And, also as in Stevenson's cases, all the subjects seemed to be equally innocent of any systematic knowledge of the languages in question. Unlike Stevenson's data, however, the data I worked with provided sufficient evidence to test the hypnotist's hypothesis that his three subjects were speaking nineteenth-century Bulgarian, fourteenth-century Gaelic, and nineteenth-century Apache, respectively. The analysis showed that the subjects did not know the basic vocabulary of their putative earlier native languages; in addition, and perhaps even more significantly, it showed that their utterances in the "previous lives' languages" were so unsystematic as to be impossible components of *any* natural human language.

The linguistic performances of these three hypnotic subjects and of Stevenson's hypnotic subjects as well, in spite of the indeterminacy that results from Stevenson's flawed methodology, all point to the same conclusion: If you want to speak a foreign language, you will need to learn it through systematic exposure to its words and structures during your current lifetime.

Postscript

Robert F. Almeder (1988), in a response to "Past Tongues Remembered?" made two main objections to my arguments in the article. First, he said, two of Stevenson's cases that I did not analyze—Jensen (Swedish) and Sharada (Bengali)—are stronger than the Gretchen case and certainly require a paranormal explanation. Second, even Gretchen produced enough appropriate responses that a paranormal explanation is required to account for them. Almeder's reasoning was not, in my opinion, persuasive.

First, Almeder is mistaken in his belief that Jensen (Stevenson 1974) and Sharada (Stevenson 1984) provide better evidence for paranormal linguistic competence than Gretchen does. Jensen is very similar to Gretchen linguistically: a tiny vocabulary, a few independent appropriate responses, and some wildly inappropriate responses—e.g., "my wife" in answer to a question about what he would pay for some item at the market. In fact, the same basic argument against Stevenson's paranormal proposal can be made (with, of course, some differences in detail) for both Jensen and Gretchen. Sharada seems relatively fluent, although Stevenson gives too little actual data to permit a detailed analysis of her linguistic performance. But his own discussion shows clearly that the subject had ample opportunity for learning a considerable amount of Bengali in a perfectly ordinary way and that she had a motive for doing so. Moreover, she spoke natively an Indic language that is very closely related to Bengali—so close that large numbers of vocabulary items and grammatical constructions are similar or identical. And she had studied Sanskrit, which is a close approximation to the linguistic ancestor of all the modern Indic languages. Under these circumstances, believing in Sharada's alleged paranormal linguistic ability requires a very determined anti-skeptic.

Almeder's point about the Gretchen case is that her answers to questions show "a firm capacity to understand the simple grammatical forms of the language." If, on an exam, a student of mine answered a German question about what he eats for breakfast with an erroneous German translation for "bedroom," I would not characterize his grasp of the simple grammatical (or lexical) forms of the language as firm. More seriously, Almeder overlooked my observation that a large amount of successful guesswork is predictable in a very restricted conversational framework like that of the Gretchen interviews, given a minimal prior acquaintance with a few German words and phrases. Laymen may be startled to learn that people can guess right, quite often, about the meanings of things said to them in a language they don't know. This is actually quite easy to do if the situation provides clues (as in the Gretchen case) to a questioner's intent. Interesting examples can be found, for instance, in answers to the questions that judges in American courtrooms ask non-English-speaking defendants in order to decide if an interpreter is needed for a trial; and many people who have traveled abroad have anecdotes, like the one a correspondent told me after my response to Almeder appeared in the *Skeptical Inquirer,* that confirm this.

Finally, readers may be interested in an indirect reference to "Past Tongues Remembered?" by Ian Stevenson in the *Journal of Parapsychology* (51:373, 1987). In listing reviews of his 1984 book, Stevenson dismissed one commentary because it appeared in "a magazine" rather than a scholarly journal; in the context, one can infer that he has my article in mind and that he views its publication in the *Skeptical Inquirer* as a justification for ignoring it.

Notes

1. There is also the possibility of divine intervention; this is the source that is sometimes claimed for "foreign-language speaking" in cases of *glossolalia,* or "speaking in tongues," in charismatic Christian churches. In this paper, however, I will be concentrating on reincarnation claims. For a thorough study of glossolalia, see Samarin 1972.

2. In his 1984 book Stevenson also discusses a third case, from India; but he gives very little data from this subject, so there is no way to arrive at an independent judgment on the material.

3. Among other things, this means that in these cases it seems unnecessary to deal so thoroughly with the question of possible fraud. If either subject cheated, the cheating was so

unsuccessful that it might reasonably be compared to a case in which a student cheats on an exam by copying from the failing student sitting nearby.

References

Almeder, Robert F. 1988. Response to "Past Tongues Remembered?" *Skeptical Inquirer*, 12 (Spring 1988): 321-323.

Samarin, William J. 1972. *Tongues of Men and Angels: The Religious Language of Pentecostalism.* New York: Macmillan.

Stevenson, Ian. 1974. *Xenoglossy: A Review and Report of a Case.* Charlottesville: University Press of Virginia.

———. 1984. *Unlearned Language: New Studies in Xenoglossy.* Charlottesville: University Press of Virginia.

Thomason, Sarah G. 1984. Do you remember your previous life's language in your present incarnation? *American Speech,* 59:340-350.

BARRY BEYERSTEIN

The Myth of Alpha Consciousness

Psychophysiology seeks to understand how mechanisms in the nervous system mediate consciousness and behavior. Over the years this hybrid field has seen many newly discovered brain processes reportedly linked with unique psychological states only to have further research reveal that the relationship is far more complex than suspected. Correcting these misinterpretations takes time, even in the professional literature. Beyond the lab, it is even more difficult to retire obsolete notions about brain–behavior relationships when the popular press, profit motives, and a host of quasitheological beliefs conspire to perpetuate them. Alpha brain-waves and biofeedback are two areas in which such misapprehensions are legion.

In the late 1960s a reawakened interest in altered states of consciousness was buoyed by claims that patterns in the electroencephalogram (EEG) called "alpha waves" were indicators of meditative or psychic states. This, plus the understandable attraction of anything offering quick relief from anxiety and stress, spawned a multimillion-dollar industry aimed at teaching people to maximize EEG alpha through a technique called biofeedback. This occurred despite a growing realization among psychophysiologists that the alleged benefits were based upon unsupported assumptions about alpha and despite growing reservations about the efficacy of biofeedback in general.

Alpha waves (Figure 1) are rhythmic pulsations in an EEG record produced under certain conditions by electrochemical activity in cells of the brain. They range in frequency from 8 to 12 Hz (hertz = cycles per second). The precise meaning of the alpha rhythm continues to be debated among brain researchers, though one would scarcely know it from most popular articles or the advertisements of the alpha-conditioning industry. Purveyors of biofeedback devices assert that enhancing alpha production brings about a special state of the nervous system that is both subjectively pleasant and psychologically, medically, and, some say, psychically beneficial. This putative state is known as "alpha consciousness," or simply "the alpha state."

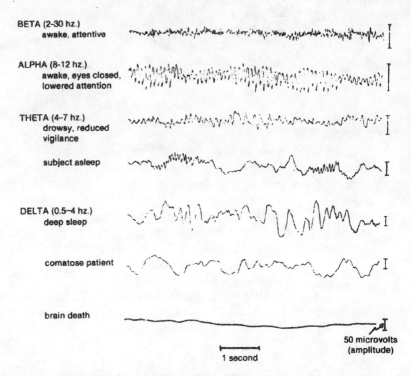

FIGURE 1. Typical electroencephalographs at different levels of arousal, in coma, and at death.

Biofeedback employs electronic sensors to inform people of variations in physiological processes whose activities are not normally accessible to consciousness (e.g., brain or muscle electrical activity, blood pressure, etc.). Pioneers in biofeedback hoped that, by receiving immediate feedback about these unfelt bodily changes, people could bring them under voluntary control. This was touted by some as a shortcut to higher states of awareness and by others as an antidote to stress. Early positive reports achieved wide currency, but later disconfirmation from better-controlled studies have tended to remain buried in technical journals. I shall concentrate on claims surrounding EEG biofeedback. Those interested in a critical assessment of other forms of biofeedback will find informative a recent review by Simkins (1982).

Early History of the Alpha Wave

The wave-forms we now call "alpha" were, among others, apparent when Richard Caton discovered in 1875 that weak pulsating electrical currents could be recorded from the exposed surface of animals' brains. A decade later, Adolph Beck noticed that the large rhythmic oscillations in the brains of awake but resting animals disappeared when they attended to stimuli—a phenomenon

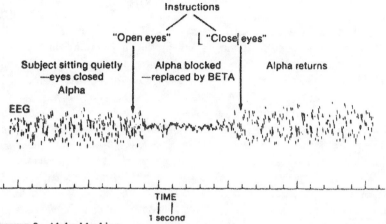

FIGURE 2. Alpha blocking.

we now know as "blocking" of the alpha rhythm. (See Figure 2.) The slow alpha waves were supplanted by smaller, higher frequency activity now known as *beta waves* (13 to 30 Hz). Beta waves predominate in the EEG during activities requiring attention or mental effort. In 1925, a German psychiatrist, Hans Berger, discovered that the electrical activity of the human brain could be recorded from electrodes placed in the scalp (Berger 1929).

The prominence of alpha over the visual areas of the brain and the fact that alpha tends to be blocked by opening the eyes led Berger to suggest that it was inversely related to attention paid to visual information, a notion that has continued to receive support (e.g., Adrian and Mathews 1934; Morrell 1967; Mulholland 1968; Plotkin 1976; Beyerstein 1977; Ray and Cole 1985).

Given this background, one can see why many brain researchers were astonished to see alpha suddenly identified with transcendent or psychic states and offered as evidence for a unique and beneficial state of consciousness.

Origins of the Alpha-Meditation Link

The first suggestions that alpha is special and desirable arose from observations in Japan and India that experienced Zen and Yoga meditators showed much alpha in their EEGs while meditating (Bagchi and Wenger 1957; Anand et al. 1961; Kasamatsu and Hirai 1966). They also produced more than usual amounts of alpha with their eyes open and were less likely than untrained persons to show alpha-blocking in response to distracting stimuli.

As far as they go, these findings are reliable and I have confirmed them in my own lab. Writers of the early reports were careful not to make the logical mistake of assuming that, because two things are correlated, one must necessarily cause the other. Later interpreters were not so cautious. It became axiomatic within the consciousness-expansion fraternity that alpha and meditation were necessarily linked. The error in assuming this without the appropriate experimental controls is apparent in the following. Suppose we noticed that finger movements decrease during meditation. Can we conclude that quieting of the hands is an index of meditation, or that by immobilizing our hands we can propel ourselves into the same state as the person whose hands became still when he began to meditate? Just as there are many reasons for reduced finger mobility, there are many states of consciousness in which alpha might appear. Meditation is one of several states in which people tend not to process much visual information, hence the preponderance of alpha in their EEGs. Trained meditators have developed the ability to ignore stimuli that usually block alpha, but alpha by itself does not guarantee someone is meditating.

There are other problems in equating alpha with meditation. One is that, although the procedures and the subjective states associated with different meditative disciplines vary considerably, all seem about equally related to alpha production, even some to which adepts refuse to grant the status of meditation at all. Another problem is that, as Richard Caton knew more than 100 years ago, lower animals produce alpha, but most people are reluctant to conclude that their pets meditate. Finally, most (but not all) people produce alpha when they simply close their eyes and refrain from active thinking or remembering. Is that all meditation is?

Despite these shortcomings, it came to be widely believed that because highly trained individuals in a self-reported pleasant state of consciousness produce a lot of alpha, alpha must be responsible. By extension, it was argued, if nonmeditators could be taught to lengthen their alpha periods they could achieve the same benefits with great savings in time and effort. A major new growth industry was born.

In 1969, Joe Kamiya reported that ordinary people could learn to enhance alpha output with feedback. Kamiya presented subjects with a signal whenever an electronic filter detected alpha in their EEGs. (Figure 3). They were simply told to do whatever they wanted in order to keep the alpha feedback tone on as much as possible. His subjects found the exercise enjoyable, strengthening the presumptive link between increased alpha and transcendent states. We shall return to the question of whether learned control of EEG alpha has in fact been demonstrated or whether the pleasurable reports might not have been the result of subjects' prior expectations. First, a few speculations as to why the initial claims failed to receive the critical scrutiny they deserved.

Alpha and the "New Age"

The enthusiastic reception accorded the notion of alpha consciousness seems to be yet another manifestation of the change in *Zeitgeist* or world-view that overtook many Western societies during the 1960s and 1970s. On the heels of social, political, and economic upheavals came increasing disillusionment with many conventional beliefs and goals and their philosophical underpinnings

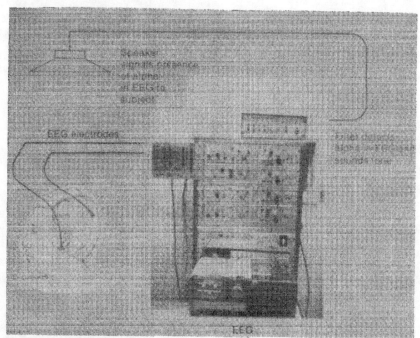

FIGURE 3. Alpha biofeedback.

(Singer and Benassi 1981). Rising popular interest in alternatives offered by Eastern religions and altered states of consciousness coincided with a loosening of the behaviorist hold on academic psychology, making the exploration of conscious contents once again a legitimate field of research. Meanwhile, many people who had sought enlightenment through consciousness-affecting drugs began to realize that this path is not without its costs as well. Nonchemical means to similar ends rose in value accordingly. Suggestions that alpha feedback was an electronic shortcut to benefits that cost others much time and effort added to its marketability.

The medical establishment was not spared its share of criticism during this time of reassessment either. Critics who saw modern medicine as increasingly mechanistic, impersonal, and preoccupied with cure rather than prevention were quick to see in biofeedback, especially alpha biofeedback, an alternative to the surgeon's knife, the psychiatrist's pills, and even the before-dinner cocktail. Thus the allegedly relaxing and curative powers of alpha consciousness were eagerly welcomed by the "holistic healing" movement. The putative link between alpha and mystical states long claimed to have healing properties enhanced its attractiveness to this constituency. Accumulating scientific evidence for psychological contributions to the onset and remission of certain diseases also helped open the doors of many establishment clinics to this fledgling therapy. Amid the exuberances of this era, many claims for alpha biofeedback were granted without the supporting data normally required of an experimental therapy.

Those of us who advocated a wait-and-see position regarding biofeed-back and cautioned that earlier panaceas claiming such sweeping benefits had invariably proved disappointing were frequently dismissed as apologists for the status quo and holdovers of the outmoded "linear thinking" the New Age was seeking to transcend. However, the data that ought to have been gathered prior to the public promotions gradually accumulated. They strongly suggested that the initial zeal was premature.

The Scientific Evidence

Three major issues ought to have been settled before selling alpha feedback to the public: (1) How good is the evidence that increases in alpha during feedback sessions are in fact due to learned enhancement? (2) Is there really any such thing as "alpha consciousness"? That is, are there any mental contents that are invariably present when, and only when, alpha waves predominate in the EEG? (3) The "truth in advertising" criterion. How reliable are the psychological, medical, and mystical dividends alleged to accrue to those who cultivate "alpha consciousness," if indeed it exists?

Has learned enhancement of alpha really been demonstrated?

Given the extent of the promotional efforts, it may surprise many to learn that it is questionable whether EEG control as touted by the alpha-conditioning industry has ever been satisfactorily demonstrated. But what then of the many journal articles reporting a steady rise in alpha over the course of the feedback training? To assess these data, we must first rule out alternatives to direct learned control that could produce similar results. Remember that most normal people exhibit alpha when they simply close their eyes and refrain from intense mental effort. Eyes-closed (EC) alpha production is measured at the outset of a feedback session as a baseline for later comparison. The task in alpha biofeedback is to do whatever is necessary (usually unspecified) to (1) produce more EC alpha at the end of the session than during the EC baseline and (2) eventually be able to meet or exceed the EC baseline with eyes open. For these comparisons to be meaningful, we must first be sure that the initial baseline was not artificially suppressed; for, if it were, a rise in alpha production due to dissipation of the suppressor could be mistaken for a learned increase due to the feedback.

Since novelty, excitement, and anxiety tend to diminish alpha in the EEG (see later exceptions), it seems several attributes of first exposure to a feedback session could initially depress alpha baselines—e.g., newness of surroundings and apparatus, anticipation of something pleasurable or even mystical, and apprehensiveness about succeeding at something that in some circles has become an index of personal worth (non-alpha-producers are not children of the New Age, according to many popular writers). Since the typical alpha-feedback

Equipment Problems: Alpha's Saturday Night Specials

If it cannot be demonstrated that enhancing alpha promotes health and well-being, it is almost gratuitous to point out that most inexpensive portable alpha-feedback machines are technically inadequate anyway. Several companies sprang up to cash in on the alpha-consciousness fad and even some of the most reputable scientific supply houses were quick to start selling "Saturday Night Special" alpha units.

Electrical signals in the brain are many times smaller than the electromagnetic noise that pervades modern buildings. Furthermore, the EEG can be swamped by bio-electrical activity of muscles, skin, heart, and eye movements. Simply jiggling the electrode cables can give rise to spurious signals. Selecting the real EEG out of this maze of artifacts is no small task, one for which units that sold for as little as $50 are clearly unfit.

Reliable EEG recording requires meticulous care in preparing the skin to receive the electrodes, minimizing electrode impedance, shielding and grounding subject and apparatus, filtering signals, etc. Even labs costing millions of dollars are occasionally plagued by interference and artifacts.

An artifact-resistant EEG electrode alone costs about $25—one of the "Specials" someone brought into our lab used for electrodes the chrome-plated disks intended for plugging holes in automotive body-work! The instructions suggested placing the electrodes in a position that maximized contamination from eye movements, but this hardly mattered because the wires carelessly soldered to the "electrodes" were left unshielded, forming a perfect aerial for interference.

Occasionally individuals bring their home alpha-apparatus to our lab to prove they can achieve results without our shielded recording suites and expensive apparatus. Never has anyone come to us out of dissatisfaction with their machines, but rarely has anyone failed to leave disappointed when side-by-side tests with our equipment showed what their devices were actually recording. Without adequate equipment and trained technicians, the majority of home alpha conditioners and alpha-parlor patrons have probably been "blissing out" on a symphony of eye movement and 60-cycle wall-main interference. Alpha conditioning cannot guarantee to lighten your spirits but it can easily lighten your pocketbook. *Caveat emptor!—B.B.*

setting is quiet, monotonous, and undemanding, this by itself tends to promote relaxation and, thereby, the dissipation of factors spuriously lowering the EC baseline. Relaxation does tend to enhance alpha output, but the converse is not necessarily true. When these situational variables are adequately controlled for, there is little evidence that increases in EC alpha-production with feedback are due to learned enhancement. Is there a better case to be made for the reports of eyes-open enhancement?

Several lines of evidence suggest that the rise in production of alpha during eyes-open (EO) feedback does represent a kind of learning, but not of the sort the feedback entrepreneurs had believed. What people learn with the aid of EO feedback (and probably with meditation training, too) are eye-movement, focusing, and attentional strategies that gradually overcome blocking of alpha that occurred before training. Thus it would seem that the apparent control over the EEG is mediated by learned behaviors that have long been known to affect EEG patterns. While this is of interest to psychophysiology, its relevance to therapy and spiritual enlightenment is questionable.

Alpha production goes up when active visual processing is minimized. Paskewitz and Orne (1973), for instance, showed that the alpha "enhancement" reported by others does not occur if subjects are given eyes-open feedback in darkness. They argue that, for the slow rise in alpha output to be seen, there must first be suppressing factors present (such as visual input to scan and assimilate) that the subject learns to ignore.

In 1977, I demonstrated a related phenomenon in subjects capable of hypnotically suggested blindness. (Note: Hypnosis, by itself, has no unique effect on EEG records.) When these subjects were told during eyes-open alpha-feedback that they could no longer see anything, their alpha output shot up immediately to almost equal their eyes-closed baseline. Output of matched subjects not given the blindness suggestion gradually rose to the eyes-closed baseline by the end of the feedback session.

Of course simply removing visual input does not mean that visual processing stops completely—visual images can be attended to from memory. Work in our lab has shown that attending to these subjective images also attenuates alpha (McBain 1983). Since attention is likely to flag over a period of alpha feedback, this too could contribute to the apparent enhancement effect.

Attention paid to the internal or external visual milieu affects both the kind and amount of eye movements. Plotkin (1976) compared the effects of instructions designed to promote feelings reported to accompany an "alpha experience" (relaxed serenity, loss of body and time awareness, diminution of thought, egolessness, etc.) to instructions that demanded specific visual strategies. He found that learning visual control had a far greater effect on the alpha output.

In reviewing the relevant research, Johnson (1977) concluded that there has never been a conclusive demonstration of learned enhancement of alpha in excess of the true (i.e., nondepressed) eyes-closed baseline.

Does alpha indicate a unique (higher?) state of consciousness?

It is a common error to mistake correlation for causation. As we have seen, alpha tends to be present during meditation or relaxed, dreamy wakefulness, but it also appears when one sits quietly with eyes closed. It seems the real correlation is with some underlying variable—probably similar visual activity rather than common subjective experience—that these states share. If this is true, we certainly could not guarantee access to any particular state of mind by teaching people to enhance alpha (assuming this were possible).

If alpha is compatible with a variety of subjective states, can it be found in states less desirable than those the alpha industry alleges are produced by their devices? The best alpha producer I ever encountered was a ten-year-old hyperactive child—hyperactivity hardly being a condition intuitively associated with serene contemplation. But hyperactivity *is* linked to difficulty in focusing attention, and such focusing tends to block alpha. On the other hand, approximately 10 to 15 percent of the normal population produces little or no alpha under any circumstances. In my experience they are not, as a group, any more anxious, "uptight," or less able to experience reverie than alpha producers. They do, however, seem to differ in the degree to which they employ visual imagery in their thinking and problem solving.

My initial mistrust of suggestions that alpha could be a quick and easy route to relaxation or "higher states" was reinforced by an informal experiment we carried out several years ago at the behest of a local physician. He had several extremely anxious patients maintained on high doses of tranquilizers. He wondered if alpha biofeedback could help reduce or eliminate their prescriptions. The first of several surprises appeared in the (drug-free) pre-training alpha baselines—several patients already produced abundant alpha at the outset and never surpassed their baseline in the ensuing sessions. Obviously some very anxious people can produce alpha. When the experiment was over, there was little to suggest that initial alpha output, or success or failure in augmenting it, bore any systematic relationship to the patients' evaluation of the training. Several showed no increase in alpha but nonetheless reported substantial relief. An equal number showed large increases in alpha output but complained they felt as anxious as ever.

Perhaps the most dramatic proof that there is no such thing as a unique "alpha state" comes from a study by Orne and Paskewitz (1974). Half their subjects received the calm, relaxing instructions of the typical alpha-feedback experiment, while the other half was treated brusquely and threatened with (but did not actually receive) painful electric shocks if they failed to increase their alpha production. Although the latter group, not surprisingly, reported anger, fear, and frustration (substantiated by physiological indications of agitation and high arousal), they nonetheless produced as much alpha as those who underwent the procedure designed to encourage relaxation and enjoyment. Thus, whatever strategies people acquire with the aid of alpha feedback, they can be learned in stressful conditions and have no automatic power to dispel unpleasant subjective states. Nevertheless, many people do report alpha conditioning to be pleasant and relaxing, and one might wonder why.

How did the widely held association between alpha waves and therapeutic or pleasurable states become established?

Critics of biofeedback, like those skeptical of psychic surgery, faith healing, or the "pop psychology" of EST and Scientology, are quickly inundated with testimonials from satisfied customers. But it has been shown repeatedly that such affirmations, while not to be dismissed out of hand, are nonetheless a weak currency (Nolen 1974; Randi 1980, Chap. 9; Gardner 1957; Rosen 1978).

There are several reasons why recipients of dubious "therapies" may honestly report improvements that cannot be objectively supported or why they may show real improvements that cannot legitimately be attributed to the "therapy" per se.

Most physical and psychological complaints are self-limiting. Many ineffectual treatments therefore are beneficial in the recipients' estimation because they coincide with recovery by natural restorative processes. In addition, these patients may simultaneously be receiving other, proved therapies or altering dietary, exercise, rest, and drug-use habits in salutary ways. And, of course, many complaints are essentially hypochondriacal to begin with and are amenable to simple assurance. Before a putative treatment can claim success, contributions of these other factors must be excluded.

Furthermore, as the connection between anxiety and stress and psychosomatic illness has been documented, it has also been found that psychological processes can aid in recovery. They do so directly by affecting body chemistry and indirectly by promoting beneficial changes in life-style. To the extent that any physiologically inert treatment instills the belief that it will work, it is likely to have these positive spinoffs. They are known as "placebo effects." Virtually none of the commercial biofeedback establishments and relatively few published research studies have included the placebo controls necessary to separate specific curative effects from these secondary factors that mimic them (Simkins 1982).

The reasons people seek alpha biofeedback fall into two broad categories. One group desires relief from stress-related complaints like anxiety, sleep disorders, drug abuse, tension headaches, hypertension, and chronic pain. The other is primarily interested in altered states of consciousness. Both present fertile ground for placebo effects. In evaluating the effectiveness of any treatment it is essential to know two things in advance: the proportion of people with a specific complaint who typically improve with no treatment at all (the spontaneous remission rate), and the proportion who respond favorably to a placebo. The placebo and its administration must as closely as possible resemble the treatment being evaluated, save withholding the suspected active principle or ingredient. Recipients should be carefully matched for kind and severity of symptoms and randomly assigned to an active or a placebo group, and neither they nor the evaluators should be aware who is in which group (the "double-blind" control). Only if these conditions have been met and the response to the active treatment reliably exceeds both the placebo response rate and the spontaneous recovery rate should the effectiveness of a candidate therapy be acknowledged.

When sufferers arrive at the faith-healer's stage, the shrine, or the alpha-conditioning parlor, they harbor many beliefs and expectations conducive to a placebo effect. Of course, if subjective or objective relief is forthcoming, this is not a bad thing (unless exorbitant sums are extracted under false pretenses or sufferers are prevented from receiving other, more effective treatments), but realizing that many putative treatments may actually be capitalizing on placebo effects might allow them to be used more efficiently, stripped of their complicated, expensive, and often mystical trappings.

So how did alpha feedback stack up when controls for expectancy and placebo effect were belatedly put in place? Several reviews (e.g., Johnson 1977; Plotkin 1979; Simkins 1982) concur that the effectiveness of biofeedback for physical ailments has not been adequately established. Reviewing the literature on applications to pain control, Melzack and Wall (1982) conclude that bio-feedback was "not found to be superior to less expensive, less instrument-oriented treatments such as relaxation and coping-skills training." Its limited value is seen as a distractor, essentially a mechanical placebo.

As for the claims that alpha feedback necessarily produces an enjoya-ble altered state of consciousness, Plotkin (1979) convincingly argues that, while this may occur, it can be accounted for by a combination of expectancy effects and factors related to the feedback situation rather than by any inherent connection with alpha per se. Plotkin examined studies where, unlike Kamiya's early reports, care was taken to find subjects who had no preconceived notions about the "alpha state," or where subjects' expectations were manipulated by overt or implicit suggestions from the experimenter. The inescapable conclusion is that to achieve "alpha consciousness" one must first be inclined to do so, by self-motivation, suggestion, or a number of situational variables common to most biofeedback settings.

The typical feedback setting is a complex social milieu including many psychological demands that amplify effects of preexisting desires and expec-tations. Furthermore, there are several aspects of the feedback routine that are known to affect consciousness in ways that bear a weak resemblance to meditative or hypnotic experiences. Among these are the effects of reduced sensory input and prolonged narrowing of awareness to the feedback signal, and mild elation at apparently succeeding in altering a physiological process the subject thinks will produce a desirable state of consciousness. It would seem that the reason there are so many satisfied alpha-feedback customers is that the process offers a restful temporary escape from the hustle of daily living under conditions that are themselves conducive to mild alterations in consciousness. In the absence of pleasant expectations, people are about equally

Theta: The Wave of the Future

If, as now seems incontrovertible, alpha is not the hasty man's short-cut to nirvana or even a credible alternative to a warm bath for reducing stress, there are other candidates waiting to take its place. The top contender at the moment is another frequency band of the EEG, theta (4 to 7 Hz). Many benefits once attributed to the "alpha state" are reappearing as purported consequences of high theta states. Like alpha, the theta rhythm has long been known to EEG researchers and several different theories exist to account for it. Theta has been extensively studied in relation to a brain structure called the hippocampus, which is involved in such diverse functions as arousal, attention, voluntary movements, and learning and memory. (See Bennett 1977, Chap. 11.) Although many biofeedback enthusiasts and occultists are embracing theta as the vehicle of alpha's lost promise, they seem likely to fall victim to the same logical and empirical inconsistencies that have plagued alpha research. Consider the Thomas Bennett statement (1982, 106) that "theta waves . . . can sometimes be observed during emotional stress in adults, particularly if the stress is produced by disappointment or frustration." In review-ing his own and others' research on theta, L. Johnson (1977) concludes that theta in the human EEG is "an artifact of general lowering of arousal level." Evaluating reports by Elmer Green and others of a link between theta and meditative and hallucinatory states, N. Birbaumer (1977) asserts that ". . . no sufficiently controlled study exists which supports any of these speculations" and that such claims are founded upon unsystematic observations lacking proper experimental designs. He goes on to agree with Johnson that there is no sound basis at the present time for attributing any benefits to theta biofeedback.

likely to find the experience euphoric, unpleasant, or neutral. That the presence of alpha is incidental if predisposition and setting are right is apparent from an incident where a new owner of a feedback device had a transcendent "alpha experience" only to find that his faulty apparatus was giving feedback unrelated to his EEG (Alcock 1979). I have had people report transcendent experiences when, unbeknownst to them, I had actually been giving them feedback to *suppress* their alpha! These are examples of accidental and intentional placebo control groups, respectively.

As Andrew Neher (1980) points out, we need not deny the existence of transcendent experiences to question popular mystical explanations for them. Neher shows how a quasi-meditative state like "alpha consciousness" is the expected outcome of a number of psychological manipulations and how to experience it without expensive apparatus or training.

Alpha and the Publishing Industry

Pandering to the narcissism of the "Me Generation" and playing on the legitimate concerns of those caught in high-tension life-styles has enriched authors and publishers, too. Best-sellers extolling the curative and mystical powers of "alpha thinking" run the gamut from the generally responsible but erroneous to the demonstrably absurd. On the one hand, we have books like Barbara Brown's (1974), which promotes her own views and warns against unscrupulous purveyors of the expensive, essentially useless alpha-conditioning apparatus. At the other extreme, we have laughable efforts like Jess Stearn's (1976) that are little more than advertising for seminars that offer paranormal powers under the guise of "alpha thinking" but never go near an EEG machine.

The middle ground is occupied by works like Lawrence's (1972) and Pines's (1973). They avoid much of the sensationalism of the genre but accept unquestioningly that alpha is incompatible with tension and anxiety and somehow related to ESP. However, for the most part they are only echoing claims by such researchers as Barbara Brown, Joe Kamiya, and Elmer Green. They fail to appreciate that these views were, and are, in the minority position among psychophysiologists. Lawrence's and Pines's books came out before some of the wilder assertions about alpha were overtaken by disconfirming data, but it is unfortunate that their misconceptions continue to be widely quoted.

Alpha and Psi Phenomena

Believers in the paranormal differ concerning whether or not psi will eventually be explained by the mechanisms of conventional science. Those in the affirmative were heartened by early demonstrations that biologically generated electrical fields could be detected outside the body. At last it was possible to suggest a mechanism whereby the emanations of one mind could conceivably interact with objects or other minds at a distance. Hans Berger's initial interest in the electrical activity of the brain stemmed from just such a desire to find a physical basis for psychic phenomena. Eight years before his classic report of the first human EEG recording, he had already published a monograph on psychic phenomena and he had devoted part of his rectoral address to the University of Jena to the subject. In the last publication of his life he propounded his brain-wave-propagation theory of telepathy (Brazier 1961, 112).

Subsequent demonstrations that the brain's electromagnetic fields obey the inverse square law, dropping to infinitesimal strength only millimeters from the scalp, and the failure of researchers to find (or even suggest) plausible brain mechanisms for receiving such emanations, if they could somehow traverse the required distances, has dampened but not eliminated theorizing along Berger's lines. A more prevalent recent trend among psi proponents has been to look to the EEG as a possible indicator of psi-conducive states (see, e.g.,

McCreery 1967, Part 2; Morris 1976; Honorton, 1977). Interest in alpha in this regard stems from two main sources. First, there was the presumed connection between alpha and meditative states, which, in turn, have been associated with psychic powers at least as far back as the *Vedas* of ancient India (Honorton 1977, 437). Second, it is a prevalent theme in psi research that the probability of psi is enhanced if the percipient is in a state of " 'detachment,' 'abstraction,' 'relaxation' and the like" (J. B. Rhine, quoted in McCreery 1967, 92). Honorton's (1977) description of psi-conducive "internal attention states" wherein external distractions and striving are minimized resembles many authors' recipes for "alpha consciousness." However, this runs counter to another strong theme in parapsychology (e.g., see Alcock 1981) that says highly motivated subjects perform better on psi tasks. Supporters of the latter position would presumably not expect alpha to be a concomitant of psi. The waters are further muddied by the realization, contrary to earlier belief, that EEG alpha does not necessarily denote a subjective state of effortless tranquility. It is not surprising, therefore, that when one tries to correlate two such elusive (and some would say nonexistent) phenomena as psi and alpha consciousness the resulting literature is, even to sympathetic reviewers like Morris (1976), "confusing."

Probably the staunchest defender of the alpha-psi link is Elmer Green. One of the most sanguine proponents of the curative potentials of biofeedback, he was brought to work at the Menninger Clinic by the noted parapsychologist Gardner Murphy. Green was recently seen on a nationally televised "documentary" on psychic phenomena claiming that some unremarkable EEG tracings were indicative that their producer was in telepathic communication. The loose procedure and lack of scientific controls displayed in this instance suggest such claims cannot be taken at face value.

Barbara Brown, though generally more responsible than most writers on the suggested alpha-psi link, nonetheless exhibits a generally credulous attitude toward claims in the area (Brown 1974, 405-407). She believes that not only will EEG research eventually succeed in identifying the brain correlates of psi states but through biofeedback people will be trained to enter them at will. To her credit, Brown acknowledges that other researchers had disputed her conceptions, but she attempts to cast doubt on their position somewhat curiously as arising from ". . . scientists . . . already at work to strengthen the dichotomy between science and spiritual development." In voicing her suspicions of a hidden agenda among critics who fail to see anything mystical in alpha waves, does she perhaps reveal one of her own?

Conclusion

Three of the main concerns of CSICOP are the demarcation of pseudoscience, evaluation of dubious health remedies, and examination of the evidence for alleged psychic powers. This article has touched upon aspects of alpha biofeedback relevant to each of these areas. I have tried to show how recent social trends set the stage for popular and professional espousal of a largely mistaken conception of the alpha rhythm. Its compatibility with the Zeitgeist perhaps accounts for the behavior of some respected scientists vis-à-vis alpha feedback that, if not pseudoscientific, at least falls short of the ideals of scientific verification. That much evidence of the previous 70 years made the revisionist notions about alpha improbable failed to impede the bandwagon effect. In the popular arena, pseudoscientific pronouncements on the curative effects of alpha abound, and the success of numerous marketing schemes fueled by throngs of satisfied customers demonstrates once again the difficulty of disputing such claims.

Alpha biofeedback seems to be yet another in a long series of putative treatments to benefit from the ubiquitous placebo effect—a placebo to appeal to those whose faith runs more along technological than along theological lines. Finally, the debatable status of psychic powers has not been enhanced by their alleged association with "alpha consciousness," a putative state whose own existence is in grave doubt. We are reminded once again of Bertrand Russell's observation that there will always be more defenders of popular falsehoods than unpopular truths.

References

Adrian, E., and B. Mathews. 1934. The Berger rhythm: Potential changes from the occipital lobes in man. *Brain*, 57:355-385.

Alcock, J. 1979. Psychology and near-death experiences. *Skeptical Inquirer*, 3, no. 3: 25-41.

———. 1981. *Parapsychology: Science or Magic?* Oxford: Pergamon Press.

Anand, B., G. Chhina, and B. Singh. 1961. Some aspects of electroencephalographic studies in Yogis. *EEG & Clinical Neurophysiology*, 13:452-456.

Bagchi, B., and M. Wenger. 1957. Electrophysiological correlates of some Yogi exercises. *EEG & Clinical Neurophysiology*. Suppl. No. 7:132-149.

Bennett, T. 1977. *Brain and Behavior*. Monterey, Calif.: Brooks-Cole.

———. 1982. *Introduction to Physiological Psychology*. Monterey, Calif.: Brooks-Cole.

Berger, H. 1929. Uber das Elektrenkephalogramm des Menchen. *Arch. f. Psychiat.*, 87:527-570.

———. 1930. Uber das Elektrenkephalogramm des Menchen. *J. Psych. Neurol.*, 40:160-179.

Beyerstein, B. 1977. Effects of visual stimulation and attentional levels on EEG alpha production in hypnotized subjects. Paper presented to the Western Psychological Assn., Seattle, Wash., April.

Birbaumer, N. 1977. Operant enhancement of the EEG-theta activity. In J. Beatty and H. Legewie, eds., *Biofeedback and Behavior*, 135-146. New York: Plenum Press.

Brazier, M. 1961. *A History of the Electrical Activity of the Brain*. London: Pitman.

Brown, B. 1974. *New Mind, New Body*. New York: Harper & Row.

Gardner, M. 1957. *Fads and Fallacies in the Name of Science*. New York: Dover.

Honorton, C. 1977. Psi and internal attention states. In B. Wolman, Ed., *Handbook of Parapsychology*, 435-472. New York: Van Nostrand-Reinhold.

Johnson, L. 1977. Learned control of brain activity. In J. Beatty and H. Legewie, eds., *Biofeedback and Behavior*, 73-94. New York: Plenum Press.

Kamiya, J. 1969. Operant control of the EEG alpha rhythm and some of its reported effects on consciousness. In C. Targ, ed., *Altered States of Consciousness*, 519-529. New York:

Anchor Books.

Kasamatsu, A., and T. Hirai. 1966. An electroencephalographic study on the Zen meditation (Zezen), *Folio Psychiat. & Neurolog.* Japonica, 20:315-336.

Lawrence, J. 1972. *Alpha Brain Waves.* New York: Avon Books.

McBain, I. 1983. Occipital Alpha Rhythm and Visual Processing. Unpublished honors thesis, Dept. of Psychology, Simon Fraser University, Burnaby, B.C.

McCreery, C. 1967. *Science, Philosophy, and E.S.P.* London: Faber and Faber.

Melzack, R., and P. Wall. 1982. *The Challenge of Pain.* Harmondsworth: Penguin.

Morrell, F. 1967. Electrical signs of sensory coding. In G. Quarton, T. Melnechuk, and F. Schmitt, eds., *The Neurosciences: A Study Program,* 452-468. New York: Rockefeller.

Morris, R. 1976. Biology and psychical research. In G. Schmeidler, ed. *Parapsychology: Its Relation to Physics, Biology, Psychology, and Psychiatry.* Metuchen, N.J.: Scarecrow Press.

Mulholland, T. 1968. Feedback electroencephalography. *Activities Nervosa Superior* (Prague), 10:410-438.

Neher, A. 1980. *The Psychology of Transcendence.* Englewood Cliffs, N.J.: Prentice-Hall.

Nolen, W. 1974. *Healing: A Doctor in Search of a Miracle.* New York: Random House.

Orne, M., and D. Paskewitz. 1974. Aversive situational effects on alpha feedback training. *Science,* 186:458-460.

Paskewitz, D., and M. Orne. 1973. Visual effects on alpha feedback training. *Science,* 181:361-363.

Pines, M. 1973. *The Brain Changers: Scientists and the New Mind Control.* New York: Harcourt, Brace, Jovanovich.

Plotkin, W. 1976. On the self-regulation of the occipital alpha rhythm: Control strategies and the role of physiological feedback. *J. Experimental Psychol: General,* 105:66-69.

———. 1979. The alpha experience revisited: Biofeedback in the transformation of psychological state. *Psychological Bulletin,* 86:1132-1148.

Randi, James. 1982. *Flim-Flam.* Buffalo, N.Y.: Prometheus.

Ray, W. J., and H. W. Cole. 1985. EEG Alpha activity reflects attentional demands, and Beta activity reflects emotional and cognitive processes. *Science,* 228:750-752.

Rosen, R. D. 1978. *Psychobabble: Fast Talk and Quick Cure in the Era of Feeling.* New York: Atheneum.

Simkins, L. 1982. Biofeedback: Clinically valid or oversold. *Psychological Record,* 32:3-17.

Singer, B., and V. Benassi. 1981. Occult Beliefs. *American Scientist,* Jan-Feb.:49-55.

Stearn, J. 1976. *The Power of Alpha Thinking: Miracle of the Mind.* New York: Signet.

Zaffuto, A. (with M. Zaffuto). 1974. *Alphagenics: How to Use Your Brain Waves to Improve Your Life.* New York: Warner Paperback Library.

DONALD D. JENSEN

Pathologies of Science, Precognition, and Modern Psychophysics

In philosophy there are two different views as to the source of knowledge: (1) rationalism (the view that what is true is what is logical, rational, and self-consistent) and (2) empiricism (the view that what is true is what is consistent with the evidence of the senses, with what can be observed). Usually these two views are seen as competitors or opposites, but they can also be seen as independent and potentially complementary views. Mysticism is an example of reliance upon neither logic nor observation, but upon revelation or intuition; while science involves the use of both logic and experience, both reason and observation, to guide opinion.

DATA USED	Empiricism	Science
DATA NOT USED	Mysticism	Rationalism
	LOGIC NOT USED	LOGIC USED

Four Pathologies of Science

The combining of data and logic is not always easily or appropriately accomplished, as Stephen Jay Gould (1986) noted at the 1986 CSICOP conference in Boulder when he described four pathologies of science. The four pathologies he described can be labeled fraud, finagle, propaganda, and prejudice. *Fraud* refers to the manufacture of evidence or data "from whole cloth," to fiction masquerading as fact, to hoaxes of the grosser sort—like those effectively exposed several decades ago by Houdini (1924) and more recently by James Randi (1982). *Finagle* refers to minor hoaxes performed by "massaging the data" and to intentional systematic errors of observation, data description, or data recording that lead to misrepresentation of phenomena. Gould's discussion (1981) of early investigations of the brain weights of different races gives an excellent example of finagled data, but many scientists know of more recent cases in their own fields where data have been slightly modified, sometimes simply by the discarding of a few troublesome observations, to enhance the statistical significance of the data and to ensure publication of the research report. See Barber (1976) for discussion of a number of different frauds and finagles in human psychological research. See Klass (1983; 1988) for frauds and finagles in UFO reports and Kusche (1975) for those relevant to the Bermuda Triangle.

It is important not simply to describe these four pathologies of science but to consider how to detect, correct, and prevent them. For the first two—fraud and finagle—this is usually accomplished in science by peer scrutiny and replication. Peer scrutiny is the critical and skeptical reading of research reports by other scientists. It may include reanalysis of data made available in the research reports or in archives. Replication is the process of repeating an investigation by another researcher, using another sample of subjects or another research method; a research finding is said to be replicated if substantially the same conclusions are reached in the repeat investigation as in the original one.

Any curious or unexpected finding is normally accepted only after having been scrutinized by other scientists without discovery of major flaws or omissions in research design, data gathering, data analysis, or argument. What is considered acceptable changes as research methods and statistical techniques develop and become standard practice; much research that was acceptable 25 or 50 years ago would be unpublishable today because some standard practices of that era were found to be flawed and have been supplanted by better methods. The publication of research reports and the reading of papers at scientific meetings are important because they facilitate this process of peer scrutiny, which has become easier and more effective as modern technology has become widely used in data collection and analysis. Modern researchers usually have a rich record verifying their data-gathering and analysis—data sheets, computer printouts, automatically produced charts and data reduction documents, and so on. Often it would be more work to fake such documents convincingly than to do the research that would generate them. The scrutiny of skeptical peers may include examination of such data records, and investigators can protect themselves from charges of fraud and finagle by making detailed and original documents available to peers. See Roman (1988) for a popular account of a case where peer scrutiny uncovered fraud in drug testing.

In the final analysis, the ultimate test of a scientific finding is its capacity for replication—the ability of others to do the same kind of research and get substantially the same data. Published reports of replication are enhanced by the same characteristics that make peer scrutiny effective: clear and detailed description of procedures for gathering data, of the data obtained, of the data analyses performed, and of the arguments made from those analyses. Peer scrutiny and replication are means of identifying frauds and finagles, and the threat of their occurrence almost certainly serves to discourage and limit the first two pathologies of science. But there are times when actual attempts at replication are necessary to identify frauds or finagles. One such example from my own research area is a report of learning in planaria (Griffard and Peirce 1964); letters requesting additional details to support replication of the study were not answered, and attempts to replicate the study produced data suggesting that the study would have been impossible to perform as described (Reynierse, Larson, and Jensen 1966). Here the threat of scrutiny and replication did not suffice, and an attempt to replicate was essential for the detection of a case of fraud or finagle.

The other two pathologies of science—propaganda and prejudice—are less easily detected, treated, or prevented. Propaganda refers to a pattern of selective presentation or marshaling of evidence for a preestablished point of view; ignoring or minimizing disconfirming data; and "accentuating the positive and eliminating the negative" in the discussion of and argument from the data gathered. Propaganda or biased presentation may be "good salesmanship" and acceptable in commercial affairs and certain legal realms, but it is considered inappropriate in science. A number of authors have recognized this as a problem in science (Chamberlin 1890; Platt 1984). Several decades ago, I encountered the practice while reviewing research on learning in paramecia and planaria (Jensen 1965). In one flagrant case, aspects of the data that contradicted the hypothesis that planaria can be classically conditioned to respond to light were hidden by the graphs and tables presented. Regraphing of the data clearly indicated that some other process was at work.

Scientific *prejudice* is the passive equivalent of scientific propaganda. It involves the acceptance of argument and data for a preestablished or favored point of view that would never be acceptable for a contrary opinion. It demands more of alternative ideas than is demanded of one's favorite ideas. It plays the game of science on an uneven field that favors one's own views.

Most readers of the *Skeptical Inquirer* see believers in the paranormal as examples of prejudiced evaluators who are much more responsive to evidence for the reality and mysterious nature of paranormal phenomena than to evidence of their spurious or trivial nature. It may surprise some readers to hear that skeptics are held to be equally guilty of the sin of scientific prejudice. It has been argued that scientists have shown prejudice against extrasensory perception (ESP) by requiring greater amounts and a higher quality of evidence for ESP than for other findings. There may be more than a grain of truth in this view, but the demand for greater scrutiny and more and stronger data is triggered by *any* exceptional and unexpected finding; it is not just the paranormal that is given "extra-high hurdles" to jump.

Consider the literature on poison-aversion learning in experimental psy-

chology. Extremely close scrutiny and unusually high demands for less ambiguous data were made of research that contradicted widely accepted views.
Poison-aversion learning, also called the "Garcia effect," refers to very rapid
learning of decreased preference for the novel flavors associated with gastrointestinal illness. This learning occurs with very long delays (hours rather than
seconds) between taste and illness and with "backward associations" (i.e., when
the flavor is tasted after the illness has begun); both conditioning with long
delays and with backward pairing were deemed impossible by the theories
of learning in vogue at the time the Garcia effect was first reported. Garcia's
data were so unexpected and challenging that he had difficulty getting papers
on this research published in major journals and in obtaining funds to continue
his work. But the research did eventually pass peer scrutiny, it was soon and
often replicated, and it is now a central part of the psychology of learning
(Seligman and Hager 1972; Barker, Best, and Domjan 1977). Detailed and
replicable data convinced the doubters, silenced the skeptics, and encouraged
a host of other researchers to work on the topic.

Any novel and unexpected phenomenon or view must be evaluated especially critically and subjected to attempted replication. The purported phenomenon is accepted when more and better data accumulate; or, if that is
not the case, it is shelved, becoming part of the history of a field rather than
an area of active research. The search for less ambiguous data characterized
the controversies over learning in planaria and the Garcia effect. In one case
(learning in planaria) this evidence was not generated, but in the other case
(the Garcia effect) it was. The data showing one-trial learning of taste-aversions
with long time intervals between taste and illness in many species and settings
accumulated. Controversy was replaced by eager acceptance of the phenomenon
and rapid development of a research literature.

It appears, then, that scientific investigation of a controversial topic involves the search for more, better, less ambiguous, and more interpretable
data and use of the best available methods of research. While the search
for *more* data can be subsumed under replication, the search for *better* data
(data that are more interpretable, less ambiguous, more convincing, more nearly
irrefutable, etc.) goes beyond replication and involves the improvement of
research methods. Better data may be obtained by improving observational
methods (improving reliability and providing checks on authenticity and
accuracy), increasing the power of experimental manipulations and controls,
improving research designs (adding placebo controls, using double-blind procedures, etc.), and by improving the methods by which data are summarized
and interpreted.

One of the advantages of statistical testing of data is that it provides
clear and relatively neutral or unbiased methods of summarizing and evaluating data. Standard descriptive statistics provide useful tools for understanding large data sets and "seeing the forest rather than the trees." Standard
tests of statistical significance ask whether the data depart from chance
expectation sufficiently to make chance an unreasonable explanation. Most

scientists now routinely ask this question of all data sets used as evidence, and only those that pass this criterion of statistical significance are normally accepted as evidence for or against any theory.

Given that better data and better methods of data analysis can protect against scientific propaganda and prejudice, what kind of data would be the best possible evidence for clairvoyance or precognition? What kind of data would an unbiased modern investigator find acceptable and worthy of serious consideration?

Needed: Unambiguous Data for Precognition

The evidence normally offered in the popular press for paranormal perception is the individual case of apparent precognition. This is the report of a hunch or a feeling or fear that an event would occur, followed by a verified occurrence of the feared event. Such events are frequently reported to bridge great distances to signal catastrophes occurring to close relatives or friends. A number of alternative explanations have been offered for such reports, including retrospective falsification and memory distortion (a kind of finagle) as well as fraud. These are important possibilities, but the point of this article is a very different one—that even in the absence of fraud and finagle, such cases are not and never can be acceptable evidence for clairvoyance. The reason for this has to do with the inevitable ambiguity of evidence gathered by a research method in which a stimulus is always presented.

Early Psychophysical Methods

The "method of limits" is one of the standard research techniques of classical psychophysics. One of its uses involves giving stimuli that vary in intensity and asking the subject to indicate whenever the stimulus or signal is detected. The data are typically analyzed by grouping responses by stimulus intensity and computing the proportion or percentage of responses given to each stimulus intensity. The data are then graphed with percentage of responses displayed on the vertical axis and intensity of the stimulus on the horizontal axis. (See Figure 1.)

Sometimes this evidence for stimulus detection is relatively unambiguous, as when a subject never responds to weaker intensities of stimulation, sometimes responds to middle intensities, and always responds to stronger intensities. Here a simple rule works well to provide a measure of sensitivity to stimulation. The rule is to determine the 50-percent point, the amount of stimulation expected to produce response 50 percent of the time.

The data can be ambiguous if certain other patterns of responding occur. (See Figure 2.) Consider several different sets of data. Curve A represents the "good subject" described in the last paragraph. Curve B represents a subject who says yes occasionally at all intensities of stimulation, but more often at higher intensities. Curve C represents a subject who says yes about half the time at lower intensities, with the response rate increasing at higher intensities. Curves A, B, and C represent subjects whose responses are related to stimulus intensity, but differ in general tendency to say yes, in what is called "response bias." This general tendency undermines the validity of the standard method of computing the absolute threshold (determining the 50-percent point), since curve B has less far to go to get to the 50-percent level, and curve C starts there!

The other three patterns of data show another complication. Curves D, E, and F (dotted horizontal lines) represent subjects who say yes at different rates (seldom, half the time, and always) but do so in a way unrelated to the intensity of the stimulation given. Curve E can be produced by someone who tries to do what is asked, and so says yes part of the time, but doesn't seem to be able to detect stimuli at any of the intensities used; this subject shows a moderate response bias and no ability to detect the stimulus. But what about curves D and F? These two patterns of data are ambiguous. The top dotted line (curve F) could represent either a person who was sensitive to all of the stimulus intensities sampled or a person who simply said yes

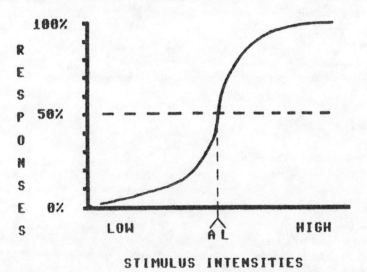

FIGURE 1. The absolute threshold of stimulation is usually identified as that intensity of stimulation expected to produce a response 50 percent of the time; the absolute threshold was the standard measure of sensitivity to stimulation for almost 100 years. The method of limits and reports of clairvoyance differ in many regards (i.e., only the first involves systematically varying intensity of stimulation), but they resemble each other in that both involve only stimulus-present situations or trials. In both cases, there is always a stimulus or signal present, and the question is whether or not the person was sensitive to or detected the signal. In both, a response is considered to be evidence of detection of a signal.

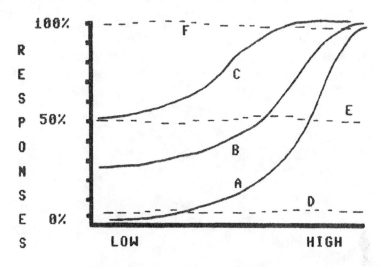

STIMULUS INTENSITIES

FIGURE 2

all the time. The bottom dotted line (curve D) could represent either someone who was deaf or blind to the stimuli being presented or someone who was uncooperative and never said yes.

To tell the difference between indiscriminate responding (response bias) and detection of a signal or stimulus, trials in which no signal or stimulus is given are necessary. If the subject responds on trials when the stimulus is present but not on trials when the stimulus is absent, then there is relatively unambiguous evidence that the subject is detecting the stimulus. On the other hand, if responses are just as likely in the presence of a stimulus as in its absence, then the subject shows a tendency to respond (response bias) but no ability to detect the stimulus. Mixtures of response bias and signal detection are evident when responses occur in the absence of the stimulus but are more likely in the presence of the stimulus.

A standard terminology exists for discussing data from research with both stimulus-present and stimulus-absent trials. If stimulus-absent trials occur only occasionally, they are called "catch" trials, since they are intended to catch or identify subjects who show indiscriminate responding. If stimulus-absent trials occur frequently, the following nomenclature is used to describe the four possible combinations of two kinds of response (yes and no) with two kinds of trials (stimulus-present and stimulus-absent).

STIMULUS PRESENT	Hit (Correct Positive)	Miss (Incorrect Negative)
STIMULUS ABSENT	False Alarm (Incorrect Positive)	Correct Rejection (Correct Negative)
	YES	NO

RESPONSE

A subject who responds when a signal is present scores a "hit" or makes a correct positive response. A subject who responds when a stimulus is absent gives a "false alarm" or makes an incorrect positive response. A subject who does not respond when a stimulus is present is scored for a "miss" or an incorrect negative response: A subject who does not respond when the stimulus is absent is scored for a correct rejection or a correct negative response. Data from one level of stimulation are evaluated, not by percentage of "yes" responses given at that level of stimulation, but by the set of four percentages obtained at that level. The data from a modern psychophysical experiment would include one table of four percentages for every level of stimulus investigated.

Perfect detection of a stimulus intensity that is presented on half of the trials is shown by the following data:

50% hits	0% misses
0% false alarms	50% correct rejections

Perfect detection is shown by "yes" responses only in the presence of the signal; all such responses are hits or correct rejections.

No detection is shown when there are equal numbers of hits and false alarms, even though the percentage of hits may vary. No detection is shown by the two data sets shown below, but they differ in the amount of response bias shown. The first data set shows a low general tendency to say yes, or a low response bias; the second set shows maximal response bias (*always* saying yes).

Low Response Bias Without Signal Detection

10% hits	40% misses
10% false alarms	40% correct rejections

Maximal Response Bias Without Signal Detection

50% hits	0% misses
50% false alarms	0% correct rejections

Both of the above sets of data represent 50 percent correct responses (the sum of hits and correct rejections) but the meaning of 50 percent correct is very different in the two cases, since the second set of data shows much greater response bias.

A mixture of detection and response bias is shown by the following data in which there are more hits ("yes" when stimulus present) than false alarms ("yes" when stimulus absent):

Moderate Response Bias and Moderate Signal Detection

40% hits	10% misses
25% false alarms	25% correct rejections

Graphing Detection and Bias: ROC Curves

All possible data sets of the type just discussed can be shown on a diagram produced by representing each set of data by a single point on a graph. (See Figure 3.) The vertical axis represents the percentage of hits and the horizontal axis represents the percentage of false alarms. A single point can represent the set of four values if equal numbers of signal-present and signal-absent trials are given, since the other two can be obtained by subtraction. This kind of diagram is called an ROC diagram; ROC stands for Receiver Operating Characteristic, a term that, along with response bias, signal detection, hits, misses, false alarms, and correct rejections, is a legacy from electrical engineering, where the data-analysis methods we have been discussing were developed to aid the evaluation of radar devices (Green and Swets 1966).

Any set of data with equal percentages of hits and false alarms and corresponding to no detection would lie on the diagonal running from the lower left to the upper right. The ROC diagram graphically separates signal detection from response bias, and data obtained using signal-detection methods are unambiguous. The person who says yes only when the signal is present (perfect signal detection) is located in the upper left corner of the ROC diagram

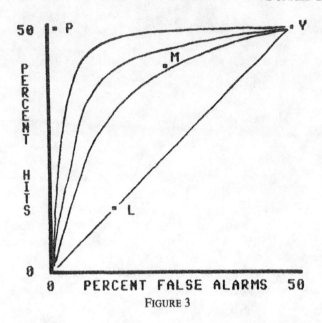

FIGURE 3

(point P): The person who says yes on every trial (maximal response bias) is in the upper right corner on the line of bias without detection (point Y). The person who says yes on one-fifth of the trials but without regard to the presence or the absence of the signal is shown lower on the line of bias without detection (point L). The person who says yes most of the time when the stimulus is present and half the time when the stimulus is absent is shown above the diagonal of bias without detection (point M).

Mixtures of different amounts of signal detection and response bias are represented by points on the ROC diagram located above and to the left of the diagonal of no detection. Equal amounts of detectability are shown along contours running in bow-shaped curves from the lower left to the upper right corners of the ROC diagram. The standard way to represent stimulus detectability and response bias is by use of two different numerical values, d' (d-prime, which reflects detectability of signals) and beta (which reflects response bias). It is d' that normally varies as a function of stimulus intensity and replaces absolute threshold as an index of performance in modern psychophysical research. The exact procedures by which values of d' and beta are calculated is of little interest here, but the difference between signal detectability and response bias is crucial to our discussion.

Application to Precognition

Signal-detection methods have been widely used in modern psychology. Their rapid acceptance into psychology is understandable because they separate detectability (of interest to psychologists studying sensory and simple perceptual processes) from response bias (of interest to psychologists studying motivation, personality, and psychopathology). Modern sensory psychology and neurophysiology provide explanations for the remarkable capacities that exist for signal detection; other kinds of psychology provide multiple alternative and supplementary explanations for response bias. For example, various kinds of epileptic conditions (Beyerstein 1988) and schizophrenia are characterized by hallucinations (perceptions without external referents, sensory responses without signals, "false alarms").

Drugs, sleeplessness, and sensory isolation produce mild and brief sensory disturbances in normal individuals. High arousal and sustained stress tend to increase false alarms, as when deer hunters shoot at cows and other hunters as well as deer, or when sentries shoot at nonexistent intruders. Ellson (1941) demonstrated that hallucinatory responses could be produced in normal subjects by classical conditioning. This was accomplished by following a "ready" signal by a tone that increased and then decreased in intensity but did not have an abrupt onset or offset. After a number of trials, the "ready" signal was given without being followed by any tone at all. Subjects pushed buttons to report hearing the tone and were adamant that the tone had in fact occurred. The behavior of Ellson's subjects was evidence that they had received the "ready" signal, but not that the tone had sounded.

While all of these varied explanations must be considered in analyzing evidence for detection of distant and future events (i.e., clairvoyance and precognition), they are all explanations for response bias, not for stimulus detectability.

Now let us apply the ideas of response bias to a real-life situation. It is graduation night at the local high school. Several hundred seniors are leaving the ceremony for parties and social activities. Mothers are worried: Will there be terrible auto accidents on this night-of-nights for high school seniors? If we substitute "Mother's Worry" for "Yes Response" and substitute "Catastrophe Occurs" for "Signal Present," we can use the now-familiar terms of hits, misses, false alarms, and correct rejections.

CATASTROPHE OCCURS	Hit	Miss
NO CATASTROPHE	False Alarm	Correct Rejection
	MOTHER'S WORRY	NONE

When a mother worries, and something terrible happens to her child, the popular press and an unsophisticated observer would say: "A case of precognition, clairvoyance, ESP, mysterious power of the human mind, etc." The more sophisticated observer would note the hit but would ask as well about the number of false alarms. It is plausible that most mothers of graduating seniors will worry, and most will feel relief when their darlings return unharmed if tipsy from their graduation parties; false alarms are common on graduation nights. But let a catastrophe occur, and the same response bias that produced many false alarms can produce a hit by chance and chance alone. It is important to note that only the "hits" are likely to be remembered or publicized. Unfortunately, human memory is very selective; we remember hits (cases of fear followed by catastrophe) and misses (cases of unexpected catastrophe) much better than false alarms (fear followed by no catastrophe); false alarms (fear without catastrophe) are normal daily events that we usually neither name nor count. Further, only "hits" are newsworthy, and published accounts of "hits" represent only one of the four kinds of events. Newspapers do not report misses, false alarms, or correct rejections.

To interpret hits effectively, one needs information on all four classes of events, and not just information about hits. Given such data, and evidence that other classes of events are as faithfully reported as are hits, one could obtain relatively unambiguous evidence of signal detection (i.e., statistically, a significantly greater percentage of hits than false alarms). Such information on false alarms, misses, and correct rejections, as well as on hits, is difficult but not impossible to obtain. One could, for example, ask students in large classes to predict, each time they come to class, whether a major domestic airline crash (more than 10 fatalities within the United States) is to occur between that and the next class meeting. Over several semesters, this procedure should produce a number of hits (students saying yes when a catastrophe does occur before the next class meeting), misses (students saying no when a catastrophe does occur), false alarms (yes when a catastrophe does not occur), and correct rejections (no when no catastrophe occurs), and these data could show signal detection as well as response bias. But note that no single case of fear followed by catastrophe, no matter how detailed and remarkable, is acceptable evidence for precognition or clairvoyance, because all four kinds of events must be observed to obtain relatively unambiguous data for signal detection.

Modern psychophysics, with its separation of response bias and signal detection, offers an improved paradigm for investigating paranormal perceptions. At an informal level, that paradigm is already being used. The editors of the *Skeptical Inquirer* frequently report more than the hits that professional psychics report; they count misses (Chernobyl, the Challenger disaster, the volcanic eruption in Colombia, etc.) and false alarms (predictions that fail to occur). There are, of course, other ways of dealing with response bias as a contaminating influence upon data; formal tests of paranormal abilities can be designed so that the effects of response bias are controlled or prevented.

A dowser, for example, might be tested by having half of a sample of pipes containing flowing water or half a set of cans containing precious metal, with the dowser instructed to identify only half of the stimuli as positive cases. Response bias is prevented in many psychophysical procedures by presenting stimuli intervals in pairs, one with and one without the signal, and allowing the subject being tested to select one and only one of each pair as the positive case. This has the effect of requiring simultaneous pairs of yes and no responses and preventing response bias; data of this kind, like that obtained when half the trials involve stimulus-absent conditions, can show stimulus detection (more correct choices than incorrect), zero detection (equal numbers of correct and incorrect choices), and "worse than chance" responding (fewer correct than incorrect choices). Statistical significance of differences from chance responding then can be estimated by a variety of statistical tests.

In general, the pathologies of propaganda and prejudice can be prevented by use of modern research methods, such as those for separating response bias and signal detection, and by formal testing that prevents response bias. While modern research methods do not guarantee freedom from fraud, finagle, propaganda, or prejudiced evaluation of data and argument, they can make these pathologies of science less likely to occur in the investigation of normal and of purportedly paranormal phenomena.

Summary

Modern psychophsyics has methods for producing relatively unambiguous data relevant to precognition and clairvoyance. These methods make possible the separation of signal detection and response bias by the study of hits, misses, false alarms, and correct rejections. Experimental procedures can also be used to eliminate response bias in tests of paranormal perception. Contemporary psychology provides many explanations for response bias that may masquerade as detection in situations in which only stimulus-present trials are considered. Modern psychophysical methods are one of the antidotes available for the pathologies of science (fraud, finagle, propaganda, and prejudiced evaluation). These methods require attention to all relevant data, and not just to "hits," which may result from either signal detection or response bias.

References

Barber, T. X. 1976. *Pitfalls of Human Research.* New York: Pergamon Press.
Barker, L. M., M. R. Best, and M. Domjan. 1977. *Learning Mechanisms in Food Selection.* Waco, Tex.: Baylor University Press.
Beyerstein, B. L. 1988. Neuropathology and the legacy of spiritual possession. *Skeptical Inquirer,* 12:248-262.
Chamberlin, T. C. 1890. The method of multiple working hypotheses. Reprinted in *Science,* May 7, 1965.

Ellson, D. 1941. Hallucinations produced by sensory conditioning. *Journal of Experimental Psychology,* 28:1-20.

Gould, S. J. 1981. *The Mismeasure of Man.* New York: Norton.

————. 1986. Keynote address, CSICOP Conference, Boulder, Colorado, April 25, 1986.

Green, D. M., and J. A. Swets. 1966. *Signal Detection Theory and Psychophysics.* New York: Wiley.

Griffard, C. D., and J. T. Peirce. 1964. Conditioned discrimination in the planarian. *Science,* 144:1472-1743.

Houdini, H. 1924. *A Magician Among the Spirits.* New York: Arno Press.

Jensen, D. D. 1965. Paramecia, planaria, and pseudolearning. *Animal Behavior Supplement,* 1:9-20.

Klass, P. J. 1983. *UFOs: The Public Deceived.* Buffalo, N.Y.: Prometheus Books.

————. 1988. *UFO-Abductions: A Dangerous Game.* Buffalo, N.Y.: Prometheus Books.

Kusche, Larry. 1975. *The Bermuda Triangle Mystery—Solved.* Reprinted in paperback by Prometheus Books. Buffalo, N.Y., 1984.

Platt, J. R. 1984. Strong inference. *Science,* 146:347-353.

Randi, J. 1982. *Flim-Flam.* Buffalo, N.Y.: Prometheus Books.

Reynierse, J. H., D. Larson, and D. D. Jensen. 1966. Shock duration and adhesion in the planarian, Phagocata gracilis. *Psychonomic Science,* 8:269-270.

Roman, M. B. 1988. When good scientists go bad. *Discover,* 9(4):50-58.

Seligman, M. E. P., and J. L. Hager. 1972. *Biological Boundaries of Learning.* New York: Appleton, Century, and Crofts.

SUSAN BLACKMORE

The Elusive Open Mind: Ten Years of Negative Research in Parapsychology

Everyone thinks they are open-minded. Scientists in particular like to think they have open minds, but we know from psychology that this is just one of those attributes that people like to apply to themselves. We shouldn't perhaps have to worry about it at all, except that parapsychology forces one to ask, "Do I believe in this, do I disbelieve in this, or do I have an open mind?"

The research I have done during the past ten or twelve years serves as well as any other research to show up some of parapsychology's peculiar problems and even, perhaps, some possible solutions.

I became hooked on the subject when I first went up to Oxford to read physiology and psychology. I began running the Oxford University Society for Psychical Research (OUSPR), finding witches, druids, psychics, clairvoyants, and even a few real live psychical researchers to come to talk to us. We had Ouija board sessions, went exploring in graveyards, and did some experiments on ESP and psychokinesis (PK).

Within a few weeks I had not only learned a lot about the occult and the paranormal, but I had an experience that was to have a lasting effect on me—an out-of-body experience (OBE). It happened while I was wide awake, sitting talking to friends. It lasted about three hours and included everything from a typical "astral projection," complete with silver cord and duplicate body, to free-floating flying, and finally to a mystical experience.

It was clear to me that the doctrine of astral projection, with its astral bodies floating about on astral planes, was intellectually unsatisfactory. But to dismiss the experience as "just imagination" would be impossible without being dishonest about how it had felt at the time. It had felt quite real. Everything looked clear and vivid, and I was able to think and speak quite clearly.

You can imagine the intellectual conflict I experienced (and of course I had no idea it was only a prelude to far worse mental conflicts!). The psychologists and physiologists who were teaching me made quite different assumptions about human nature from those made by the people I met through the OUSPR. The latter, for the most part, assume that there is "another dimension" to man, that we can communicate directly mind to mind, that there are "other worlds" waiting to be explored in altered states of consciousness, and even that consciousness is separable from its physical home and might survive the death of its body. The conflict was a challenge to me and I conceived the objective (I think naively, rather than purely arrogantly) of proving my teachers wrong, or at least showing that psychologists were closed-minded in ignoring the most important of human potentials—the paranormal.

Even at that very early stage I made a crucial mistake—or a series of crucial and related mistakes. First, I assumed that all these odd and inexplicable things—ESP, PK, OBEs, mystical experiences, ghosts, poltergeists, and near-death experiences—were related and that one explanation would do for all. Second, I assumed that there had to be a paranormal explanation—that we were looking for psi. Third (and I don't know whether this was just cowardice or an attempt at being sensible for a change), rather than launching straight into what really interested me—the OBE—I thought it was more "scientific" to begin with psi. After all, there had been research done on ESP and PK and, though generally rejected, it had some basis in scientific research. It seemed far easier, and safer, to start there. I didn't notice what I was doing. I can only point it out with the benefit of hindsight. I just took psi to be the key to the mysteries and wanted to study parapsychology.

The first thing I did was to develop my own theory of psi. This theory involved the notion that psi and memory are aspects of the same process, that memory is a specific instance of the more general process of ESP. Eventually I got a place at Surrey University to do a Ph.D., and it was then that I set about testing my theory.

While I was at Surrey I was lucky enough to be given the chance to teach a parapsychology class. It attracted more than a hundred students, so I had plenty of subjects for my experiments. I began three kinds of tests. First, I predicted a positive correlation between ESP and memory. That is, if memory and ESP are aspects of the same process, then the same people should be good at both of them. I did many tests of this kind (Blackmore 1980a). Second, I predicted that the best target materials for ESP should not be those that are easy to perceive, but those that are easy to remember. I did a series of experiments with different target materials (Blackmore 1981a). Third, I predicted that the errors and confusions made in ESP should more closely resemble those made in memory than those made in perception. I had high hopes for this method since the study of errors has always been so useful in psychology, for example, in the study of visual illusions. I also did many experiments to test this (Blackmore 1981b). However, the only noteworthy thing about all of the results was the number that were not significant.

After a long series of experiments I had no replicable findings and only a large collection of negative results. Clearly they could not answer my original questions, nor test my special theory. Some of you may already be protesting: What an idiot. Why didn't she just give up and do something useful instead? But I would have responded: This could be useful! If ESP exists, it could be one of the most important findings for science; and in any case you can never tell in advance what research will be useful in the end. You may also be thinking, as many people said at the time: "Oh but this is just what you'd expect. She has only shown that there is no psi." But of course I hadn't done that, and couldn't do that. No amount of negative results can prove the nonexistence of psi. Psi might always be right around the next corner, and there were plenty of corners to look around.

There were also plenty of parapsychologists eager to suggest corners I had not yet turned and reasons why my experiments had not worked. And I was eager to carry on the search. Some said it might be the subjects; students are notoriously not the best ones. So, instead of testing my class, I tested people who came to me with claims of special powers. I tried to design experiments that would test what they claimed to be able to do and that would allow me to impose sufficient controls. In some ways this upset me more than anything, because I met lots of genuine and well-meaning people who were convinced they could communicate by telepathy, or find underground pipes or hidden water, until they tried to do it under conditions that ruled out normal sensory information. Then they, and I, were always disappointed.

Then I tried using young children as subjects. At that time, Ernesto Spinelli was getting outstandingly good results with preschool children in ESP tests (Spinelli 1983). So I set about designing experiments to use a method similar to his (though not a direct replication) to test my memory theory. It was much harder work than the previous experiments, but much more fun. The children were three- to five-year-olds in playgroups, and they thoroughly entered into the whole idea, being convinced they could transmit pictures to one another. But the results were quite clear. The proportion that were "nonsignificant" was as high as before. The overall results were nonsignificant and so were the correlations with age (Blackmore 1980b).

Why? Spinelli had many suggestions. It could have been that I used colored pictures, while his were black and white; or that the sweets I used as a reward (based on someone else's previously successful experiments) were too well liked by the children and were disruptive; or that I simply didn't have the right personality and rapport with the children. I could only say that I seemed to get on well with the children, but perhaps this was not well enough.

Another suggestion was that the problem was not the subjects themselves, but the state of mind they were in during the experiments. At that time, the ganzfeld experiments were the "latest thing," and the results from Carl Sargent (1980) at Cambridge, and Chuck Honorton (1977) at Princeton, seemed impressive. So I set about doing a ganzfeld study. My subjects each had half of a ping-pong ball covering each eye, lay on a reclining chair, and heard only white noise fed through headphones. I wrote down everything they said. Then they had to look at four pictures and choose which one they thought the agent had been looking at.

I had for some months led an imagery training group, in which we practiced relaxation, guided imagery, and many imagery tasks adapted from Buddhist training techniques. For my ganzfeld study I chose ten test subjects from this group and ten control subjects.

This study taught me a lot. Being in ganzfeld is in itself an interesting experience. Images come pouring in, and it is tempting to imagine that you are picking them up from somewhere outside of yourself. I also had one very impressive experience in which I was subject and my brother was agent. I "saw" people fishing and lakes, mountains, and swiss chalets, and when I saw the targets I picked the correct one right away. It was an amazingly close hit. It set me to wondering whether I had at last found the key! However, in the course of the experiment I saw many equally amazing correspondences, but to the wrong pictures. My remarkable hit rapidly disappeared among the chance scores.

This should have taught me something important, something I should have known all along; that is, one should not rely on subjective estimations of probability (see Blackmore and Troscianko 1985). One should rely only on the statistics, and they were telling me that there was nothing there. Of course I tried it again with my brother, but the second time it did not work. Overall the results were close to chance expectation.

Why did this study also fail? I had used trained subjects in psi-conducive conditions and a method others had found successful. The ultimate suggestion of most parapsychologists was that it was an experimenter effect—more than that, it was a psi-mediated experimenter effect. That is, either I was using my own negative psi or I had some kind of personality defect, or defect in belief, that suppressed the psi of other people. I was a psi-inhibitory experimenter, so that whatever I did I would always get negative results. I began to get the feeling that I had some creeping sickness. I was a failure, a reject; there was something in me that suppressed the true spiritual nature of other people. I tried not to let it upset me, but I must admit that there is something terribly unflattering about being labeled "psi-inhibitory"!

Well, what could I do about it? It is not entirely an untestable idea. But Sargent had already tested the personalities of successful and unsuccessful experimeters and found the successful ones to be extroverted, confident, non-neurotic, and so on. In fact I fitted the description quite well—except for my results.

The other key to my failures seemed to be belief. I was told that I didn't get results because I didn't believe strongly enough in psi, because I didn't have an open mind! But what could I do about that? I couldn't just change my beliefs overnight or test ten subjects while believing and another ten while not! I argued that in the beginning I had believed in psi and still had got no results, but I couldn't prove this against the counter-argument that I had never *really* believed at all.

However, I did have an idea. There were still things in which I did believe.

I could test the Tarot. I had, in my preoccupation with everything occult, been reading Tarot cards for about eight or nine years. They really did seem to work. People told me that I could accurately describe them using the cards, and this was, naturally, gratifying. I even thought it might have a paranormal basis. So I set about testing the cards, doing readings for ten people, keeping the procedure as close as possible to a normal Tarot reading, but isolating myself, as the reader, from the subjects. They then had to rank all ten readings to see whether they picked their own more often than chance would predict (Blackmore 1983).

It worked! The results were actually significant. You can imagine my excitement—perhaps I had at last found something. Perhaps there was no psi to be found in the standard laboratory experiments, but something paranormal could appear when the conditions were closer to real life. But then I talked to Carl Sargent. He pointed out that all my subjects knew one another, and if they knew one another their ratings and rankings could not be independent. So I had violated an assumption of the statistical test I was using.

This seemed so trivial. Their knowing one another could not help them pick the right reading, could it? No it couldn't; but this meant that the estimate of probability was inaccurate—and, after all, the results were only marginally significant. So I repeated the experiment twice more with subjects who did not know one another. I expect you can predict the results I obtained—entirely nonsignificant.

You may choose to interpret these results in different ways. Some parapsychologists have claimed that the first experiment found genuine psi and that the later ones didn't summon the same attitude, the same novelty, the same enthusiasm, that made psi possible—or even that psi itself doesn't like being replicated. But I think I had finally reached a stage where I no longer felt it was worth pursuing such arguments. I chose this point to say: "I think that, however many more experiments I do on psi, I am probably not going to find it."

Now we finally come to the question: "What do these negative results tell us?" Of course the one thing they do not tell us is that psi does not exist. However long I went on looking for psi and not finding it they could not tell us that. But I found myself simply not believing in psi anymore. I really had become a disbeliever. Like one of those doors with a heavy spring that keeps it closed, my mind seemed to have changed from closed belief to closed disbelief.

But either way I suffered. There was mental conflict whether I believed or disbelieved. I had many questions. One was this: How far could I generalize these negative results? The situation was the converse of the normal situation in science when one gets positive results and has to ask how far they can be generalized. Here I had to ask whether my negative results applied only to those experiments carried out by me, at those particular times, or whether they applied to the whole of parapsychology. There is no obvious answer to that question. If one had replicability one could answer the question as one does in other areas of science. But without replicability it is impossible.

The next question was: How could I weigh my own results against the results of other people, bearing in mind that mine tended to be negative ones while everyone else's seemed to be positive ones? I had to find some kind of balance here. At one extreme I could not just believe my own results and ignore everyone else's. That would make science impossible. Science cannot operate unless people generally believe other people's results. Science is, and has to be, a collective enterprise.

At the other extreme I could not believe everyone else's results and ignore my own. That would be even more pointless. There would have been no point in all those years of experiments if I didn't take my own results seriously. Indeed, it is a fundamental principle in science that one has to take notice of the results one finds.

So there is no right answer to how to weigh them up. And these problems are only aspects of the basic dilemma of parapsychology, which is whether to believe or disbelieve in the existence of psi. Either way, I suggest, one meets conflict.

In the believer's position one is saying: "I believe there is something negatively defined, defined as communication without the use of the recognized senses, or action without the use of the muscles of the body. I have faith that future experiments will find this thing, even though so far they have failed to produce a replicable effect." If one takes this position, then one not only has to accept the open-ended nature of the search but also has to face up to the mounting negative results.

But what about the disbeliever's position? The disbeliever is only saying: "I do *not* believe there is this negatively defined thing. I do not believe the search will be successful. I have faith that all experiments with positive results could be successfully debunked." So the disbeliever is in a kind of mirror-image of the believer's position. But of course one can never debunk all the experiments, and there will always be more in the future. So the search is equally open-ended. And the disbeliever has to take notice of those positive results. I am thinking particularly of the results of Carl Sargent, Charles Honorton, Helmut Schmidt, and Robert Jahn. I suggest that if we think these can easily be dismissed then we are only deluding ourselves. One cannot offer simplistic counterexplanations and throw all these results away. I am not saying that these results may not, in the future, succumb to some normal explanation; they may well do so. But at the moment we do not have such an explanation.

Whether you are a believer or a disbeliever you will suffer mental conflict and anguish. So what is the solution? Easy, isn't it? Have an open mind. But human beings are not built to have open minds. If they try to have open minds they experience *cognitive dissonance*. Leon Festinger (1957) first used this term. He argued that people strive to make their beliefs and actions consistent and when there is inconsistency they experience this unpleasant state of "cognitive dissonance," and they then use lots of ploys to reduce it. I have to admit I have become rather familiar with some of them.

First there is premature closure. You can just pick one theory and stick to it against all odds. But I could not do that after all those years. What I could do was only slightly more subtle; that is, I could prefer one theory and ignore the evidence that goes against it. In this way the believer can dismiss negative results by using all the old arguments: The time, the place, the emotional state, or the "vibes" weren't right. Or the disbeliever can refuse to look at the positive results. You may think I wouldn't refuse, but I have to admit that when the *Journal of Parapsychology* arrives with reports of Helmut Schmidt's positive findings I begin to feel uncomfortable and am quite apt to put it away "to read tomorrow."

Alternatively one can jump on a simple counterexplanation, such as "It's all fraud and delusion." Well, maybe it is, but that too creates dissonance of its own. To go around thinking that all parapsychologists are cheating, or deluding themselves, can turn one into a permanently suspicious and miserable sort of person, and it can damage one's self-esteem. Suspecting that some effect is fraudulent and tracking that down systematically is one thing, but approaching everything one hears about as though it must be fraud is destructive.

Then there are other cheap ploys. You can decrease the perceived attractiveness of the opposition. The believer can find it easy to put down one famous critic as a dried-up old professor with no real contact with the field anymore, or a more recent one as having shifty eyes and too bushy a beard! Or the disbeliever can dismiss research on the grounds that all parapsychologists are Scientologists, or are too committed to religious beliefs, or are too dreamy eyed and vague to be doing "real science." But none of this will really wash. And most of us know it won't. Nevertheless, we go on doing it because it is so very hard to have an open mind.

I have said rather a lot about what negative results do not tell us, but is there anything they do tell us? I think we are now in a position to see that there is. I suggest that, wherever you start in parapsychology, if you base your research on the psi hypothesis then you will be forced to do ever more and more restricted research, to back up into ever less and less testable positions, and to produce ever more feeble and flimsy buttresses to hold your theory together. In the end, whatever the questions you started with, you are forced to ask more and more boring questions until there is only one question left: Does psi exist? That question, I submit, is unanswerable.

This process is not restricted to those who get negative results. Helmut Schmidt is among the best researchers in parapsychology, and he has been forced to ask the question "Does psi exist?" Charles Honorton is another example. He is working on fraud-proof, fully automated procedures, even though he might prefer, as most people in parapsychology, to do process-oriented research, as I did when I started with my question "Is ESP like memory?"

I think that is the problem with parapsychology, and it is a problem that starts from the very hypothesis of psi. The structure and definitions of parapsychology are to blame. The negative definition of psi, the hundred years

of bolstering failing theories, and the powerful will to find something are
at fault. They not only force us to ask, "Does psi exist?" but force us to
answer in terms of belief. Where there is no rational and convincing answer,
belief takes over, and that is why there are two sides, and such misunder-
standing.

Here, it seems to me, lies the crux. All those negative results teach us
only one thing, that we have been asking the wrong question. And the whole
history of parapsychology looks like a string of wrong questions. Parapsychology is, if it is based on the psi hypothesis, a magnificent failure; not because
psi doesn't exist, but because it asks unanswerable questions.

An entirely different aspect of my research was prompted by my personal
out-of-body experience. I never entirely forgot it. I went on wanting to understand it and eventually tried to tackle it directly.

The first question I asked was the obvious one: "Does anything leave
the body in an OBE?" This question may seem close to the unanswerable
"Does psi exist?" but I think it is different enough, or perhaps I was just
more ruthless in trying to answer it. From experiments of my own, and from
reading the literature, I concluded that we do have an answer. And it is "No."
You may have heard about an isolated incident of an OBE when someone
correctly read a five-digit number (Tart 1968), or when a cat responded to
its owner's out-of-body presence (Morris et al. 1978), but I prefer to look
at the whole body of evidence (see Blackmore 1982). I concluded that these
were unreplicable and that in general we have enough evidence to answer
that there is no real evidence for psi in OBEs, there is no evidence of anything
leaving the body, and there is no evidence of effects caused by out-of-body
persons.

The next question I asked was "Why does the OBE seem so real?" To
someone who has not experienced an OBE this might seem a silly starting
point, but those of you who have will probably understand why I asked it.
That then set me to ask, "Why does anything seem real?" Here I provided
myself an answer that seemed to account for the OBE (Blackmore 1984).

Very briefly, I argued that the cognitive system cannot make its decision
about what is "real" or "out there" at the low level of chunks of input. Rather,
it makes its decisions at the higher level of global models of the world. That
is, it constructs models of the world, and chooses one, and only one, as
representing "the world out there."

I next had to ask, "Can this decision go wrong?" And the answer is
obviously "Yes." When there is inadequate input—damage to the system, drugs,
trauma, or any of the many things that can precipitate OBEs—then it might.
But what would happen if it goes wrong, the system loses contact with reality?
I would say that a sensible strategy would be to try to replace the lost input
model with the next best approximation—one built from memory. And we
know a lot about memory models. For example, as Ronald Siegel (1977)
has pointed out, they are often built on a bird's-eye view. We know they
are schematized, simplified, and often plain wrong. Indeed, they are just like

the OBE world.

I proposed that the OBE comes about very simply when the system loses input control and replaces its normal "model of reality" with one constructed from memory. It seems real because it is the best model the system has at the time, and it is therefore chosen to represent "out there."

This answered a lot of questions about the OBE; especially about the phenomenology of the experience. It also led to some predictions I have successfully tested. For example, if the OBE occurs when the normal model of reality is replaced by a bird's-eye view constructed from memory, then the people who have OBEs should be better able to use such views in memory and in imagery. In several experiments I found that OBEers were better at switching viewpoints, were especially good at imagining scenes from a position above their heads, and were more likely to recall dreams in a bird's-eye perspective. I actually had some positive results at last (Blackmore 1986a)!

This theory also led to a new approach to altered states of consciousness in general. To that persistent question "What is altered in an altered state of consciousness?" I could now answer that a person's "model of reality" is altered. I could look at changes induced by meditation, drugs, hypnosis, or a mystical experience, in terms of the changing models of reality (Blackmore 1986b). The OBE could then be seen as only one of a variety of experiences that become possible when the input-driven model of reality is lost.

Interestingly, this theory treats the OBE as a kind of error of reality modeling. And so once again the error can be used to throw light on the normal process at work. But I was only able to come back to this insight once I had abandoned looking for psi. It wasn't that I had rejected the possibility of psi, I had simply ignored it.

I mention my OBE research only to contrast it with my previous work based on psi. In my early work, starting from the psi hypothesis, I was forced to ask, "Does psi exist?" In this research I never had to ask it. The other difference is that I no longer had to worry about having an open mind. That makes me wonder what it is like in other sciences. Of course it is always important to have a potentially open mind. If one's results show that one's hypothesis is wrong, then one has to be prepared to change it; but that need not happen very often—at least if one's hypotheses are any good it shouldn't. One doesn't have to have a permanent open mind. And so it was with the OBE research—and what a relief!

I can conclude that all my negative results did teach me something. Or am I perhaps only trying to get my 50-cents worth? A few years ago I read an article in the *British Psychological Society Bulletin* about the "Royal None-such of Parapsychology." The author, H. B. Gibson (1979), described Mark Twain's wonderful story of cognitive dissonance, about the show that never was. Many people were lured into paying 50 cents to see a nonexistent show, but instead of decrying the fraud they went out and persuaded others to see it and pay their 50 cents too. Gibson was reminded of this tale, he said, by a conference paper given by a woman who had spent two years in fruitless research on parapsychology. He suggested that parapsychology is only kept going by the "very human tendency to try to get one's 50-cents worth after one has been misled . . . by an unkind fate which has led one into an immense expense of effort in a blind alley."

I fought back in print (Blackmore 1979a), arguing that I was not just trying to get my 50-cents worth, that I was after the truth and an understanding of the Nature of Life, the Universe, and Everything. But the problem is that it is very hard to understand the nature of life, the universe, and everything, if you start with the psi hypothesis.

In the end I think my negative results told me that the psi hypothesis leads only to unrepeatability (Blackmore 1985). It forces us to ask ever more boring questions, culminating in the question "Does psi exist?" and to that question there is no obviously right answer. Where there is no right answer, we are in ignorance; and, where we are in ignorance, we should do only one thing—have an open mind. But that is too difficult. After all these years of research, I can only conclude that I don't know which is more elusive—psi or an open mind.

Postscript

It has been several years since I wrote this article. I'm still studying the paranormal and, after several fruitless attempts to "give it up," have finally "given in." Either Gibson was right after all, or it is just too interesting for me to leave alone.

In the meantime, the issue of what parapsychology is or should be has progressed. The idea of a "psychical research without psi" (Blackmore 1989) met much resistance. It has become clear that for most parapsychologists the interest lies only in the claimed anomalies, not in understanding strange or even life-transforming experiences in themselves. Whether this is because the anomalies undermine the current world-view or because the researchers are secretly dualists in search of the soul, as Alcock (1987) claims, is a moot point. But this whole issue has received sophisticated discussion in *The Behavioral and Brain Sciences.*

An exception is Rhea White (1990), who has made a plea for an "experience-centered approach to parapsychology." She and I agree in starting from the experiences themselves. The difference is that she thinks the psi hypothesis is essential. I do not. Of course if our research is any good, and we keep hold of that elusive open mind, time alone will tell.

References

Alcock, J. E. 1987. Parapsychology: Science of the anomalous or search for the soul? *Behavioral and Brain Sciences,* 10:553-565 (reprinted in Alcock's *Science and Supernature,* Prometheus

Books, 1989); see also, in the same issue of that journal, K. R. Rao and J. Palmer, pp. 539-551, and Open Peer Commentary, pp. 566-643.

Blackmore, S. J. 1979. Correspondence. *Bulletin of the British Psychological Society*, 32:225.

———. 1980a. Correlations between ESP and memory. *European Journal of Parapsychology*, 3:127-147.

———. 1980b. A study of memory and ESP in young children. *Journal of the Society for Psychical Research*, 50:501-520.

———. 1981a. The effect of variations in target material on ESP and memory. *Research Letter* (Parapsychology Laboratory, Utrecht); 11:1-26.

———. 1981b. Errors and confusions in ESP. *European Journal of Parapsychology*, 4:49-70.

———. 1982. *Beyond the Body*. London: Heinemann.

———. 1983. Divination with Tarot cards: An empirical study. *Journal of the Society for Psychical Research*, 52:97-101.

———. 1984. A psychological theory of the out-of-body experience. *Journal of Parapsychology*, 48:201-218.

———. 1985. Unrepeatability: Parapsychology's only finding. In *The Repeatability Problem in Parapsychology*, edited by B. Shapin and L. Coly, 183-206. New York: Parapsychology Foundation.

———. 1986a. Where am I? Viewpoints in imagery and the out-of-body experience. *Journal of Mental Imagery*.

———. 1986b. Who am I? Changing models of reality in meditation. In *Beyond Therapy*, edited by G. Claxton, 71-85. London: Wisdom.

———. 1989. Do we need a new psychical research? *Journal of the Society for Psychical Research*, 55:49-59.

Blackmore, S. J., and Troscianko, T. S. 1985. Belief in the paranormal: Probability judgements, illusory control and the "Chance Baseline Shift," *British Journal of Psychology*, 76:459-468.

Festinger, L. 1957. *A Theory of Cognitive Dissonance*. Evanston: Row Press.

Gibson, H. B. 1979. The "Royal Nonesuch" of parapsychology. *Bulletin of the British Psychological Society*, 32:65-67.

Honorton, C. 1977. Psi and internal attention states. In *Handbook of Parapsychology*, edited by B. B. Wolman, 435-472. New York: Van Nostrand Reinhold.

Morris, R. L., Harary, S. B., Janis, J., Hartwell, J., and Roll, W. G. 1978. Studies of communication during out-of-body experiences. *Journal of the American Society for Psychical Research*, 72:1-22.

Sargent, C. L. 1980. Exploring psi in the ganzfeld. *Parapsychological Monographs* No. 17. New York: Parapsychology Foundation.

Siegel, R. K. 1977. Hallucinations. *Scientific American*, 237:132-140.

Spinelli, E. 1983. Paranormal cognition: Its summary and implications. *Parapsychology Review*, 14(5):5-8.

Tart, C. T. 1968. A psychological study of out-of-the-body experiences in a selected subject. *Journal of the American Society for Psychical Research*, 62:3-27.

White, R. 1990. An experience-centered approach to parapsychology presented at conference "Parapsychology, Philosophy and Religion." Santa Barbara, Calif., August 1990.

ANTONY FLEW

Parapsychology, Miracles, and Repeatability

David Hume (1711-1776), the first of the two great philosophers of the eighteenth-century Age of Enlightenment, was also the first thinker of the modern period to develop systematically a world outlook that was thoroughly skeptical, this-worldly, and human-centered. A friend and admirer of Benjamin Franklin, living just long enough to hear of and to welcome the American Declaration of Independence, Hume was the philosophical founding father of what is in the United States today so widely and so fiercely denounced as "secular humanism."

Our immediate concern, however, is with only twenty or thirty pages of all of Hume's writings. These few pages, the treatment "Of Miracles" in *An Inquiry Concerning Human Understanding*, provoked in his own lifetime more protest and controversy than most of the rest of Hume's published work put together. He believed that he had "discovered an argument . . . which, if just, will, with the wise and learned, be an everlasting check to all kinds of superstitious delusion, and, consequently, will be useful as long as the world endures. For so long, I presume, will the accounts of miracles and prodigies be found in all history, sacred and profane."

Hume himself, like his contemporary critics, was most interested in "the accounts of miracles and prodigies" found in what in those days people still distinguished as "sacred history." Above all, both he and his critics were concerned with the application of his "everlasting check" to accounts of the resurrection of Jesus. For then all parties agreed that this allegedly well-evidenced alleged event constituted both the best reason for accepting that the Christian candidate is a genuine revelation and one essential element in that revelation.

Here my own primary concern is with the phenomena, or putative phenomena, of parapsychology as they appear, or appear to appear, in an entirely secular context. So far, no one seems to have appreciated the full significance for parapsychology of Hume's argument. For those who prefer the big words, it is an epistemological rather than an ontological argument. It is directed not at the question of whether miracles occur but at the question of whether—and, if so, how—we could know that they do, and when and where they have.

Hume's argument takes off from an observation about "the very nature of the fact" or, better, the logic of the concept. Since a miracle must essentially involve an overriding of the ordinary order of Nature, presumably by some supernatural power, there is bound to be an irresolvable conflict of evidence. Since all evidence for insisting that some conceivable occurrence (were it in fact to have occurred) constituted such an overriding of the natural order must at the same time and by the same token be evidence against the contention that the particular principle precluding occurrences of this particular kind is in fact an element in that order; and, of course, also the other way about.

It is unfortunate that Hume disqualified himself from exploiting the full potentialities of this most promising gambit. In his zeal to defend his great negative insight about causation—"If we reason *a priori*, anything may appear able to produce anything"[1]—he denied the crucial notions of physical necessity and physical impossibility.[2] In his official view, the relation of cause to effect is no more than a regularity of observed precedence and succession; there is no explicit and explained reference to the need for efforts either experimentally to break or actively to exploit such merely observed correlations. No combination of statements—each expressible as what is misleadingly called a "material implication"—can be got formally and vigorously to entail any contrary-to-fact conditional (such as, "If this were to have happened—though it did not—that would have followed). Yet every causal proposition as well as every nomological proposition—every proposition expressing a law of nature—does carry contrary-to-fact entailments. If you say, to take a homely illustration, that the cause of the trouble was that there was no gas in the tank, then your statement implies that, had the tank been full, all other things being equal the engine would have started.

Hume's accounts both of causation and of laws of nature are therefore grossly inadequate. Another way of bringing out both the nature and the severity of these deficiencies is to say that Humian causes, unlike real causes, do not bring about their effects. They do not, by themselves, make the occurrence of their effects physically necessary and all alternative eventualities, in precisely that existing situation, physically impossible. Nevertheless, and very understandably, in his discussion "Of Miracles" Hume wants to make much of a distinction for which he cannot himself find room. Thus he argues: "But in order to increase the probability against the testimony of witnesses, let us suppose that the fact, which they affirm, instead of being only marvelous, is really miraculous."

To be "really miraculous," as opposed to being "only marvelous," is to be physically (or practically or contingently) impossible, as opposed to being an event merely very rare, unusual, or surprising. It is precisely and only because (it is believed that) it would be in this everyday sense impossible for any power within the universe, human or nonhuman, to bring about a "really miraculous" event that religious people would say that, were it to occur, it would have to be the work of some supernatural power.

That is not a question for us to pursue here. What is, however, both relevant and important is that psi phenomena, the putative subject-matter of parapsychology, are, or would be, phenomena whose occurrence all of us—including most of the time the believing and practicing parapsychologists themselves—would with complete confidence rule out as physically (or practically or contingently) impossible. This point has in a way been recognized by all those who have insisted that psi phenomena are (or would be) inconsistent with (what are currently believed to be) the laws of physics. This is, I believe, most of what J. B. Rhine and others have had in mind when they have claimed that psi phenomena are (or would be) *non*physical.

But the truth, as C. D. Broad argued long ago in a landmark paper,[3] is that the Basic Limiting Principles that rule out such goings on as, in this sense, impossible are less sophisticated than the development of physical science. They also have been, and remain, largely independent of its development. Broad originally stated these Basic Limiting Principles in a highly abstract way. Both their nature and their importance will come out more clearly in the concrete.

Suppose, for instance, that there has been yet another security leak in Washington, or Bonn, or London. Then everyone, or almost everyone, assumes that some hostile agent has had some form of direct or indirect sensory access to the Top Secret material that is now secret no longer. It never seriously enters most people's heads that that material might have been telepathically or clairvoyantly "read" by an agent who at no time came within normal sensory range. Extrasensory perception (ESP) is thus in practice ruled out as impossible. That information can be acquired without employment of the normal senses is precluded by a Basic Limiting Principle (BLP).

Suppose that there had actually been an explosion in the nuclear power station at Three Mile Island. No one, or almost no one, would have suggested that this might have been a case of sabotage by psychokinesis (PK). That too is precluded as practically impossible by another BLP.

A second thing to notice about these BLPs, in addition to the fact that they are both familiar and more fundamental than any of the named laws of physics, is that to appeal to them as reasons for dismissing some alleged occurrence as physically (or practically or contingently) impossible is not—any more than to appeal in a similar context to some named law of established physics—to dismiss such allegations dogmatically and apriori.

Many contributors to the *Skeptical Inquirer,* including some fellows of CSICOP, are quite unnecessarily embarrassed by, while making dreadfully heavy weather of, such charges of apriori dogmatism. Certainly, since none of us is infallible, we ought to be always ready to consider any strong evidence suggesting that some proposition we had believed expressed a true BLP or a true law of nature is, after all, false. Yet it is simply grotesque to complain, in the absence of any such decisive falsifying evidence, that these appeals to the BLPs and the named laws of established physics are exercises in apriori dogmatism. For what "apriori" means is: prior to and independent of experience. But in both of these kinds of cases we have an enormous mass of experience supporting our present beliefs and our present incredulities.

So, now, what sort of evidence should we demand as sufficient to show that we have been mistaken in dismissing all alleged psi phenomena (ESP and PK) as physically impossible? When, back in 1955, G. R. Price made the first attempt to deploy Hume's argument "Of Miracles" as a challenge to parapsychology, Price called not for a demonstration *type* but a demonstration *token*.[4] He demanded not an algorithm for producing psi phenomena at will, whenever and wherever required, but rather a single, once-and-for-all decisive, knock-down falsification of one or all of the precluding BLPs. In this, Price revealed that he had not appreciated the full richness and strength of Hume's argument.

What needs to be remembered is that already, when publishing the first *Inquiry,* Hume had for some time been intending to devote his future to history. The Catalogue of the British Library still puts our greatest philosopher down as "Hume, David, the historian." The section "Of Miracles" is thus, among other things, an examination of the presuppositions of critical history.

In effect, Hume's thesis is that the detritus of the past can only be interpreted as historical evidence—and, as such, employed to tell us what actually happened—by applying to it everything we know, or think we know, about what is probable or improbable, possible or impossible. Confronted with a story of a miracle, or of any other story that he knows, or believes he knows, to be impossible, the critical historian is therefore required to reject it as a fiction. Hume gives as an example of sound historical practice the reaction of the famous physician De Sylva to the case of Mademoiselle Thibaut: "The physician declares, that it was impossible she could have been so ill as was proved by witnesses; because it was impossible she could, in so short a time, have recovered so perfectly as he found her. He reasoned, like a man of sense, from natural causes."

What, regrettably, Hume did not allow for was the possibility that later historians, following the same sound methodological principles but having the advantage of further scientific findings, might have to admit that some of the stories in question had after all been true—although the events thus truly recorded were not miraculous. For instance, the stories of supposedly miraculous cures wrought by the Roman Emperor Vespasian in Egypt, stories ridiculed by Hume and all like-minded contemporaries, would, in the light of advancing knowledge of psychosomatic possibility, appear to have been true.[5]

The moral for us is that any supposedly once-and-for-all-decisive yet not-in-practice-repeatable demonstration of the reality of psi phenomena has to be rejected. It has to be rejected in the same emphatic way, and for the same excellent reasons, that critical historians reject stories of what they know, or believe they know, to be physically impossible. So to the objection that there are some rare phenomena that, though not repeatable at will, are admitted by science, the correct and properly crushing reply should be that these are not phenomena for which we have the strongest or indeed any experimental reasons for thinking impossible.

There are three further reinforcing reasons that we have to demand full repeatability and to refuse to accept any substitute.

1. In the first place, parapsychology is by now a fairly old subject. The (original, British) Society for Psychical Research was founded in 1882. Serious work has been going on for more than a century, while the amount done each year appears still to be increasing. Nevertheless, the long sought repeatable demonstration of any psi phenomenon seems to be as far away as ever. It is still stubbornly the case that those best-informed about the field automatically assume that anyone claiming to demonstrate psi capacities with night-after-night regularity must be some sort of fraud, achieving their effects by mere conjuring tricks. So long as this situation continues, there will every year be better and better reason to close the books, concluding that the whole business was a wild-goose chase up a blind alley.

Another dampening and damaging feature of the history of the subject is the ever lengthening succession of shameful, shabby cases—cases that at one time and to many people had seemed to constitute knock-down demonstrations of the reality of these putative phenomena but have since been definitively discredited as fraudulent. For instance, this applies to every one of the cases commended by the various contributors to J. Ludwig's *Philosophy and Parapsychology* (Prometheus, Buffalo, 1978). In particular, it is true of the once famous and now notorious work of S. G. Soal on Gloria Stewart and Basil Shackleton.[6] Soal's sharply righteous reply to G. R. Price falls now upon disillusioned ears.

2. The second reason for viewing the whole business with the deepest suspicion, and the second reason reinforcing the demand for repeatability or nothing, is the fact that no one has been able to think up any halfway plausible theory accounting for the occurrence of any psi phenomenon. This is important, because a plausible theory relating these putative phenomena to something that undoubtedly does occur would tend both to explain and to probabilify their actual occurrence.

3. Third, and finally, there are the reasons arising from the fact that all the psi concepts are negatively defined. This important truth is often overlooked because such expressions as "by telepathy" and "by psychokinesis" sound like the expressions "by telephony" and "by psychoanalysis." But the fact, of course, is that all the psi terms refer rather to the absence of any means or mechanism, or at any rate to the absence of any normal and understood means or mechanism.

One consequence is that no sense has been given to a distinction between single hits achieved by ESP and single hits due to chance alone. Only when, after a series of guesses (or whatever) has been made and has been scored up against the targets, it turns out that there have been significantly more hits than we could have expected by chance alone are we entitled to begin to talk of psi, or of a psi factor. The phenomenon, therefore, is—so far, at least—defined as essentially statistical. Furthermore, and despite some protests to the contrary, the same applies not only to the experimental work but also to the supposed spontaneous or sporadic phenomena. If, for instance, someone has a dream of a maritime disaster "on the night when the great ship went down," then there is no way of identifying this dream as a psi phenomenon save by summing single items of correspondence between dream and reality and arguing that there are too many correspondences, and too few noncorrespondences, to be put down to chance alone.

PK, on the other hand, should not be similarly statistical. Nor would it have been had the evidence actually offered been what we ought to have expected. For, if people really were able to exert some force at a distance on other objects at will, then we should have expected this to be demonstrated by the use of some extremely delicate and very carefully shielded apparatus. If the subject's willings were always followed by the occurrences of the willed movement, and that movement was one that we had taken every care to ensure would not otherwise occur, then we would be home and dry; and, presumably, we should in this have a repeatable demonstration.

But the actual "dice work" has been different. In fact, it is once again essentially statistical. A batch of dice are tossed mechanically, and the subject is told to will them all to come up on one particular side. The procedure is repeated ad nauseam, and well beyond. The experimenter's hope is that he will find significantly more willed sides turning up than chance alone would lead us to expect. If that hope is fulfilled, the experimenter reports a PK effect. So, once again, no operational sense is in fact given to the notion of a single PK hit, as opposed to a run of falls suggesting the operation of a PK factor.

The second and further consequence of all this is that there is no way of decisively identifying even a single run in which a psi factor was operating. Since no identifiable means or mechanism is being employed, it must remain always possible to say that any single run was no more than a statistical freak—however improbable, not impossible. There is therefore once again no substitute for what there is ever less reason for expecting we shall in fact get—namely, a repeatable demonstration, showing psi phenomena being produced and inhibited at the will of the experimenters and/or their subjects. Only this would really demonstrate that the targets actually are causing the subjects to come up with correct guesses and/or that subjects actually are influencing the fall of the dice.[7]

Notes

1. Hume, of course, rendered the expression *a priori* as two words, printed in italics. Since the first employment recorded in the big *Oxford English Dictionary* was in 1710, in Berkeley's

Principles, it is surely more than time to grant it citizenship in the English language. We shall render it, as I do below, as a single unitalicized word.

2. See, for a more adequate treatment of the immediately relevant points both of philosophy and of Hume interpretation, "Another Idea of Necessary Connection," in *Philosophy,* 57(1982):487-494.

3. "The Relevance of Psychical Research to Philosophy," in *Philosophy,* 24(1949):291-309.

4. See "Science and the Supernatural," in *Science,* 131 (1955): 359-367.

5. For further discussion, compare either Chapter 8 of my *Hume's Philosophy of Belief* (London and New York: Routledge and Kegan Paul, and Humanities, 1961), or the Introduction to the separate printing of the section "Of Miracles," which should, by the time the present article appears, have been published by Open Court of LaSalle, Ill.

6. The Amazing Randi loves to say that academics, and especially philosophers, find it hard to say either "I don't know" or "I was wrong." So let me say here and now that in my first book, *A New Approach to Psychical Research* (London: C. A. Watts, 1953), I was totally wrong about, among other things, this now wholly discredited research.

7. Readers wanting to pursue somewhat further the questions raised in this essay are referred to Antony Flew (Ed.), *Readings in the Philosophical Problems of Parapsychology* (Buffalo: Prometheus Books).

KENDRICK FRAZIER

Ganzfeld Studies: First Detailed Appraisal Finds Serious Flaws, No Evidence of Psi

In the years since publication of the first "ganzfeld" ESP experiment in 1974, reporting pro-psi results, a series of similar experiments have been published in the parapsychological literature. Parapsychologists and others have considered these to be among the strongest scientific evidence for the existence of extrasensory perception.

Ganzfeld experiments are based on the idea that sensory deprivation is conducive to the manifestation of psi abilities. The research subject is generally isolated from visual and other sensory contact. Then various experiments are carried out to test the subject's ability to perceive outside information.

For instance, a person undergoing perceptual deprivation might be asked to "receive" an image from a photograph randomly selected from four photos and being concentrated on by a "sender" in another location. While doing so he might be requested to verbalize his thoughts, feelings, and images. Afterward, the subject and, in some cases, independent judges would be asked to assess the degree of correspondence between the picture and the subject's imagery. Positive results from a number of such experiments have been presented as evidence for psi.

The first detailed scholarly evaluation of the ganzfeld studies has now been published. The critique, prepared over a period of several years by psychologist (and CSICOP Executive Council member) Ray Hyman, of the University of Oregon, gives little comfort to proponents of the ganzfeld experiments as the best hope for proving psi abilities exist.

". . . I believe that the ganzfeld psi data base, despite initial impressions, is inadequate either to support the contention of a repeatable study or to demonstrate the reality of psi," Hyman concludes in his 47-page critical appraisal, published in the March 1985 *Journal of Parapsychology*. "Whatever other value these studies may have for the parapsychological community, they have too many weaknesses to serve as the basis for confronting the rest of the scientific community. Indeed, parapsychologists and others may be doing themselves and their cause a disservice by attempting to use these studies as examples of the current state of their field."

The journal follows Hyman's analysis with an equally detailed response by parapsychologist Charles Honorton disputing many of his conclusions. Honorton carried out the first reported ganzfeld ESP experiment in 1974, and his subsequent ganzfeld studies include some of the most positive (pro-ESP) results.

Hyman had been asked to prepare a critical appraisal of parapsychology. Rather than attempting to take on the whole field, he looked for a systematic research program that parapsychologists considered especially promising. It needed to consist of a series of studies carried out by a variety of researchers. He chose the ganzfeld psi studies. Respected investigators had conducted them. Hyman was intrigued by their claims that significant psi scores had been achieved in more than half of the experiments and that the studies had been conducted with a high level of sophistication and rigor.

Hyman sought Honorton's cooperation. Honorton felt it important to have an outside critic like Hyman assess the ganzfeld literature. He supplied Hyman with a copy of every reported ganzfeld study he knew of.

As a result, Hyman evaluated all 42 studies reported from 1974 through 1981. Honorton classified 23 of them as having achieved significance as evidence of psi. This amounts to a claimed replication rate of 55 percent.

Hyman prepared a preliminary critique, which he presented at the combined meetings of the Society for Psychical Research and the Parapsychological Association in Cambridge, England, in August 1982 (*Skeptical Inquirer*, Winter 1982-83). As a result of comments on that paper by Honorton and others, he prepared a new and more systematic analysis of the data.

Hyman focused on two questions: (1) Does the data base, taken as a whole, supply evidence for psi? (2) Does the ganzfeld psi study yield evidence for psi that is replicable?

The basic index for these questions is some measure of hitting or target-matching compared with a chance baseline. This, Hyman noted, creates special problems; assumptions about chance levels and probability distributions take on a great burden.

He divided his critique into four phases:

—Rechecking the "vote count"

—Assessing the actual opposed to the assumed level of significance

—Assigning procedural flaws to studies

—Analyzing correlations among flaws, positive effects, and significance.

The "vote count" check assessed whether the studies claimed to be successful really amount to 55 percent of the total. Hyman found a lot depends on how studies containing multiple conditions are divided up. He found, for instance, that a study counted as one "successful" replication could be viewed "with equal justification" as adding one successful and 11 unsuccessful replications to the total.

Then he considered the "file-drawer" problem. How many ganzfeld studies have been conducted but not reported? Surveys have identified other studies, and their inclusion tends to lower the success rate.

The important question here was whether there was evidence for biased reporting—specifically, is there a possibility that only those experiments that begin with a string of successes end up being reported? "It is [easy] to imagine that a large number of experimenters . . . might have begun conducting some trials and then abandoned the study when the first few trials turned out to be unpromising. On the other hand, a few of these exploratory ventures might have started with initially successful trials, encouraging the experimenter either to continue or to stop and write up the result as a successful replication."

Is there any evidence for such a suggestion? Yes, Hyman says. He found a tendency for the studies with the fewer trials to have a higher proportion of significant outcomes. "The most obvious conclusion is that such a strange relationship is due to selective bias. It suggests a tendency to report studies with a small sample only if they have significant results."

Another, related bias Hyman calls a "retrospective bias." "This is the tendency to decide to treat a pilot or exploratory series of trials as a study if it turns out that the outcome happens to be significant or noteworthy." He found two studies in the data base that were clearly retrospective and strong circumstantial evidence of four others.

Next Hyman considered whether the chances of getting successful results in ganzfeld ESP studies without invoking psi are really as low as psi proponents suggest. The studies varied widely in variables and in the questions being asked. Notes Hyman, with some understatement, "Many confusing questions arise about what probability levels to assign to the various tests of significance."

Generally, a ganzfeld experiment is taken to show evidence of psi if the statistics indicate there is no more than a .05 probability that the results are due to chance. Hyman's various analyses found that the probability of obtaining at least one significant outcome per experiment was instead .24—"over four times the assumed level of .05."

The discrepancy results from the use of multiple indices—the availability of several different ways of getting "hits" without their being included in the probability estimates. Hyman found that more than half the studies he evaluated "clearly used multiple indices without taking this into account in computing their statistical significance." Multiple indices was but one of six categories of multiple testing he checked the studies against. Forty percent of the studies, for instance, used multiple baselines; 64 percent used multiple groupings.

Using this kind of analysis, Hyman found one study that had increased the probability of getting successful results "almost surely beyond .50." In other words there was a better than 50-50 chance of getting a positive, "pro-psi" result just by chance. "Indeed, if we consider the eight intervening practice conditions, the chances of coming up with a significant outcome are well over .80! And this is just one of the many studies in this data base that exhibit such complex options either explicitly or implicitly."

Hyman found that the actual rate of successful replication is less than 30 percent. "And the arguments in this section strongly suggest that this rate of 'successful' replication is probably very close to what should be expected by chance given the various options for multiple testing exhibited in this data base."

Hyman then turned to procedural flaws in the studies. He found that 36 percent of them used improper randomization procedures, 55 percent used only a single target (which allows various chances for sensory leakage), 24 percent allowed contaminating feedback, 38 percent of the published studies (and 81 percent of the unpublished ones) gave inadequate documentation, 24 percent had inadequate security, and 29 percent appeared to use erroneous statistical procedures.

Hyman says he was very conservative in assigning these flaws. Those that were not too common or depended on suspicions or hard-to-objectify criteria he did not count. "In any case the existence of so many elementary defects in this data base is both disturbing and surprising. Only two studies were entirely free of the six procedural flaws. And if we include multiple-testing errors, not a single study in this data base was flawless.

"It is important to realize that the defects being discussed are not obscure or subtle. Rather, I suspect that a typical parapsychologist would spontaneously list them as being unacceptable in a psi experiment."

Are these defects important? Yes, Hyman believes, in two ways. First, they are a symptom of something seriously wrong. When studies put forth as among the field's strongest evidence of psi have so many elementary deficiencies, it is a sign that quality control is lacking. There are so many problems with these studies in fact that Hyman told the *Skeptical Inquirer* he believes most were informal exploratory studies that were reported only because they gave positive results. No one knows how many exploratory studies giving null results have been carried out for each one giving a positive result. This is an old problem in parapsychology. Until it is solved, it makes the statistical case for psi almost meaningless.

Second, the tendency to get stronger results correlates with greater presence of experimental deficiencies. Hyman examined the pattern of relationships among indices of success and various flaws. He compared the presence of flaws in each study with its outcome. The flaws concerning randomization, feedback, documentation, and statistics seemed to correlate with three different measures of significance. "The more likely a study was to be assigned any of these flaws, the more likely it was to be classified as significant." A similar but weaker pattern was found in the size of the effect reported by the study.

Hyman told the *Skeptical Inquirer* he doesn't contend that there is a one-to-one correspondence between a flaw and a positive outcome. "I argue that it's not any one defect alone. Probably they work in combination."

"Whatever the reasons, the 42 studies in the present data base cannot by any stretch of the imagination be characterized as flawless, and I suspect that most of them were not well planned," Hyman concludes. "The current data base has too many problems to be seriously put before outsiders as evidence of psi."

In the concluding section, Hyman offers a number of suggestions by which the Parapsychological Association and others could establish guidelines for what should constitute an adequate confirmatory study. (The Council of the Parapsychological Association has now commissioned a study group to develop just such guidelines. Hyman has accepted an offer to serve on the committee.) Only if a large body of studies meeting such rigorous guidelines can be accumulated should the scientific community have any obligation to take notice, Hyman concludes.

In Honorton's 41-page response to Hyman's analysis, he presents his own meta-analysis of ganzfeld research, which he says eliminates the multiple-analysis problems Hyman criticized. He disputes the view that selective-reporting bias has anything to do with the positive results reported in the literature. He contends that, contrary to Hyman's assessment, no significant relationship is found between study outcomes and measures of study quality. He also disagrees with some of Hyman's assignments of flaws. "Is there a significant psi ganzfeld effect?" Honorton asks. "I believe my evaluation of direct-hits studies justifies an affirmative answer to the question."

Hyman will prepare a rebuttal to Honorton's response for a future issue of the *Journal of Parapsychology*. He points out that Honorton, in his response, concentrates only on Hyman's correlation analysis while ignoring the larger issue. "He ignores the fact that *all* the experiments are flawed. In ten years, why hasn't anyone done them right? He hasn't faced up to that." Hyman's published evaluation included all ganzfeld studies published through 1981, but he says he hasn't seen one reported since then that avoids the problems he cites.

Hyman's is the most detailed critical analysis of the ganzfeld ESP studies ever done. Constructive in intent, it is nevertheless sobering, even damning, in result. It is clear from Honorton's spirited rebuttal that the controversy will continue. But parapsychologists who hoped that ganzfeld ESP studies would at last achieve the scientific demonstration of psi have been given a good dose of the kind of critical scrutiny their claims will have to face and overcome. If future studies can pass that kind of test, then they would deserve the attention of outside scientists; if not, it appears that oblivion is their destiny.

Postscript

The ganzfeld papers described above provoked intense, sometimes bitter, controversy. Nevertheless, psychologist Ray Hyman and parapsychologist Charles Honorton met at the 1986 meeting of the Parapsychological Association and made an unprecedented decision. Rather than another round of debate on the psi ganzfeld experiments, they would collaborate on a joint communique. The earlier debate emphasized their differences; the communique emphasized points of agreement. The 15-page joint paper was published in the December 1986 *Journal of Parapsychology* (50:351-364), and it has since been reprinted in Hyman's *The Elusive Quarry* (Prometheus, 1989). "To the best of our knowledge," they wrote, "this is the first time a parapsychologist and a critic have collaborated on a joint statement of this type."

"We agree that there is an overall significant effect in this database that cannot reasonably be explained by selective reporting or multiple analysis," the statement said. "We continue to differ over the degree to which the effect constitutes evidence of psi, but we agree that the final verdict awaits the outcome of future experiments conducted by a broader range of investigators and according to more stringent standards." They made recommendations about how such experiments should be conducted and reported, including many specifics on such things as randomization, judging, feedback, statistics, and so on. They concluded that "psi researchers and their critics share many common goals" and said they hoped their joint communique "will encourage future cooperation to further these goals."

Despite the amiable tone of the joint communique, there remain strong differences between Hyman and Honorton in their interpretation of ganzfeld experimental results, and the controversy continues. In the June 1990 *Journal of Parapsychology* (54:99-139), Honorton and colleagues (now all with the Psychological Research Laboratories, Plainsboro, N.J.) described 11 new experiments that they claimed were done according to the Hyman-Honorton guidelines. They said the subjects correctly identified randomly selected and remotely viewed targets to a statistically significant degree. No independent evaluation of these experiments is yet available.

KENDRICK FRAZIER

Improving Human Performance: What About Parapsychology?

In 1984 the Army Research Institute asked the National Academy of Sciences to form a committee to examine the value of various techniques claimed to improve human performance. Most of these techniques had been developed outside the mainstream of the human sciences and most made quite extraordinary claims. Many of them grew out of the human-potential movement of the 1960s. They included guided imagery, meditation, biofeedback, neuro-linguistic programming, sleep learning, accelerated learning, split-brain learning, and a variety of techniques claimed to reduce stress and improve concentration. The Army was also interested in whether parapsychology had discovered helpful mental skills.

Many of these claims were regularly publicized in the media and gained considerable acceptance from the public. The promoters of these claims used the language of science but for the most part were not trained in science. They did appeal to the basic human drive to improve performance, however, and the U.S. Army understandably has a great interest in any legitimate techniques that can make its troops and support personnel more effective.

The Army asked the NAS committee to recommend general policy and criteria for future evaluation of enhancement techniques. The National Academy of Sciences is a private organization of the nation's most distinguished scientists. It is officially chartered by Congress to provide scientific advice to the U.S. government. Through its operating branch, the National Research Council (NRD), it can call upon scientific experts nationwide to address issues and problems of interest to the government.

The NRC formed the Committee on Techniques for the Enhancement of Human Performance, chaired by John A. Swets of Bolt Beranek and Newman, Inc., in Cambridge, Massachusetts, and consisting of 13 members—psychologists, neurologists, training experts, and other scholars.[1] It also formed several subcommittees. The committee met with representatives of the Army, conducted site visits, commissioned ten analytical and survey papers, and examined state-of-the-art reviews of the relevant literature as well as unpublished documents. The result is a valuable 299-page report, *Enhancing Human Performance*, of wide general interest and available to the public.[2]

It is a significant study, and an unusual one for the Academy, which only on rare occasions has been asked to evaluate claims residing along the fuzzy fringes of science. In fact, the section on paranormal phenomena may represent the first time the Academy has ever addressed this controversial and emotional subject.

What follows is an overview of the report, a summary of its conclusions, and a summary of its longest section (the one that is perhaps of the most interest to our readers), on claims of paranormal phenomena.

Techniques Evaluated

The study quotes an estimate that American companies are spending $30 billion a year on formal courses and training programs for their employees. Even so, this is only the tip of the iceberg. The courses are actively promoted by entrepreneurs who probably realize there is a goldmine in selling self-improvement techniques.

The results of the study provided answers to several questions on how best to improve human performance.

There were some positive findings. It appears it may be possible to prime future learning by presenting material to a subject during certain stages of sleep (although not deep sleep). Learning can be improved by integrating certain instructional elements. Skilled performance can be improved through particular combinations of mental and physical practice. Stress can be reduced by providing information to the subject that increases his or her sense of control. Group performance can be improved by using organizational cultures to transmit positive values. Nothing too surprising here.

There were some negative findings. The committee found a lack of supporting evidence for such techniques as visual training exercisers, hemispheric synchronization, and neurolinguistic programming. It found a lack of scientific justification for the parapsychological phenomena it examined. It found ambiguous evidence for the effectiveness of a suggestive accelerated learning package.

Throughout its report, the committee emphasizes the importance of having adequate scientific evidence or compelling theoretical argument, or both, in support of any techniques proposed for consideration by the Army. And it comes down hard on the utility of testimonials as evidence. "Personal experiences and testimonials cited on behalf of a technique are not regarded as an acceptable alternative to rigorous scientific evidence. Even when they have high face validity, such personal beliefs are not trustworthy as evidence." Recent research on how people arrive at their beliefs "indicate that many sources of bias operate and that they can lead to personal knowledge that is invalid despite its often being associated with high levels of conviction."

Some specific findings and conclusions:

Learning During Sleep. The committee found no evidence to suggest that learning occurs during verified sleep. However, it found some evidence that waking perception and interpretation of verbal material could be enhanced by presenting the material during the lighter stages of sleep.

Accelerated Learning. The committee found little scientific evidence that so-called superlearning programs derive their instructional benefits from elements outside mainstream research and methods. Effective instruction comes from quality teaching, practice, study, motivation, and matching of training to job demands. "Programs that integrate all these factors would be desirable."

Improving Motor Skills. Motor skills can be improved by mental practice. Programs claiming to enhance cognitive and behavioral skills by visual concentration have not been shown to be effective and are not worth further evaluation. The effects of biofeedback on skilled performance have yet to be determined.

Altering Mental States. The committee was not able in the time allotted to evaluate self-induced hypnotic states or other techniques claimed to improve concentration and performance. It did review literature on brain hemispheres; this review "refutes claims that link differential use of the brain hemispheres to performance." The committee found no scientifically acceptable evidence to support claimed effects of techniques intended to integrate hemispheric activity. Attempts to increase information-processing capacity by presenting material separately to the two hemispheres do not appear to be useful.

Stress Management. Stress is reduced by giving an individual as much knowledge and understanding as possible regarding expected events. Giving the individual an effective sense of control is effective. Biofeedback can reduce muscle tension, but "it does not reduce stress effectively."

Influence Strategies. "The committee finds no scientific evidence to support the claim that neurolinguistic programming is an effective strategy for exerting influence." (See box.)

Parapsychology. "The committee finds no scientific justification from research conducted over a period of 130 years for the existence of parapsychological phenomena." This strongly worded conclusion is followed by the statement that "there is no reason for direct involvement by the Army at this time." The committee does recommend monitoring certain areas, such as the work being done in the Soviet Union and the "best work" in the United States. The latter includes research being carried out at Princeton University by Robert Jahn; at Maimonides Medical Center in Brooklyn by Charles Honorton, now in Princeton; at San Antonio by Helmut Schmidt; and at SRI International by Edward May. It suggests site visits by both proponents and skeptics. As for future studies, it recommends that a common research protocol be agreed upon; that this protocol be used by "both proponents

Neurolinguistic Programming: No Evidence

The committee found no evidence to support claims for the effectiveness of neurolinguistic programming (NLP)—a widely touted system of procedures and models that purports to enable people to be more influential and better communicators.

"In brief, the NLP system of eye, posture, tone, and language patterns as indexing representational patterns is not derived or derivable from known scientific work. Furthermore, there is no internal evidence or documentation to support the system. . . . Overall there is little or no empirical evidence to date to support either NLP assumptions or NLP effectiveness. Different critics may attach different values to the quality of these studies [testing one or another aspect of NLP], but the fact remains that none supports the effectiveness of NLP in improving influence or skilled motor performance."

and skeptics" in any research they conduct; and that practical applications be looked for.

Examination of Parapsychology

The report's largest single section is devoted to an examination of parapsychological techniques and claims of paranormal phenomena. A parapsychology subcommittee chaired by psychologist Ray Hyman of the University of Oregon, also a member of the overall committee, assisted with this part of the report.

The committee examined a range of claimed parapsychological phenomena from scientifically serious to near-trivial matters that people nevertheless sometimes take seriously.

Since the study was done for the Army it was inevitable that some claims that had been heard from parts of the military in recent years would be examined. These claims first surfaced in newspaper columns and later in several books.

Some of the military officers who had made these claims were invited to make presentations to the committee. They gave details of experiments at SRI International in which subjects were said to more or less accurately describe a distant geographical location by means of "remote viewing." The examples appeared to indicate some striking correspondences between the subjects' descriptions and the target sites.

The presentations included anecdotal descriptions of psychic mind-altering techniques, the levitation claims of Transcendental Meditation groups, psychotronic weapons, psychic metal-bending, dowsing, thought photography, and bioenergy transfer. The officers maintained that the Soviet Union is far ahead of the United States in developing applications of such paranormal phenomena. They gave personal accounts of spoon-bending parties in which participants believed they had bent cutlery by mind-power alone, as well as instances of walking barefoot on hot coals, leaving one's body at will, and bursting clouds by psychic means.

The committee examined these claims and those of "psychic warfare," often in the news in the 1980s. "The claimed phenomena and applications range from the incredible to the outrageously incredible," says the report. "The 'antimissile time warp,' for example, is supposed to somehow deflect attack by nuclear warheads so that they will transcend time and explode among the ancient dinosaurs, thereby leaving us unharmed but destroying many dinosaurs (and, presumably, some of our evolutionary ancestors). Other psychotronic weapons, such as the 'hyperspatial nuclear howitzer,' are claimed to have equally bizarre capabilities. Many of the sources cite claims that Soviet psychotronic weapons were responsible for the 1976 outbreak of Legionnaires' disease, as well as the 1963 sinking of the nuclear submarine *Thresher*."

The committee observed that some people, including some military decisionmakers, have imagined a variety of potential military applications of the two broad categories of psychic phenomena—extrasensory perception (ESP), which includes telepathy, precognition, and clairvoyance, all alleged methods of gathering information about objects or thoughts without the intervention of known sensory mechanisms; and psychokinesis (PK), the alleged influence of thoughts upon objects without the intervention of known physical processes, popularly called "mind over matter." ESP, if real, for example, could be used to gather intelligence and to anticipate the enemy's actions. PK, if real, might be used to jam enemy computers, prematurely trigger nuclear weapons, and induce sickness. One might plant thoughts in people's minds, erect psychic shields, and make psychotronic weapons. The committee noted that one suggested application was to form a "First Earth Batallion" of "warrior monks," who would have mastered the techniques the committee was considering, including ESP, leaving one's body at will, levitating, psychic healing, and walking through walls.

The committee refers to these as "colorful examples" of claims. The question is whether they have any validity.

The committee says the cumulative body of data in the discipline of parapsychology—with reports since 1882—"enables us to judge the degree to which paranormal claims should be taken seriously." It notes that, as scientists, their inclination is to restrict themselves to evidence that purports to be scientific. But it recognizes the substantial appeal, and biasing influences, of the public's strong interest and beliefs.

"The alleged phenomena that have apparently gained most attention and that have apparently convinced many proponents do not come from the parapsychological laboratory. Nothing approaching scientific literature supports the claims of psychotronic weaponry, psychic metal-bending, out-of-body experiences, and other potential applications supported by many proponents.

"The phenomena are real and important in the minds of proponents, so we attempt to evaluate them fairly. Although we cannot rely solely on a scientific data base to evaluate the claims, their credibility ultimately must stand or fall on the basis of data from scientific research that is subject to adequate control and is potentially replicable."

So the committee examined "the best scientific arguments for the reality of psychic phenomena." (These turn out to be experiments on remote-viewing and research on the ganzfeld, or whole visual field.) It then also considered the arguments of proponents who rely on "qualitative" or subjective as opposed to "quantitative" or objective evidence for the paranormal. The committee rightly recognized the compelling power of subjective experience in forming paranormal beliefs.

"Such evidence depends on personal experience or the testimony of others who have had such experiences. Most, if not all, of this evidence cannot be evaluated by scientific standards, yet it has created compelling beliefs among many who having encountered it. Witnessing or having an anomalous experience can be more powerful than large accumulations of quantitative, scientific data as a method of creating and reinforcing beliefs."

To evaluate the best *scientific* evidence for the existence of psi, the committee conducted visits to two noted parapsychological laboratories: Robert Jahn's Engineering Anomalies Research Laboratory at Princeton University and Helmut Schmidt's laboratory at the Mind Science Foundation in San Antonio, Texas. The chair of the parapsychology subcommittee also visited SRI International, another major laboratory, in California.

The committee gathered what insights it could from demonstrations of experiments and talks at these laboratories. It says it was impressed with the dedication and sincerity of these investigators. But it found many unresolved problems, and no standardized, easily replicable procedures. For making scientific judgments the committee relied, "as we would in other fields of science, on a careful survey of the literature."

The claims of remote-viewing, especially those by physicists Harold Puthoff and Russell Targ when they were at SRI International, were carefully examined and found to be so severely deficient as to be almost totally dismissable: "After 15 years of claims and sometimes bitter controversy, the literature on remote viewing has managed to produce only one possibly successful experiment that is not seriously flawed in its methodology—and that one experiment provides only marginal evidence for the existence of ESP. By both scientific and parapsychological standards, then, the case for remote viewing is not just very weak, but virtually nonexistent."

"It seems that the preeminent position that remote viewing occupies in the minds of many results from the highly exaggerated claims made for the early experiments, as well as the subjectively compelling, but illusory, correspondences that experiments and participants find between components of the descriptions and the target sites."

Tests of Micro-PK

Random-number generators for parapsychological research next came under the group's scrutiny. A random-number generator (RNG) is simply an electronic device that uses either radioactive decay or electronic noise to generate a random sequence of symbols. It becomes essentially an electronic coin-flipper. The best-known and most widespread use of RNGs is in what proponents call micropsychokinesis, or micro-PK. A subject attempts to mentally bias the output of the random-number generator to produce a nonrandom sequence. Some departures from randomness have been reported. The question is what that means.

The committee examined Helmut Schmidt's experiments at San Antonio (averaging approximately 50.5 percent hits over the years) and Robert Jahn and colleagues' more than 200 times greater number of trials. In 78 million trials, the percentage of hits in the intended direction was only 50.02 percent, an average of 2 extra hits every 2,500 trials.

The committee says it looks as if all the success of Jahn's huge data base can be attributed to the results from one individual, who has produced 25 percent of the data. This individual was presumably familiar with the equipment. The Princeton experiments are faulted for such things as failing to randomize the sequence of groups of trials at each session, inadequate documentation on precautions against data tampering, and possibilities of data selection. Similar criticisms can be directed at Schmidt's experiments.

What would it take to conduct an adequate RNG experiment? The committee notes that one group, E. C. May, B. S. Humphrey, and G. S. Hubbard, in a project summarized in a 1980 SRI technical report, set out to do one. They reviewed all previous experiments, including their deficiencies, and devised a careful research protocol to overcome them. This included setting out in advance the precise criteria by which their test could be judged a success. They obtained successful results, and then subjected their equipment to all sorts of extremes to see if an artifact might have accounted for the results.

"It is unfortunate, therefore," notes the committee, "that this carefully thought out experiment was conducted only once. After the one successful series, using seven subjects, the equipment was dismantled, and the authors have no intention of trying to replicate it. It is unfortunate because this appears to be the only near-flawless RNG experiment known to us, and the results were just barely significant. Only two of the seven subjects produced significant results, and the test of overall significance for the total formal series yielded a probability of 0.029."

Even this experiment still had some problems. It was never reported in the peer-reviewed scientific literature. Despite the authors' equipment tests, a physicist with several years of experience in constructing and testing random-number generators told the committee it is quite possible for the human body to act as an antenna in some circumstances, possibly biasing the output.

The committee notes that May and his colleagues, in their technical report,

PK Parties and Self-Delusion

Another example of beliefs generated in circumstances that are known to create cognitive illusions is macro-PK, which is practiced at spoon-bending, or PK [psychokinesis] parties. The 15 or more participants in a PK party, who usually pay a fee to attend and bring their own silverware, are guided through various rituals and encouraged to believe that, by cooperating with the leader, they can achieve a mental state in which their spoons and forks will apparently soften and bend through the agency of their minds.

Since 1981, although thousands of participants have apparently bent metal objects successfully, not one scientifically documented case of paranormal metal bending has been presented to the scientific community. Yet participants in the PK parties are convinced that they have both witnessed and personally produced paranormal metal bending. Over and over again we have been told by participants that they know that metal became paranormally deformed in their presence. This situation gives the distinct impression that the proponents of macro-PK, having consistently failed to produce scientific evidence, have forsaken the scientific method and undertaken a campaign to convince themselves and others on the basis of clearly nonscientific data based on personal experience and testimony obtained under emotionally charged conditions.

Consider the conditions that leaders and participants agree facilitate spoon bending. Efforts are made to exclude critics because, it is

surveyed all the RNG experiments up through 1979. They found all incomplete in at least one of four areas: (1) No control tests were reported in 44 percent of the cases. (2) Necessary details were not given about the physics and construction of the electronic equipment. (3) Raw data was not saved for later independent analysis. (4) None of the experiments reported controlled and limited access to the experimental apparatus.

Concludes the committee: "As far as we can tell, the same four points can be made with respect to the RNG experiments that have been conducted since 1980. The situation for RNG experiments thus seems to be the same as that for remote-viewing: Over a period of approximately 15 years of research, only one successful experiment can be found that appears to meet most of the minimal criteria of scientific acceptability, and that one successful experiment yielded results that are just marginally significant."

A variety of similar problems were noted with regard to the data base for ganzfeld experiments. (See previous chapter, this volume.)

asserted, skepticism and attempts to make objective observations can hinder or prevent the phenomena from appearing. As Houck, the originator of the PK party, describes it, the objective is to create in the participants a peak emotional experience. To this end, various exercises involving relaxation, guided imagery, concentration, and chanting are performed. The participants are encouraged to shout at the silverware and to "disconnect" by deliberately avoiding looking at what their hands are doing. They are encouraged to shout "Bend!" throughout the party. "To help with the release of the initial concentration, people are encouraged to jump up or scream that theirs is bending, so that others can observe." Houck makes it clear that the objective is to create a state of emotional chaos. "Shouting at the silverware has also been added as a means of helping to enhance the emotional level of the group. This procedure adds to the intensity of the command to bend and helps create pandemonium throughout the party.

A PK party obviously is not the ideal situation for obtaining reliable observations. The conditions are just those which psychologists and others have described as creating states of heightened suggestibility and implanting compelling beliefs that may be unrelated to reality. It is beliefs acquired in this fashion that seem to motivate persons who urge us to take macro-PK seriously. Complete absence of any scientific evidence does not discourage the proponents; they have acquired their beliefs under circumstances that instill zeal and subjective certainty. Unfortunately, it is just these circumstances that foster false beliefs.

—From *Enhancing Human Performance*

The Question of Subjective Evidence

The committee noted that it "continually encountered the distinction between qualitative and quantitative evidence" for the existence of the paranormal. Even scientists who believe in the paranormal in some way use or exploit the distinction. Often, the committee noted, proponents acknowledged that they themselves rely on subjective evidence for their own beliefs but referred the committee to the experiments at Princeton or SRI for supposed supporting quantitative data.

Most proponents, the committee noted, "seem impatient with the request for scientific evidence." As observers of paranormal belief systems have long known, most people are convinced through their experiences or the vivid testimonies of those they trust. Many even argue that qualitative evidence is superior to quantitative, and they offer a variety of holistic arguments.

The study addressed these questions directly.

"We see two problems regarding qualitative evidence. First, personal observation and testimony are subject to a variety of strong biases of which

Conclusions on Psychic Phenomena

In drawing conclusions from our review of evidence and other considerations related to psychic phenomena, we note that the large body of research completed to date does not present a clear picture. Overall, the experimental designs are of insufficient quality to arbitrate between the claims made for and against the existence of the phenomena. While the best research is of higher quality than many critics assume, the bulk of the work does not meet the standards necessary to contribute to the knowledge base of science. Definitive conclusions must depend on evidence derived from stronger research designs. The points below summarize key arguments in this chapter.

1. Although proponents of ESP have made sweeping claims, not only for its existence but also for its potential applications, an evaluation of the best available evidence does not justify such optimism. The strongest claims have been made for remote viewing and the ganzfeld experiments. The scientific case for remote viewing is based on a relatively small number of experiments, almost all of which have serious methodological defects. Although the first experiments of this type were begun in 1972, the existence of remote viewing still has not been established. Furthermore, although success rates varying from 30 to 60 percent have been claimed for the ganzfeld experiments, the evidence remains problematic because all the experiments deviate in one or more respects from accepted scientific procedures. In the committee's view, the best scientific evidence does not justify the conclusion that ESP—that is, gathering information about objects or thoughts without the intervention of known sensory mechanisms—exists.

2. Nor does scientific evidence offer support for the existence of psychokinesis—that is, the influence of thoughts upon objects without intervention of known physical processes. In the experiments using random number generators, the reported size of effects is very small, a hit rate of no more than 50.5 percent compared with the chance expectancy of 50 percent. Although analysis indicates that overall sig-

most of us are unaware. When such observations and testimony emerge from circumstances that are emotional and personal, the biases and distortions are greatly enhanced. Psychologists and others have found that the circumstances under which such evidence is obtained are just those that foster a variety of human biases and erroneous beliefs. Second, beliefs formed under such circumstances tend to carry a high degree of subjective certainty and often resist alteration by later, more reliable confirming data. Such beliefs become self-sealing, in that when new information comes along that would ordinarily nificance for the experiments, with their unusually large number of trials, is probably not due to a statistical fluke, virtually all the studies depart from good scientific practice in a variety of ways; furthermore, it is not clear that the pattern of results is consistent across laboratories. In the committee's view, any conclusions favoring the existence of an effect so small must at least await the results of experiments conducted according to more adequate protocols.

3. Should the Army be interested in evaluating further experiments, the following procedures are recommended: first, the Army and outside scientists should arrive at a common protocol; second, the research should be conducted according to that protocol by both proponents and skeptics; and third, attention should be given to the manipulability and practical application of any effects found. Even if psi phenomena are determined to exist in some sense, this does not guarantee that they will have any practical utility, let alone military applications. For this to be possible, the phenomena would have to obey causal laws and be manipulable.

4. The committee is aware of the discrepancy between the lack of scientific evidence and the strength of many individuals' beliefs in paranormal phenomena. This is cause for concern. Historically, many of the world's most prominent scientists have concluded that such phenomena exist and that they have been scientifically verified. Yet in just about all these cases, subsequent information has revealed that their convictions were misguided. We also are aware that many proponents believe that the scientific method may not be the only, or the most appropriate, method for establishing the reality of paranormal phenomena. Unfortunately, the alternative methods that have been used to demonstrate the existence of the paranormal create just those conditions that psychologists have found enhance human tendencies toward self-deception and suggestibility. Concerns about making the experimental situation comfortable for the alleged psychic or conducive to paranormal phenomena frequently result in practices that also increase opportunities for deception and error.

—From "Paranormal Phenomena,"
Enhancing Human Performance, NRC

contradict them, the believers find ways to turn the apparent contradictions into additional confirmation."

The study provides two extended examples of such "problematic beliefs" formed "under conditions known to generate cognitive illusions and strong delusional beliefs." One involves the tests in 1974 at the University of London by John Hasted and a group of distinguished physicists on the apparent metal-bending and other supposed psychic powers of Uri Geller. The experimenters allowed Geller to dictate the conditions for the test, saying that the best results come when everyone is in a relaxed state, all present sincerely want him to succeed, and the experimental arrangement is "aesthetically or imaginatively appealing" to the subject being tested for PK. Mutual trust should be encouraged; the slightest hint of suspicion is said to stifle the appearance of the powers.

The committee chided Hasted and colleagues for their naivete. "In their quest for psi-conducive conditions, they have created guidelines that play into the hands of anyone intent on deceiving them. The very conditions that are specified as being conducive to the appearance of paranormal phenomena are almost always precisely those that are conducive to the successful performance of conjuring tricks."

PK parties, touted by some of the military officers who made presentations to the committee, are another example. (See box "PK Parties . . .") "When proponents encounter a new phenomenon or psychic, they are strongly motivated to create conditions that will not drive the phenomenon away. The special atmosphere of PK parties and the suggestions of the British physicists are just two examples of attempts to generate psi-conducive conditions that also seem to be deception-conducive and bias-conducive."

Claims Lack Support

The parapsychology section of the NRC report concludes that despite sweeping claims, the best available evidence does not support claims for the existence of ESP or for its applications. (See box "Conclusions.") In fact it concludes that the best scientific evidence "does not justify the conclusion that ESP . . . exists." As for psychokinesis, the other major category of alleged psychic phenomena, the conclusion is similar: "Nor does scientific evidence offer support for the existence of psychokinesis. . . ."

It notes that these conclusions go counter to many individuals' beliefs in the reality of the paranormal. But it cautions against conclusions based on "alternatives" to the scientific method. "Alternative methods . . . create just those conditions that psychologists have found enhance human tendencies toward self-deception and suggestibility."

Postscript

Parapsychologists, grievously wounded by the National Research Council report, quickly fired back. Such devastating criticism from such a prestigious scientific institution was a severe blow to the future funding prospects of parapsychology. Parapsychologists complained strongly to the NRC, alleging bias. The NRC defended its procedures. Dean I. Radin, then president of the Parapsychological Association, said a somewhat negative report had been expected but felt the NRC committee had gone to extremes in its criticism. "Reports like the one by the National Research Council tend to influence people who might be interested in funding this work," he told the *Chronicle of Higher Education*. A Pittsburgh parapsychologist, Robert McConnell, sent out a mass mailing of materials that he said "show the inner workings of science in a dying civilization." It included a seven-page letter he had sent to the NRC questioning the study's "propriety." Colonel John Alexander, U.S. Army (retired), wrote an article in *New Realities* alleging that the NRC operated in a "biased and heavy-handed manner" with no channel for appeal. "What, we may ask, are they afraid of?" Congressman Claiborne Pell and his aide Scott Jones, a staunch proponent of the paranormal, even organized a one-sided government hearing attacking the study.

The most serious response was a 24-page booklet published in 1988 by the Parapsychological Association, "Reply to the National Research Council Study on Parapsychology," by John A. Palmer, Charles Honorton, and Jessica Utts (available from the PA for $2.00). It disputed the study's negative findings and asserted that it "does not represent an unbiased scientific assessment of parapsychology." It called the committee's conclusion of no scientific justification for the claims of parapsychology from 130 years of research "totally unwarranted."

Psychologist Ray Hyman, chairman of the NRC study's parapsychology subcommittee, remained low-key in his comments. He said parapsychologists were pushing for acceptance by science too soon and they need "to go back to their laboratories and clean up their act." He pointed out that parapsychologists should be gratified that the NRC took parapsychology seriously enough to do a formal study and that the study did conclude that research in certain areas should continue to be monitored. As for the overall criticism, he told the 1990 CSICOP conference during a discussion on these matters that he considered the report "very fair."

Notes

1. Members of the Committee were: John A. Swets, *Chair*, Bolt Beranek and Newman Inc., Cambridge, Mass.; Robert A. Bjork, Department of Psychology, University of California, Los Angeles; Thomas D. Cook, Department of Psychology, Northwestern University; Gerald C. Davison, Department of Psychology, University of Southern California; Lloyd G. Humphreys, Department of Psychology, University of Illinois; Ray Hyman, Department of Psychology, University of Oregon; Daniel M. Landers, Department of Physical Education, Arizona State University; Sandra A. Mobley, Director of Training and Development, the Wyatt Company, Washington, D.C.; Lyman W. Porter, Graduate School of Management, University of California, Irvine; Michael I. Posner, Department of Neurology, Washington University; Walter Schneider, Department of Psychology, University of Pittsburgh; Jerome E. Singer, Department of Medical Psychology, Uniformed Services University of Health Sciences, Bethesda, Md.; Sally P. Springer, Department of Psychology, State University of New York, Stony Brook; Richard F. Thompson, Department of Psychology, Stanford University. Daniel Druckman, *Study Director*.

2. Committee on Techniques for the Enhancement of Human Performance, Commission on Behavioral and Social Sciences and Education, National Research Council, *Enhancing Human Performance: Issues, Theories, and Techniques*, edited by Daniel Druckman and John A. Swets (Washington, D.C.: National Academy Press, 1987); available from National Academy Press, 2101 Constitution Ave., N.W., Washington, D.C. 20418, cloth $32.50; paper $22.50.

MARTIN GARDNER

Psi Researchers' Inattention to Conjuring

In 1985 the McDonnell Foundation, which funded the McDonnell Laboratory for Psychical Research at Washington University, St. Louis, announced it had withdrawn this funding. A wise decision. The lab had become an albatross around the university's neck after James Randi's notorious Alpha experiment made clear that Peter Phillips, the lab's director, though a competent physicist, had no comprehension of how to test supposed psychics.

When psychics start bending metal, rotating motors, moving objects, and performing other feats that imitate conjuring, there are only two sensible ways to conduct an investigation. Either have a knowledgeable magician present during the testing, or take a few years off to learn the art of close-up magic.

One of the dreariest aspects of psi history is the failure of otherwise intelligent researchers to understand this simple fact. Over and over again researchers and writers, ignorant of conjuring, have made fools of themselves by declaring their belief in metal-bending. Professor John Taylor, a British mathematical physicist, was duped into writing a preposterous book about metal-bending before he discovered he had been hoodwinked. Physicist John Hasted produced an even funnier book about the wonders of metal-bending. Neither Taylor nor Hasted deemed it worthwhile to seek the help of conjurors before starting their amateur investigations. To Taylor's credit, he later rejected metal-bending, but to this day he has been too embarrassed to admit how gullible he was.

In the United States the damage done to psi research by psychic mental-bending has been equally great. Both Helmut Schmidt and E. H. Walker, the two leading proponents of the quantum-mechanical explanation of psi, were taken in. As far as I know they may still be on the fence with respect to such powers.

I could go on and on with other recent cases of parapsychologists who never grasped the fact that magicians are the only experts on close-up deception. Jule Eisenbud, as far as I know, still believes Ted Serios could project his thoughts onto Polaroid film, although magicians have explained how Ted could have faked it with a palmed optical device. I suspect that every leading parapsychologist in the country now realizes that Eisenbud was deceived, but are too timid to say so. It never occurred to Charles Honorton to ask magicians how his friend Felicia Parise moved a pill bottle across her kitchen counter before he wrote a paper about this great event. Does Honorton still think Felicia did not use an invisible nylon thread? Apparently he does.

What makes this so hilarious is that it has all happened before, in the days of the great mediums. Did Conan Doyle or William Crookes or Oliver Lodge ever take a disguised magician to a séance? If so, I never heard of it. Among British journalists, the most tireless drumbeater for spiritualism was W. T. Stead, who died in the sinking of the *Titanic*. It would be hard to decide who was the biggest mark, Stead or Doyle. Stead thought it terrible that the Society for Psychical Research would try to apply scientific methods to mediums. In 1909, he attacked the Society by picturing himself as shipwrecked and drowning. (Believers in precognition have seized on this speech as evidence for psi premonition!) Suppose, said Stead, that instead of throwing him a rope someone shouts: "Who are you? What's your name?"

"I am Stead!" he imagined himself shouting back. "I am drowning here in the sea. Throw me a rope." His rescuers continue: "How do we know you are Stead? Where were you born? Tell us the name of your grandmother."

"What are known as psychical research methods were abhorrent to him," wrote spiritualist Edith Harper in her book *Stead the Man* (1914). "He held them truly unscientific. . . . He said he would rather die in the workhouse than believe that anyone would tell him a deliberate falsehood for the mere purpose of deceiving him."

Recently I obtained a copy of one of the strangest books on spiritualism ever written, or rather ghostwritten. It is *Lights and Shadows* by that magnificent charlatan D. D. Home. Doyle was furious with Home because in this book the Scottish medium exposed the methods of rivals who produced phenomena unlike his own—slate-writing, for instance. Of course Home carefully avoided any mention of his own methods. Even when he heard it from the medium he most admired, poor gullible Doyle couldn't believe that other mediums cheated as much as Home said they did.

Over and over again Home chastises his rivals for conducting séances in darkness, always adding that his own were in the light. Is it true that Home's séances were in the light? It is not. Home always *began* his sittings in the light. There would be table vibrations, raps, singing, talking, and praying; then the gaslights would be dimmed or extinguished. The room was seldom totally dark because it was necessary to see such things as fluttering white hands.

Pause to meditate on the absurdity of such darkness. Why would friendly spirits, anxious to contact loved ones, refuse to manifest themselves in significant ways except in the dark? If Home could flit about a room near the ceiling, as he often did, why did he always do this in rooms so black that the only proof he was up there was his own voice describing these Peter

Pan flights? An article by Robert Bell in the *Cornhill Magazine* (August 1860) contains a dramatic account of Home floating around in "pitch darkness." How did the sitters know Home was really up there? As he had done hundreds of times, Home left a mark on the ceiling!

It is often said that Home was never caught cheating. Well, it all depends on what you mean by "caught." In the same *Cornhill* article, Bell tells how he broke one of Home's cardinal rules by taking his hands off the table and clutching a spirit hand. "It was palpable as any soft substance, velvet or pulp, but pressure reduced it to air." White rubber gloves that glow in the dark were the stock-in-trade of nineteenth-century mediums. There are other records that strongly suggest ways in which Home cheated. In France, Baron Morio de l'Isle looked under the table and saw an empty shoe. After a woman said a spirit had touched her, Morio saw Home's foot slip back into his shoe. It is said that this ill-fated séance was one reason for Home's abrupt departure from France.

Such incidents are rare in Home's career, and for a simple reason. Home would not perform in the presence of magicians or even skeptics unless he sized up the skeptic as simple-minded. If a sitter in one of Home's séances so much as hinted doubts, the spirits would ask the skeptic to leave. Would not such negative thoughts dampen the spirits' spirits? We hear the same rationalizations today from psi investigators who want to exclude magicians and skeptics as observers.

The result of course is that dramatic PK phenomena—metal-bending, translocations, levitations, poltergeist activity—always occur when nobody capable of detecting fraud is watching. I write at the time of Edinburgh University's announcement that Robert Morris, of Syracuse University, has been appointed to Edinburgh's new Chair of Parapsychology, endowed by half a million pounds from the late Arthur Koestler's estate. Will Morris do a better job in Scotland than Phillips did in Missouri? Or will he too find excuses for excluding magicians when he starts testing for extraordinary powers?

Morris is a firm believer in the paranormal, though more cautious than most of his colleagues. He has what he once called "a high tolerance for ambiguity." As a younger man he was not always so cautious. While getting a doctorate in psychology at Duke University—his thesis was on the mating habits of ring-necked doves—he also worked at the nearby Psychical Research Foundation. This had been set up in 1960 to investigate evidence for survival after death, with William Roll, the well-known authenticator of poltergeists, as director. Morris was Roll's research assistant.

The best known of Morris's many experiments were his investigations of the powers of Blue Harary, a "psychic" who recently teamed up with Russell Targ to form a new psi-research organization and to coauthor their book *Mind Race* (1984). Morris's tests strongly confirmed Harary's ability to go "out of body" to a nearby lab where his spirit influenced the spirit of Spirit, Blue's pet kitten.

Another notable experiment was designed to test precognition in rats. The clever scheme was this: Monitor the behavior of a group of rats, select a few animals randomly, kill them, then see if anything in their previous behavior suggested foreknowledge of their doom. According to D. Scott Rogo, who describes these experiments in his *Parapsychology: A Century of Inquiry,* the rat test was "inconclusive." Morris tried again with goldfish. This time the fish were not killed, but simply held out of water long enough to cause "stress." Success! "Those goldfish that had been removed from the tank," Rogo writes, "were the ones who had been more active in the base-line period." This, Rogo informs us, could have been due to "an ability of the animals to show anxiety because of an awareness of what would be happening to them."

As these experiments indicate, Morris has been intensely interested in "animal psi." His paper "The Psychobiology of Psi," in Edgar Mitchell's *Psychic Explorations* (1974), is a readily accessible introduction to his views. Unfortunately, this survey of outstanding results on animal psi included Walter Levy's research at Dr. Rhine's laboratory on the PK power of live chicken eggs, having been written before Levy was caught cheating.

"Evidence for psi seems obtainable from a wide range of species and central nervous system complexity levels . . . ," Morris concludes. "In many ways, animals appear to respond to psi tasks in the same way that humans do—psi missing under negative conditions, habituation, response bias effects, and so on." He adds the warning that a major difficulty in such tests is that an experimenter's PK may bias results. He cites "evidence" that researchers can influence the movements of paramecia and wood lice, and Schmidt's famous tests with cockroaches, in which the results suggest it was Schmidt who influenced the randomizer because he hates cockroaches.

In recent years Morris has moved away from animal psi to other areas. In May 1984 *Omni* reported an experiment to test the abilities of humans to influence computers. Out of 33 subjects, the computer crashed with 13. Morris reported that these 13 were significantly more skeptical of PK than the others. The crashing may not have been the result of PK, Morris admitted—his high tolerance for ambiguity coming to the fore—but he added: "Why then did it occur so consistently in relation to the attitude of the people involved?"

In brief, Morris is a believer, but more hesitant than most parapsychologists in making extraordinary claims without extraordinary evidence. It will be interesting to see what results emerge from the Edinburgh laboratory. Let us hope that the lesson taught by the St. Louis fiasco will not be forgotten and that before Morris tests a psychic who performs what looks exactly like mediocre magic he will have the foresight and the courage to have someone on the scene capable of detecting fraud.

Editor's Postscript

Some interesting and occasionally amusing correspondence between author Martin Gardner and parapsychologist John Beloff about this article ensued.

It is too lengthy to reprint here but can be found in the *Skeptical Inquirer* (Summer 1986) and also in Gardner's book *The New Age: Notes of a Fringe-Watcher* (Prometheus Books, 1988).—*K.F.*

MARTIN GARDNER

The Obligation to Disclose Fraud

It is customary among editors of scientific journals to let their readers know when a published paper is found to have been based on fraud. It is the only way to prevent the paper from continuing to mislead later researchers. Such was not the practice of Joseph Banks Rhine.

Rhine outlined his policy of secrecy, in a note titled "The Hypothesis of Deception" (*Journal of Parapsychology*, 2, 151-152, 1938) as follows: "Certain friends of the research in extra-sensory perception," he began, "have recently informed us of rumors . . . that the subjects at Duke University and at other places were practicing deception . . . and that even when caught, these deceptions were deliberately withheld from the public. . . ." Rhine goes on to say that his researchers have become so skillful in safeguarding their experiments against both willful and unwitting deception that "no magician . . . is willing to attempt to work (as a magician) under such conditions." Indeed, he continues, so stringent are the controls that "the mere possibility alone" of cheating is "sufficient to bar data from acceptance. . . ."

That subjects and experimenters occasionally cheat is to be expected, Rhine says. It is not surprising, therefore, that his laboratory "[has] encountered a number of phenomena which on closer investigation proved to be fraudulently produced." Should such evidence be made public? "We do not feel," Rhine answers, "that any good purpose could be served by the exposure, à la Houdini, of these instances. . . . In a word, a research project in ESP does not become of conclusive scientific importance until it reaches the point at which even the greatest will-to-deceive can have no effect under the conditions. This criterion is the very threshold of the research field. It leaves us under no obligation to concern ourselves either with the ethics of the subjects or with the morbid curiosity of a few individuals."

My morbid curiosity was strongly aroused when I recently read in Louisa Rhine's *Something Hidden* (1983) a dramatic account of her husband's discovery that a paper he had published in his journal was based on deliberate cheating by the author. Mrs. Rhine refers to the dishonest parapsychologist only as "Jim." He had contributed many earlier articles to Rhine's journal, and this new work was "considered one of the best of those recently reported."

Banks, as Louisa called her husband, intended to make Jim's paper the "centerpiece" of a talk he was scheduled to give at a meeting of parapsychologists in Columbus, Ohio. A few weeks before the symposium, Gardner Murphy asked Rhine for Jim's original records to consider for his own speech on record-keeping and -checking. Jim brought his records to Rhine a few days before the Columbus meeting. To Rhine's horror, when he and two of his assistants began examining the records, they found unmistakable evidence of fraud. "Jim had actually consistently falsified his records. . . ," Louisa Rhine tells us. "To produce extra hits Jim had to resort to erasures and transpositions in the records of his call series." Rhine journeyed to Columbus in great anguish. He had to scrap the paper he intended to read, and deliver instead, with visible nervousness, an entirely different talk. Jim's college professor, after seeing evidence of the cheating, was profoundly shocked and even blamed himself for not being more vigilant.

"Jim's name," Louisa Rhine writes, was never "again seen in the annals of parapsychology."

This simply isn't true. Jim (I learned from a disenchanted parapsychologist) was James D. MacFarland, then a young instructor in psychology at Tarkio College, in Tarkio, Missouri. His flawed paper, "Discrimination Shown Between Experimenters by Subjects," appeared in Rhine's journal (*JP*, 2: 160-170, Sept. 1938), the issue following the one with Rhine's piece on deception. No retraction of the paper was ever published. Did references to MacFarland's research vanish from the literature of psi? It did not. J. G. Pratt, in *Extrasensory Perception After 60 Years* (1940), refers to MacFarland's work. And Pratt was one of Rhine's two assistants who originally discovered MacFarland's fudging!

In 1974 Rhine again suffered from unfortunate timing. His paper "Security Versus Deception in Parapsychology," published in his journal (vol. 38, 1974), runs to 23 pages. In it he dismisses deception by subjects as no longer significant. Self-deception by experimenters is more widespread, but this too is limited, Rhine says, to inexperienced novices who form a "subspecies of unprepared experimenters" who "may soon be approaching extinction."

Turning to deliberate deception by parapsychologists, Rhine selects twelve sample cases of dishonest experimenters that came to his attention from 1940 to 1950, four of whom were caught "red-handed." Not a single name is mentioned. What papers did they publish, one wonders. Are their papers still being cited as evidence for psi? Rhine is convinced that such fraud diminished markedly after 1960. "We have at least got past the older phase of having to use detectives and magicians to discover or prevent trickery by the subjects." He applauds the growing use of computers; but although "machines will not lie," he warns against overoptimism about their usefulness in parapsychology. Complex apparatus, he cautions, "can sometimes also be used as a screen to conceal the trickery it was intended to prevent."

The warning proved prophetic. A few months after Rhine's paper appeared, the acting director of his laboratory and the young man he had chosen to be his successor, Dr. Walter Levy, was caught red-handed tinkering with an electronic recording machine. The tinkering had beefed up the scores of a test he was making on the PK ability of rats. Levy resigned in disgrace, though, again, references to his earlier papers (one on the PK powers of live chicken eggs) have not yet entirely vanished from psi literature. Rhine tried his best to hush up the scandal; but when it was obvious he could not do so, he wrote an apologetic article about it in his journal. As usual he did not mention Levy's name, apparently under the naive delusion that readers would not learn the flimflammer's identity.

Four years later, England's most distinguished parapsychologist, S. G. Soal, was caught having deliberately fudged the data for one of his most famous tests. I see no sign that Soal's other experiments are disappearing from the literature. J. G. Pratt, almost pathologically incapable of believing anyone would cheat, came to Soal's defense. He argued that Soal may have "used precognition when inserting digits into the columns of numbers he was copying down, unconsciously choosing numbers that would score hits on the calls the subject would make later. For me, this 'experimenter psi' explanation makes more sense, psychologically, than saying that Soal consciously falsified for his own records."

I have been told on reliable authority that the files in Rhine's laboratory contain material suggesting fraud on the part of Hubert Pearce, the most talented of all of Rhine's early psychics. Who knows how much data of this sort is buried in the Rhine archives? Let us hope that someday someone with a balanced sense of history, under no compulsion to regard Rhine as one of psi's saints, will be allowed full access to those archives and give us a biography of Banks that is not a hagiography.

Let me change the subject. Early in 1987 Random House published *Intruders*, by Budd Hopkins. It is one of the funniest and shabbiest books ever written about abductions of humans by extraterrestrials who visit Earth in flying saucers. Hopkins is easy to understand. He is a hack journalist of the occult. Harder to comprehend was a full-page advertisement that appeared in the *New York Times Book Review*. It is a long "Dear reader" letter signed by no less a personage than the then-publisher of Random House, Howard Kaminsky.

Kaminsky's letter bursts with praise for Hopkins's worthless volume. The book's events are "objectively set down." You might think the author and the publisher are "kooks," Kaminsky continues, but it is "Hopkins' calmness, objectivity, and cogency—as well as the mass of medical, physical, and psychiatric evidence he presents—that make *Intruders* so *un*kooky. He is as intelligent and thoughtful as anyone I know, and questions his own evidence as severely as any skeptic would. . . . There were moments, as I read the manuscript, when I actually got chills down the back of my neck."

Well, chills slithered down *my* neck when I read those incredible remarks by the publisher of one of our nation's most distinguished publishing houses. *Newsweek* magazine (October 26, 1987) devoted page 62 to the story of how Kaminsky had been suddenly fired from Random House by his superior, Robert Bernstein, chairman of the firm, to be replaced by Joni Evans, from Simon and Schuster. I have no inside information about the personality clashes behind what *Newsweek* called the "rumble at Random House," but I suspect and hope that Kaminsky's idiotic letter played a role in the rumble.

Part 5: Examining Popular Claims

RON AMUNDSON

The Hundredth Monkey Phenomenon

Claims of the paranormal are supported in many ways. Personal reports ("I was kidnapped by extraterrestrials"), appeals to puzzling everyday experiences ("Did you ever get a phone call from someone you had just dreamed about?"), and references to "ancient wisdom" are a few. Citations of actual scientific results are usually limited to ESP experiments and a few attempts to mystify further the already bizarre discoveries of modern physics. But the New Age is upon us (we're told) and New Age authors like Rupert Sheldrake (1981) and Lyall Watson (1979) support their new visions of reality with scientific documentation. Sheldrake has a bibliography of about 200 listings, and Watson lists exactly 600 sources. The sources cited are mostly respectable academic and scientific publications. The days of "[unnamed] scientists say" and "Fred Jones, while walking alone in the woods one day . . ." are gone. Or are they?

I teach college courses in epistemology, in the philosophy of science, and in pseudoscience and the occult. Students in these courses naturally bring to class examples of remarkable and paranormal claims. During the past few years one such claim has become especially popular, the "Hundredth Monkey Phenomenon." This phenomenon was baptized by Lyall Watson, who documents the case with references to five highly respectable articles by Japanese primatologists (Imanishi 1963; Kawai 1963 and 1965; Kawamura 1963; and Tsumori 1967). Watson's discussion of this phenomenon covers less than two pages. (Except where noted, all references to Watson are to pages 147 and 148.) But this brief report has inspired much attention. Following Watson, a book (Keyes 1982), a newsletter article (*Brain/Mind Bulletin* 1982), and a film (Hartley 1983) have each been created with the title "The Hundredth Monkey." In addition we find a journal article entitled "The 'Hundredth Monkey' and Humanity's Quest for Survival" (Stein 1983) and an article called "The Quantum Monkey" in a popular magazine (*Science Digest* 1981). Each relies on Watson as the sole source of information on the remarkable and supernatural behavior of primates.

The monkeys referred to are indeed remarkable. They are Japanese macaques (*Macaca fuscata*), which live in wild troops on several islands in Japan. They have been under observation for years. During 1952 and 1953 the primatologists began "provisioning" the troops—providing them with such foods as sweet potatoes and wheat. This kept the monkeys from raiding farms and also made them easier to observe. The food was left in open areas, often on beaches. As a result of this new economy, the monkeys developed several innovative forms of behavior. One of these was invented in 1953 by an 18-month-old female that the observers named "Imo." Imo was a member of the troop on Koshima island. She discovered that sand and grit could be removed from the sweet potatoes by washing them in a stream or in the ocean. Imo's playmates and her mother learned this trick from Imo, and it soon spread to other members of the troop. Unlike most food customs, this innovation was learned by older monkeys from younger ones. In most other matters the children learn from their parents. The potato-washing habit spread gradually, according to Watson, up until 1958. But in the fall of 1958 a remarkable event occurred on Koshima. This event formed the basis of the "Hundredth Monkey Phenomenon."

The Miracle on Koshima

According to Watson, all of the juveniles on Koshima were washing their potatoes by early 1958, but the only adult washers were those who had learned from the children. In the fall of that year something astounding happened. The exact nature of the event is unclear. Watson says:

> . . . One has to gather the rest of the story from personal anecdotes and bits of folklore among primate researchers, because most of them are still not quite sure what happened. And those who do suspect the truth are reluctant to publish it for fear of ridicule. So I am forced to improvise the details, but as near as I can tell, this is what seems to have happened. In the autumn of that year an unspecified number of monkeys on Koshima were washing sweet potatoes in the sea. . . . Let us say, for argument's sake, that the number was ninety-nine and that at eleven o'clock on a Tuesday morning, one further convert was added to the fold in the usual way. But the addition of the hundredth monkey apparently carried the number across some sort of threshold, pushing it through a kind of critical mass, because by that evening almost everyone was doing it. Not only that, but the habit seems to have jumped natural barriers and to have appeared spontaneously, like glycerine crystals in sealed laboratory jars, in colonies on other islands and on the mainland in a troop at Takasakiyama.

A sort of group consciousness had developed among the monkeys, Watson tells us. It had developed suddenly, as a result of one last monkey's learning potato washing by conventional means. The sudden learning of the rest of the Koshima troop was not attributable to the normal one-monkey-at-a-time methods of previous years. The new phenomenon of group consciousness was responsible not only for the sudden learning on Koshima but for the equally sudden acquisition of the habit by monkeys across the sea. Watson admits that he was forced to "improvise" some of the details—the time of the day, the day of the week, and the exact number of monkeys required for the "critical mass" were not specified in the scientific literature. But by evening (or at least in a very short period of time) almost everyone (or at least a large number of the remaining monkeys) in the colony had suddenly acquired the custom. This is remarkable in part because of the slow and gradual mode of acquisition that had typified the first five years after Imo's innovation. Even more remarkable was the sudden jumping of natural boundaries, apparently caused by the Koshima miracle.

Documentation

In this section I investigate the relations between Watson's description of the Hundredth Monkey Phenomenon and the scientific sources by which he validates it. To be sure, we must not expect too much from the sources. Watson has warned us that the complete story was not told and that he was "forced to improvise the details." But we should expect to find some evidence of the mysteriousness of the Koshima events of 1958. In particular, we should expect to find evidence of an episode of sudden learning within the troop at this time (though perhaps not in one afternoon) and evidence of the sudden appearance of potato washing in other troops sometime soon after the Koshima event. We also have a negative expectation of the literature; it should *fail* to report certain important details. It will not (we expect) tell us the exact number of monkeys washing potatoes prior to or after the event of 1958, nor will it provide us with an explanation of how the post-event Koshima learners were able to acquire their knowledge. After all, it is Watson's claim that the event produced *paranormal* learning of potato washing. These three expectations will be tested against the literature. Was there a sudden event at Koshima? Did acquisition at other colonies follow closely the Koshima event? Does Watson improvise details *only* when the cited literature fails to provide adequate information? The following comments will be restricted to the literature on macaques actually cited by Watson.

Almost all of the information about the Koshima troop appears in a journal article by Masao Kawai (1965); the other articles are secondary on this topic. Kawai's article is remarkably detailed in its description of the Koshima events. The troop numbered 20 in 1952 and grew to 59 by 1962. (At least in the numerical sense, there was never a "hundredth monkey" on Koshima.) Watson states that "an unspecified number" of monkeys on Koshima had acquired the potato-washing habit by 1958. Actually this number was far from unspecified. Kawai's data allowed the reader to determine the dates of acquisition of potato washing (and two other food behaviors), as well as the dates of birth and genealogical relationships, *of every monkey in the Koshima troop from 1949 to 1962* (Figure 1, pp. 2-3, and elsewhere in the paper). In March 1958, exactly 2 of 11 monkeys over 7 years old had learned potato washing, while exactly 15 of 19 monkeys between 2 and 7 had the habit (p. 3). This amounts to 17 of 30 noninfant monkeys. There is no mention in this paper (or in any other) of a sudden learning event in the fall of 1958. However, it is noted that by 1962, 36 of the 49 noninfant monkeys had acquired the habit. So both the noninfant population and the number of potato washers had increased by 19 during this four-year period. Perhaps this is what suggested to Watson that a sudden event occurred in the fall of 1958. And perhaps (since one can only surmise) this idea was reinforced in Watson's mind by the following statement by Kawai: "The acquisition of [potato washing] behavior can be divided into two periods; before and after 1958" (p. 5).

So Kawai does not give a time of year, a day of the week, or even the season for any sudden event in 1958. But he does at least identify the year. And is Kawai mystified about the difference between pre- and post-1958 acquisition? Is he "not quite sure what happened"? Is he reluctant to publish details "for fear of ridicule?" No. He publishes the whole story, in gothic detail. The post-1958 learning period was remarkable only for its normalcy. The period from 1953 to 1958 had been a period of exciting innovation. The troop encountered new food sources, and the juveniles invented ways of dealing with these sources. But by 1958 the innovative youth had become status quo adults; macaques mature faster than humans. The unusual juvenile-to-adult teaching methods reverted to the more traditional process of learning one's food manners at one's mother's knee. Imo's first child, a male named "Ika," was born in 1957 (pp. 5, 7). Imo and her former playmates brought up their children as good little potato-washers. One can only hope that Ika has been less trouble to his Mom than Imo was to hers. Kawai speaks of the innovative period from 1953 to 1958 as "individual propagation" (p. 5) and the period after 1958 as "pre-cultural propagation" (p. 8). (This latter term does not indicate anything unusual for the monkey troops. The troops under normal circumstances have behavioral idiosyncrasies and customs that are passed along within the group by "pre-cultural" means. The expression only indicates a reluctance to refer to monkey behavior as genuinely "cultural.")

So there was nothing left unsaid in Kawai's description. There was nothing mysterious, or even sudden, in the events of 1958. Nineteen fifty-eight and 1959 were the years of maturation of a group of innovative youngsters. The human hippies of the 1960s now know that feeling. In fact 1958 was a singularly poor year for habit acquisition on Koshima. Only two monkeys learned to wash potatoes during that year, young females named Zabon and Nogi. An average of three a year had learned potato washing during the previous five years (Table 1, p. 4). There is no evidence that Zabon and Nogi were psychic or in any other way unusual.

Let us try to take Watson seriously for a moment longer. Since only two monkeys learned potato washing during 1958 (according to Watson's own citation), one of them must have been the "Hundredth Monkey." Watson leaves "unspecified" which monkey it was, so I am "forced to improvise" and "say, for argument's sake" that it was Zabon. This means that poor little Nogi carries the trim metaphysical burden of being the "almost everyone in the colony" who, according to Watson, suddenly and miraculously began to wash her potatoes on that autumn afternoon.

Watson claims that the potato-washing habit "spontaneously" leaped natural barriers. Is there evidence of this? Well, two sources report that the behavior was observed off Koshima, in at least five different colonies (Kawai 1965, 23; Tsumori 1967, 219). These reports specifically state that the behavior was observed only among a few individual monkeys and that it had not spread throughout a colony. There is no report of when these behaviors occurred. They must have been observed sometime between 1953 and 1967. But there is nothing to indicate that they followed closely upon some supposed miraculous event on Koshima during the autumn of 1958, or that they occurred suddenly at any other time, or that they were in any other way remarkable.

In fact there is absolutely no reason to believe in the 1958 miracle on Koshima. There is every reason to deny it. Watson's description of the event is refuted *in great detail* by the very sources he cites to validate it. In contrast to Watson's claims of a sudden and inexplicable event, "Such behavior patterns seem to be smoothly transmitted among individuals in the troop and handed down to the next generation" (Tsumori 1967, 207).

Methodology of Pseudoscience

The factual issue ends here. Watson's claim of a "Hundredth Monkey Phenomenon" is conclusively refuted by the very sources he cites in its support. He either failed to read or misreported the information in these scientific articles. But Watson's own mode of reasoning and reporting, as well as the responses he has inspired in the popular literature, deserve attention. They exemplify the pseudoscientific tradition. Consider the following:

1. Hidden sources of information: Watson informs us that the scientific reports leave important data "unspecified." This is simply false. But, more subtly, he tells us that most of the researchers are still unsure of what happened and that those who "do suspect the truth are reluctant to publish it for fear of ridicule." In one fell swoop Watson brands himself as courageous, explains why no one else has dared report this miraculous phenomenon, and discourages us from checking the cited literature for corroboration. Watson got the real story from "personal anecdotes and bits of folklore among primate researchers. . . ." Those of us who don't hobnob with such folks must trust Watson. The technique was effective. Of the commentaries I have found on the Hundredth Monkey Phenomenon, not one shows evidence of having consulted the scientific sources cited by Watson. Nonetheless, each presents Watson's fantasy as a scientifically authenticated fact. Nor is additional information

available from Watson. I have written both to Watson and to his publishers requesting such information and have received no reply.

2. Aversion to naturalistic explanations: The fact is that potato washing was observed on different islands. Watson infers that it had traveled in some paranormal way from one location to another. Like other aficionados of the paranormal, Watson ignores two plausible explanations of the concurrence of potato washing. First, it could well have been an independent innovation— different monkeys inventing the same solution to a common problem. This process is anathema to the pseudoscientist. The natives of the Americas simply *could not have* invented the pyramid independent of the Egyptians—they just didn't have the smarts. In more extreme cases (von Däniken, for example) a *human being* is just too dumb to invent certain clever things—extraterrestrials must have done it.

Watson assumes that Imo was the only monkey capable of recognizing the usefulness of washing potatoes. In his words, Imo was "a monkey genius" and potato washing is "comparable almost to the invention of the wheel." Monkeys on other islands were too dumb for this sort of innovation. But keep in mind that these monkeys didn't even *have* potatoes to wash before 1952 or 1953, when provisioning began. Monkeys in at least five locations had learned potato washing by 1962. This suggests to me that these monkeys are clever creatures. It suggests to Watson that *one* monkey was clever and that the paranormal took care of the rest. A second neglected explanation is natural diffusion. And indeed Kawai reports that in 1960 a potato washer named "Jugo" swam from Koshima to the island on which the Takasakiyama troop lives. Jugo returned in 1964 (Kawai 1965, 17). Watson does not mention this. The Japanese monkeys are known to be both clever and mobile, and either characteristic might explain the interisland spread of potato washing. Watson ignores both explanations, preferring to invent a new paranormal power.

3. Inflation of the miracle: As myths get passed along, everyone puffs them up a bit. The following two examples come from second-generation commentaries that quote extensively from Watson. Nevertheless, even Watson's claims are beginning to bulge. First, the primatologists' reports had mentioned that only a few isolated cases of off-Koshima potato-washing were observed. Watson reports this as the habit's having "appeared spontaneously . . . in colonies on other islands. . . ." Not actually false, since the few individuals were indeed *in* other colonies (though only individuals and not whole colonies adopted the behavior). Following Watson, Ken Keyes reports that, after the hundredth Koshima monkey, "colonies of monkeys on other islands . . . began washing their sweet potatoes"! (Keyes 1982, p. 16). From Keyes, one gets the image of spontaneous mass orgies of spud-dunking. A second example: Regarding the primatologists' attitudes toward the events of 1958, Watson reports only that they are "still not quite sure what happened." But the primatological confusion quickly grows, for *Science Digest* (1981) reports "a mystery which has stumped scientists for nearly a quarter of a century." In

these two particular cases, Watson's own statements are at least modest. They're not what one would call accurate, but not exorbitantly false either. By the second generation we find that "not quite sure what happened" becomes "stumped for nearly a quarter of a century," and the habit that *appeared in* individuals within colonies of monkeys becomes a habit *of* colonies of monkeys. Please keep in mind that the second generation relies *only* on Watson for its information; even Watson's none-too-accurate report has been distorted— and not, needless to say, in the direction of accuracy.

4. The paranormal validates the paranormal: The validity of one supernatural report is strengthened by its consistency with other such reports. Watson's commentators show how this works. Keyes supports the Hundredth Monkey Phenomenon by its consistency with J. B. Rhine's work at Duke, which "demonstrated" telepathy between individual humans. "We now know that the strength of this extrasensory communication can be amplified to a powerfully effective level when the consciousness of the 'hundredth person' is added" (Keyes 1982, 18). Elda Hartley's film "The Hundredth Monkey" invokes Edgar Cayce. And in a remarkable feat of group consciousness, *four of the five* secondary sources emphasize the similarities between Watson's Hundredth Monkey Phenomenon and Rupert Sheldrake's notion of the "morphogenetic field." The spontaneous recognition of the similarities between Watson and Sheldrake seems to have leaped the natural boundaries between the four publications! Now *there's* a miracle! (Surely independent invention or natural diffusion couldn't account for such a coincidence.)

Conclusions

I must admit sympathy for some of the secondary sources on the Hundredth Monkey Phenomenon. This feeling comes from the purpose for which the phenomenon was cited. Ken Keyes's book uses the phenomenon as a theme, but the real topic of the book is nuclear disarmament. Arthur Stein's article and (to a lesser extent) the Hartley film are inspired by Keyes's hope that the Hundredth Monkey Phenomenon may help prevent nuclear war. The message is that "you may be the Hundredth Monkey" whose contribution to the collective consciousness turns the world away from nuclear holocaust. It is hard to find fault in this motive. For these very same reasons, one couldn't fault the motives of a child who wrote to Santa Claus requesting world nuclear disarmament as a Christmas present. We can only hope that Santa Claus and the Hundredth Monkey are not our best chances to avoid nuclear war.

Watson's primary concern is not prevention of war but sheer love of the paranormal. His book begins with a description of a child who, before Watson's own eyes, and with a "short implosive sound, very soft, like a cork being drawn in the dark," psychically turned a tennis ball inside out—fuzz side in, rubber side out—without losing air pressure (p. 18). Just after the Hundredth Monkey discussion, Watson makes a revealing point. He quotes with approval a statement attributed to Lawrence Blair: "When a myth is shared by large numbers of people, it becomes a reality" (p. 148). This sort of relativist epistemology is not unusual in New Age thought. I would express Blair's thought somewhat differently: "Convince enough people of a lie, and it becomes the truth." I suggest that someone who accepts this view of truth is not to be trusted as a source of knowledge. He may, of course, be a marvelous source of fantasy, rumor, and pseudoscientific best-sellers.

I prefer epistemological realism to this sort of relativism. Truth is not dependent on the numbers of believers or on the frequency of published repetition. My preferred epistemology can be expressed simply: Facts are facts. There is no Hundredth Monkey Phenomenon.

Follow-up

I began investigating the "Hundredth Monkey Phenomenon" in August 1984 with a letter to Lyall Watson, the author of the "phenomenon," addressed in care of his publisher, Simon and Schuster. I asked for more information about the group consciousness of monkeys reported by Watson in *Lifetide*. Neither this nor a later letter to the publisher has ever received a reply. My study was published in the Summer 1985 *Skeptical Inquirer*. Boyce Rensberger, a *Washington Post* science writer, and subsequently a recipient of CSICOP's 1986 Responsibility in Journalism Award, picked up the story. He also approached Simon and Schuster, who declined to put him in touch with Watson. Rensberger (1985) quoted Watson's editor as saying that Watson "is a distinguished and eminent scholar who, I have to say, does have some weird ideas." No news there.

Watson has now broken the silence. Ted Schultz, an editor for *Whole Earth Review,* managed to contact him. According to Schultz, Watson was "quite happy to respond to Amundson's analysis of his monkey tale." The response was published, in the Fall 1986 "Fringes of Reason" issue of *Whole Earth Review* (and reprinted in Schultz 1989). Although he begins with a swipe at "self-appointed committees for the suppression of curiosity," Watson deals "in good humor" with my critique of the Hundredth Monkey. My article was "lucid, amusing, and refreshingly free of the emotional dismissals" that, he says, CSICOP is prone to. I wish I could be proud of this distinction.

Watson continues: "I accept Amundson's analysis of the origin and evolution of the Hundredth Monkey without reservation. It is a metaphor of my own making, based—as he rightly suggests—on very slim evidence and a great deal of hearsay. I have never pretended otherwise. . . . I based none of my conclusions on the five sources Amundson uses to refute me. I was careful to describe the evidence for the phenomenon as strictly anecdotal and included citations in *Lifetide,* not to validate anything, but in accordance with my usual practice of providing tools, of giving access to useful background information."

It should be remembered that the "five sources" I used to "refute" him were the identical five sources that Watson provides as "tools" and "access" in his original discussion of the phenomenon.

Watson goes on to complain about my conclusion that the Hundredth Monkey Phenomenon does not exist. He still thinks the phenomenon is real but admitting that it didn't happen on Koshima. This is like saying that the "Geller Effect" is real, while claiming that Uri Geller himself has no special powers. Well, okay. Show us a *real* example.

Watson is unhappy about my description of his work as "pseudoscience." He admitted all along, he says, that the Hundredth Monkey story was anecdotal. This is approximately a half-truth. Watson did admit in *Lifetide* that he had to "gather the rest of the story from personal anecdotes and bits of folklore." (This was because, he said, the scientists were afraid to publish the truth "for fear of ridicule.") He then *specifically stated* that certain crucial details were missing from the scientific reports. He went on to describe the events on Koshima, "improvising" the details. The miraculous results were stated in two sentences, followed by a citation reference.

The details said by Watson to be missing *were not missing*. He falsely reported on the scientific evidence available—available, in fact, in his own citations.

Watson responds to my claim that his own documentation refutes him by explaining that his citation references were not meant as documentation at all, but as "tools." (Perhaps being refuted by your own tool is less painful than being refuted by your own documentation.) Here it should be noted that the citations were presented in *exactly* the format used to provide documentation for factual claims, both in scientific and in informal writing. *Lifetide* is peppered with raised reference numbers, each following a factual statement made in the text. The *Chicago Manual of Style* refers to this format as "notes documenting the text, and corresponding to reference numbers in the text." Does Watson anywhere warn us that his citations do *not* document the text—that they actually *contradict* the text? Does he warn us that they are merely "tools"? No. We are told only that the raised numbers "refer to numbered items in the bibliography."

As an "eminent scholar" and "holder of degrees in anthropology, ethology, and marine biology" (*Whole Earth Review*'s description), Watson must be assumed to understand the use of scientific citations. The meaning of a reference citation is not something each author simply invents for himself. It does not mean "documentation" for some writers and "tools" for others. Watson uses a format that implies documented support for a factual claim. He now says that he didn't really mean it that way.

I submit that this technique is *pseudoscientific* in the strictest sense. It falsely presents the appearance of science. Watson could have admitted that he made a mistake in his citations (or that he never read them in the first place). Instead he excuses himself by saying that the references were merely "tools." They just *looked* like scholarly citations.

Watson owes an apology to the thousands of people who took his claims to be reports of fact, rather than "hearsay" and "anecdotes." None of Watson's published commentators thought he was presenting "hearsay" about potato-washing monkeys. If *I* made a mistake by taking him seriously, so did everyone else. Let it be known that the hundreds of scientific-looking citations in Watson's books are not intended to support his factual claims. They are "tools." They look, for all the world, like scientific documentation. But it is all an illusion.

Postscript

My only regret in the writing of "The Hundredth Monkey Phenomenon" is that I didn't have the nerve to call it something like "Spud-Dunking Monkey Theory Debunked," Boyce Rensberger's priceless title in the *Washington Post*.

Reaction to the paper amazed me. I had underestimated the influence of the *Skeptical Inquirer*, and Rensberger's article certainly helped to spread the word. But besides that, I had no idea that the Hundredth Monkey had become such a compelling image in New Age thought, not only in the United States but around the world. The article has been reprinted in Australia and in Sweden (where it was translated into "Der Hundreden Apen"). It was discussed in the British science magazine *New Scientist* (1985), and I was interviewed on Australian Public Radio (an interview arranged by the good people of the Australian Skeptics). It has even received friendly attention from sources one would normally expect to be sympathetic to the New Age, such as *East-West Journal* (1985) and *Whole Earth Review*. (Discussion and related articles from the Fall 1986 *Whole Earth Review* were reprinted in *Fringes of Reason*, Schultz 1989). There was even a kindly word from Douglas Groothuis (1988) in a book advising conservative Christians about how to confront New Age beliefs. To my knowledge, the only negative reaction was Lyall Watson's gentle scolding of my narrow-mindedness (in Schultz 1989). The moral of the story seems to be that many of the thousands of people who had heard the Hundredth Monkey myth *were already skeptical about it*. Nevertheless, practically no one had bothered to chase down its origin and check its credentials.

The notable exception to this complacency was Maureen O'Hara, a humanistic psychologist who had independently critiqued the Hundredth Monkey (see O'Hara 1986). She was more tolerant than I of Watson's myth-making, laying most of the blame on Watson's commentators. But she eloquently exposed a crucial fallacy in the New Age acceptance of mass consciousness, a fallacy I had missed. New Age aficionados consider mass consciousness to be "empowering" to individuals, since "you may be the Hundredth Monkey." O'Hara points out the foolishness of this "empowerment." An individual whose beliefs are in the minority is *already* out-Hundredth-Monkeyed by the opinion of the majority. Moreover, the conviction that beliefs alone can affect social change provides a perfect excuse for complacency. Why bother to engage in political activism when it's just as effective to sit comfortably at home and *believe* things? I was especially gratified to see the same point recognized in a local Kansas newspaper; my refutation of Watson was celebrated by the *Wellington News* in an editorial entitled "Individually Responsible."

As I already confessed, I'm no heroic crusader for rationality. I studied the Hundredth Monkey Phenomenon beause my students forced me into it. Our complacency in the face of such nonsense simply allows the nonsense to spread. Other myths may not be as easy to burst as the Hundredth Monkey Phenomenon, but we'll never know until we try.

References

Brain/Mind Bulletin. 1982. The hundredth monkey. In "Updated Special Issue: 'A New Science of Life.' "

East-West Journal. 1985. Monkey business, November, p. 13.

Groothuis, Douglas R. 1988. *Confronting the New Age.* Downers Grove, Ill.: InterVarsity Press.

Hartley, Elda (producer). 1983. *The Hundredth Monkey* (film and videotape). Hartley Film Foundation, Inc. Cos Cob, Conn.

Imanishi, Kinji. 1963. Social behavior in Japanese monkeys. In *Primate Social Behavior,* Charles A. Southwick, ed. Toronto: Van Nostrand.

Kawai, Masao. 1963. On the Newly-acquired behaviors of the natural troop of Japanese monkeys on Koshima island. *Primates,* 4:113-115.

———. 1965. On the newly-acquired pre-cultural behavior of the natural troop of Japanese monkeys on Koshima Islet. *Primates,* 6:1-30.

Kawamura, Syunzo. 1963. Subcultural propagation among Japanese macaques. In *Primate Social Behavior,* Charles A. Southwick, ed. Toronto: Van Nostrand.

Keyes, Ken, Jr. 1982. *The Hundredth Monkey.* Coos Bay, Ore.: Vision Books.

New Scientist. 1985. Making a monkey out of Lyall Watson. July 11, p. 21.

O'Hara, Maureen. 1986. Of myths and monkeys. *Whole Earth Review,* Fall. Reprinted in Schultz, 1989.

Rensberger, Boyce. 1985. Spud-dunking monkey theory debunked. *Washington Post,* July 6.

Schultz, Ted, ed. 1989. *Fringes of Reason: A Whole Earth Catalog.* New York: Harmony Books.

Science Digest. 1981. The quantum monkey. Vol. 8: 57.

Sheldrake, Rupert. 1981. *A New Science Life.* Los Angeles: J. P. Tarcher.

Stein, Arthur. 1983. The "hundredth monkey" and humanity's quest for survival. *Phoenix Journal of Transpersonal Anthropology,* 7: 29-40.

Tsumori, Atsuo. 1967. Newly acquired behavior and social interactions of Japanese monkeys. In *Social Communication Among Primates.* Stuart Altman, ed. Chicago: University of Chicago Press.

Watson, Lyall. 1979. *Lifetide.* New York: Simon and Schuster.

———. 1986. Lyall Watson responds. *Whole Earth Review,* Fall. Reprinted in Schultz, 1989.

Wellington (Kansas) News. 1985. Individually Responsible, July 22.

BERNARD J. LEIKIND and WILLIAM J. McCARTHY

An Investigation of Firewalking

For centuries, some people in various cultures around the world have walked on hot coals without getting burned. Ordinarily, this is associated with religious rituals, and success is attributed to spiritual or mystical powers' protecting the walkers. Since firewalking is usually done in faraway places, many Americans are quite willing to give some credence to the firewalkers' claims that some sort of mysterious powers protect the walkers from harm—powers that can only be harnessed after long study and careful preparation. In the past year or so, many Americans have been walking across beds of hot embers as part of self-help seminars that purport to teach the student to overcome fears or to take command of their lives and achieve success. Because firewalking seems so mysterious, if not impossible, to most of us, the firewalk serves as a powerful persuasive tool, convincing the walker that all of the material taught in the seminar must be correct.

We have investigated American firewalking in Los Angeles as taught and practiced by Tony Robbins of the Robbins Research Institute. We participated in a firewalk in the fall of 1984. One of us (WJM) attended the seminar session; and the other (BJL) did not, since he wished to test the proposition that the training offered in the seminar was not necessary in order to walk across the coals.

How the Investigation Began?

One morning in April 1984, I (BJL) read an article in the *Los Angeles Times* headlined "Firewalking, the Curious Hotfoot It to a New Fad" (Krier 1984). It received a big play, beginning on page 1, continuing on page 3 for another half-page, and including three large photographs, one of which showed a rugged fellow in a dark suit striding boldly across a bed of glowing embers.

The article was filled with statements by the firewalk leaders, like this

Bernard Leikind walks on bed of hot coals. (Photo by Kimberly Willis, Physics Department, UCLA.)

one by Tolly Burkan, once professionally known as Tolly the Clown and now one of the nation's most renowned gurus of firewalking: "Just holding the thought in your mind that you're not going to injure your feet alters the chemistry of your body," he insisted. "Indeed, at many firewalking rituals throughout the world, belief is reportedly all that is needed."

Throughout the article, the consensus of the firewalkers was that in some way special mental powers altered the operation of normal physical processes. As it happened, I had read an article that dealt with firewalking in *Scientific American*'s "Amateur Scientist" column (Walker 1977). In fact, I had seen the article's author, Professor Jearl Walker of Cleveland State University, dip his fingers into molten lead, and I knew that he had walked on hot coals in his classes. Professor Walker attributed this ability to the Leidenfrost effect: the presence of a thin layer of water vapor—a poor heat conductor—from moisture on the feet, either from sweat or from damp material around the coals.

So I thought I knew how firewalking was done, and I certainly believed it had nothing to do with the kind of exotic powers claimed in the *Los Angeles Times* article. I called the reporter and was told to write a letter to the editor. I called a skeptical medical doctor who had been quoted in the article, and he began referring reporters who called him about firewalking to me.

Sometime later, one of the principal firewalkers, Tony Robbins, was interviewed on a local call-in radio show. When I heard about this show, I called the station to see if I could get a tape of the program. Bill Jenkins, the interviewer, had firewalked and was a believer in the mysterious mental powers of firewalkers. When I told him what I thought, he was quite upset and maintained that water wasn't necessary. He challenged me to go to a seminar to see for myself. I accepted.

So it was that one evening in November 1984 my psychologist friend, Bill (WJM), and I (BJL) drove up into the San Gabriel Mountain foothills above Burbank to attend as guests a firewalking seminar run by the Tony Robbins group.

I was plenty nervous. I had been going around telling everyone that I knew how it was done and that I could do it without the training. I was thinking, however, that I might get burned. I wasn't sure which would hurt more, red, burned feet or a red, embarrassed face. I had taken the precaution of calling my doctor to get some advice on first aid in case I needed it. He said that it wouldn't be too smart to burn my feet, even in the name of science, and said that not to do it would be the best first aid. Bill, on the other hand, intended to attend the training but not to walk— friendship only goes so far. So he was feeling pretty chipper.

What Happened at the Seminar?

While Bernie was feeling anxious about walking on hot coals, I (WJM) felt mostly the excitement of anticipation of a new adventure. I wasn't in any danger of burning my feet, since I wasn't planning to walk. I was just going to look and learn. What kind of people would pay $125 for the privilege of risking their soles? What would the training be like? Could people really walk on hot coals without hurting themselves? Did Bernie know what he was talking about? Would I have to drive his car home for him?

The flyers advertising the firewalking experience and several other seminars claimed participation in these meetings could help people overcome lifelong fears like claustrophobia, eliminate lifelong addictions like smoking and overeating, and cure people of impotence and chronic depression—all within one or two hours. They could, it was said, help students study more effectively and train people to know, instantly, the most effective ways to communicate with and persuade people. The flyer promised to increase people's confidence in their ability to accomplish any important goal and to overcome past failures and succeed at seemingly impossible tasks. The proof of these new abilities was to walk on fire. Thousands of people had already succeeded.

The audience of about 80 people was middle class, with an average age of about 35 and a fairly even split between men and women. The vast majority of the participants were white and somewhat formally dressed. They seemed nervously gregarious, the way a class buzzes with conversation before a midterm examination. Among those I talked to were lawyers, doctors, secretaries, and advertising consultants.

The seminar took place in a hotel conference-room, with folding chairs placed in a semicircle around a temporary stage. There was a sophisticated sound system and contemporary upbeat music.

Tony Robbins is a tall, powerfully built man with a lot of energy and a pleasant, forceful personality. He led the training for the entire six hours and was assisted by a small army of volunteers and staff members. Perhaps as many as one-fifth of the audience had attended the seminar before and were there for a refresher course.

Robbins told the audience they were "kindred" souls. He assured them that they could be as successful as he was simply by following the advice he was to give that night.

He warned against defeatist thinking, saying that fear of failure wipes out initiative and stops action. He claimed that stupid people can be successes while presumably smart ones may not be, that some stupid people may persist in the face of disappointment while the smart ones say it can't be done.

After about an hour, all 80 of us, clapping rhythmically and chanting, "Yes, yes, yes," filed out of the room and down to the parking lot to view the fires we would soon be walking on. The "yes" we chanted was the wishful answer to the question in all of our minds, "Can I walk the coals and not get burned?" The crowd of clapping participants encircled a bed of fresh sod in the middle of the parking lot. On the bed there were two bonfires of furiously crackling wood. The heat seemed particularly intense in the cold November night air. Robbins exhorted us to close our eyes and imagine ourselves conducting a successful firewalk. "What are you going to do," he asked, "when you have achieved success? You're going to celebrate!" He suggested we imagine we had just completed a successful firewalk, make a fist in the air, and shout with the elation we would feel upon achieving such a singular success. For several minutes, seemingly frenzied students shook their fists at the night sky and shouted "Yay!" "I did it!" and "Yahoo!" The din may have struck other hotel residents as yet more evidence that the strange things they had heard about California were true.

We returned to the seminar room where Robbins had presented himself as a "model" for us to emulate. He repeatedly told us that he was no different from us, that he had suffered the same anxieties and fears we were suffering, and that he nevertheless had succeeded in walking on coals many, many times without getting burned. He also encouraged all of us to think of past successes and to remind ourselves of all of our "untapped" power.

He said, "We're all masters" and our fears are often groundless and should be ignored. He listed five steps to get rid of any fear: identify it, analyze it to death, be willing to accept the worst, be willing to accept the best, and then take action.

Halfway though the seminar, Robbins began describing neurolinguistic programming, a technique he claimed could enable its practitioners to cure people of tumors and long-standing psychological problems in a fraction of the time required by conventional treatments. He claimed that neurolinguistic programming enabled him to read people's motives like an open book. Neurolinguistic programming gave him such power, he said, that he could, without touching her, make a woman have an orgasm involuntarily. He claimed that he had cured a man of impotence and a long-standing drug-addiction in 90 minutes and that he could bring a person who was brain-dead back to life.

Meanwhile, Bernie was waiting anxiously down by the conflagrations. He chatted with the attendants and measured the temperature with a pyrometer he had brought with him. The fires were hot, $1,500°F$ to $1,800°F$. He was sweating. Then, back in the lobby, he was nervously thumbing through Kittel's textbook, *Thermal Physics.* "Perhaps," he told me later, "I missed something doing a crossword puzzle in class when I should have been taking notes."

Finally, at about 1:00 A.M., the seminar reached its climax. Robbins gave us pointers about walking on coals. He said we could end up with stumps for legs if we didn't follow instructions. We were to walk, not run, breathe fully and deeply, stand very erect, look up at the sky, visualize a cool place, and chant, "Cool moss, cool moss," as we walked. At the end, we were to quickly and carefully wipe our feet and then celebrate our success.

We took off our shoes and socks, turned up our pant-cuffs, and filed on down to the parking lot chanting, "Yes, yes." Workers were taking apart the bonfires with shovels and spreading burning coals into thin beds 8 or 10 feet long. The heat was powerful enough to force us to close our eyes and take a step backward when we stood near the fires. Burning embers floated into the sky.

Robbins was the first person to walk across the coals. Several staff members then followed his example, one after the other. They carefully and in exaggerated fashion followed the instructions so we would get the right idea. They huffed and puffed just before setting foot on the coals and walked stiffly across with eyes fixed on a point in the sky.

News photographers' lights lit up the scene. They were allowed to photograph only the staffers. The students might be too easily distracted, Robbins said. They might lose their concentration and get burned. (One of the few times Robbins admits to getting burned while firewalking was when he walked on fire while being filmed for a TV show and was distracted during the firewalk by the talk-show host.)

Our spirits were high and the peptalk had been inspiring. People began walking across the coals and shouting in excitement, encouraging those yet to walk and congratulating those who finished. There was always someone walking on one bed or the other. The effect was to surround the firewalking experience with considerable noise and movement. It is not clear why the firewalk leaders encourage all of this distracting tumult while at the same time saying that a few photographers' flash lamps would distract the walkers. Although I hadn't planned to walk, by the end of the seminar I had been swept up by the group spirit and was one of the first to walk across the coals. I was thrilled.

More than 90 percent of the participants, or about 80 tenderfeet, walked. Very few got blisters or, at any rate, very few volunteered that they had. Bernie did see two women with blisters at another walk he attended and there have been news accounts of others.

* * *

While the seminarians were walking, I (BJL) was trying to take pictures of the footprints I could clearly see in the embers. After the jam of walkers eased somewhat, I took my place in line. The firewalk leaders made me stand and take some breaths, but as soon as I took my first step I violated their rules—I decided that it might be a good idea to look where I was going, something that my mother always urged me to do. I did follow one of their rules. I wiped my feet when I got to the end. By this time the embers had cooled quite a lot and were not glowing much any more. They felt like warm moss on my feet. I was quite disappointed, so when they brought over a new load of glowing embers I jumped at the opportunity and was the first to walk. This time they were a lot hotter and I thought that I might have sizzled my feet, but I couldn't find any damage.

How Can It Be Done?

Firewalking appears to be one of those strange phenomena that, while appearing to be difficult or impossible, are actually quite easy to do once the trick is discovered. Evidently this trick has been found out by many peoples through-out the world, although it is ordinarily associated with mystical or religious states of mind. For example, one firewalker from Sri Lanka said, "Anyone can do this if he prepares properly." That proper preparation, he went on to say, "may involve a week or two of fasting, prayer and meditation, devotional chants, frequent baths and celibacy" (Doherty 1982). For anyone who is planning a walk but finds this last requirement too extreme, I have been told by a reliable source that celibacy is not a prerequisite for a successful walk.

The secret to firewalking and many similar heat-defying stunts lies in the distinction between temperature and heat (or internal energy). This distinction is not a part of our commonsense notions, although all of us are actually familiar with it as part of our daily lives. For example, when we are baking a cake, the air in the oven, the cake, and the cake pan are all at about the same temperature. None of us would think for a moment before putting our hands into the hot oven air, but we know that we cannot touch the cake pan for more than an instant without being burned. Why is this? They really are at the same temperature. Why would the pan burn us and not the equally hot oven air?

The answer is that different materials at the same temperature contain different amounts of thermal or heat energy and also have different abilities to carry the energy from one place to another. Thus the air has a low heat capacity and a poor thermal conductivity, while the aluminum has a high heat capacity and a high thermal conductivity. Our bodies have a relatively high heat capacity, similar to water. When we put our hands in the hot oven air, energy flows from the air to our hands. As the energy leaves the air it cools and our hands warm up. But, because the air holds very little energy, it cools much more than our hands warm. Furthermore, because of the poor ability of the air to conduct heat from far away to our hands, it will take a long time for our hands to finally get baked. In contrast, the aluminum cake-pan holds a lot of thermal energy and is an excellent conductor of heat. When we touch the metal and energy flows from the pan into our hands, the metal does not drop in temperature very much and even brings energy from far away to replace its losses while our hands quickly warm. It is for these reasons that we put a potholder, a poor conductor of heat, between our hand and the pan and don't worry about the air. So just knowing the temperature is not enough to decide whether something will burn us.

Firewalking and walking on hot rocks, as is done in Fiji, are based on this same idea. The embers are light, fluffy carbon compounds. Although they may be at a fairly high temperature (1,000° to 1,200°F), they do not contain as much energy as we might expect from our commonsense notions of incandescent objects. Thus, so long as we do not spend too much time on the embers our feet will probably not get hot enough to burn. In fact, because the capacity of the embers is low and that of our feet relatively high, the embers cool off when we step on them. How do I know this? Well, the color and intensity of the light from the embers tells us their temperature;

yellow embers are hotter than orange, orange hotter than red, and so on. When I watched people walking across the bed of coals I could clearly see darkened footprints where the coals had cooled because of contact with the feet. In a couple of seconds the combustion reactions restored the embers' temperature and glow.

In my reading about firewalking and fire-handling, I have found the combination of low heat capacity and poor thermal conductivity to be the one common factor. For example, in Fiji, where people walk on hot rocks, they choose cobbles of volcanic rock, probably pumice. Pumice is that strange porous rock that floats in water. It has a low heat capacity and a poor thermal conductivity. Similarly, firehandlers can withstand flames on their bodies, for a short time, because the hot gases contain relatively small quantities of heat.

We may well ask, "Why is it that some people get burned and others do not?" The answer is that the practice of firewalking is not a controlled scientific experiment. There are many variables from one person to the next and from one moment to the next: how long we stay on the embers, how many steps we take, how tough the soles of our feet are, and whether we walk where the embers are deep or shallow, for example. It certainly is possible to get injured, especially if we believe that it is our mind that protects us and if we do not take into account the normal physical behavior of heat. *Rolling Stone* magazine (Krakauer 1984) reported that in one group of fire-walkers the average length of time on the coals was 1.5 seconds, with the longest being 1.9 seconds, except for one unfortunate woman with a brain and spine injury who, walking with canes and believing that her mind would protect her, courageously spent seven seconds on the embers before collapsing with severe burns. Another walker, a radio news reporter in San Francisco, a tough and fearless former war-correspondent, apparently strolled more slowly than the previous walkers and strayed to the side into a deep pile of embers, where she badly burned her arches. There are many such variable factors, and in the tumult and excitement it is very difficult to make careful observations.

Another scientifically based explanation for the firewalking is the Leiden-frost effect. This effect is produced by getting a thermally insulating layer (like a potholder) between our feet and the embers. This principle is actually known to some of us and used in our ordinary lives. For example, some cooks will sprinkle drops of water onto a pan to see if it is hot enough. If so, the drops evaporate relatively quickly; if hot enough the drops will dance or jump around for a surprisingly long time. Why does this happen? If the pan is sufficiently hot, a layer of water vapor forms between the drop and the skillet. This layer reduces the heat flow to the drop because vapors and gases are generally poor conductors of heat. When we wet our fingers before touching an iron to see if it is hot, or before putting out a candle, we are using this effect as well as taking advantage of the high heat capacity of water. It is also used in certain magic tricks, such as dipping fingers into molten lead, licking red-hot knives, and so on. However, moisture, while often present at firewalking, is not invariably present. I have been told by James

Randi, a magician who has investigated firewalking, that in Sri Lanka the walkers believe that moisture on their feet will cause the embers to stick, so they carefully dry their feet before they walk.

Since the Leidenfrost effect is well known and thoroughly documented, and since the walkers are often in a state of great physical excitement, their feet may be "sweaty" because of the nearness of a hot fire, and the surroundings of the bed of embers is often wet, I conclude that the Leidenfrost effect is likely to be helpful but not necessary for firewalking, provided the heat capacity, thermal conductivity, and temperature of the embers or rocks is suitably low. It is certainly true that at Robbins's firewalks the sod and ground around the fires are usually kept fairly wet.

All of the various other explanations for firewalking I have come across in my investigations begin with the assumption that you should get burned unless some special exotic effects are operating. Thus, instead of searching for ways in which normal physics or physiology might operate to reduce the likelihood of a burn, firewalk theorists search for anomalies in normal science or in areas on the frontier where scientists are still puzzled. Most of the explanations involve the necessity for "correct" beliefs on the part of the firewalker. For some, the belief alone is somehow sufficient. This is perhaps what is believed by the Greek firewalkers who carry statues of the saints as they walk. For others, the correct belief is supposed to induce physiological changes that protect the walker. For example, endorphins—chemicals found in the brain that have been associated with feelings of pain and pleasure—are imagined to increase because of the correct beliefs and to then protect the body from burns. Some believe that the physiological changes involve the "bioelectric field." As they approach the embers they can, they say, feel the electricity around them and believe that they are somehow shielded by it. Another theory is that the proper beliefs change the properties of nerves and muscles so that they can conduct the heat away from the feet. Still others believe that the ability of some people to cause small changes in the temperature of their hands and feet might somehow be utilized to a much larger extent by firewalkers.

One characteristic of all of these explanations is that they are totally unsupported by any direct experimental data. Where is the measurement of the bioelectric field before, during, and after firewalking? Where is the demonstration of electrostatic shielding of heat? Where is the evidence showing that endorphins reduce damage from injuries?

There is a simple experiment that could be done to prove whether one's mental state effects the thermal properties of one's feet. The Tony Robbins people make such a claim. They say that walkers are in a certain "state" that protects them. Why not measure the flow of heat into someone's foot as they go in and out of this state? This would be easy to do and would involve no risk, since it could be done at low temperatures.

Bill and I believe that the explanation for the lack of burns is found in the ordinary physics of heat and materials. There are, however, some interesting psychological effects that play a role in the experience of firewalking. Bill will now describe this role.

What Are the Psychological Factors?

Psychology can explain why some people feel no pain or heat, even when they have been exposed to enough heat to produce blisters. It is necessary to distinguish the concept of pain from the concept of being burned. Pain from a burn is a perception that the body has been injured. People can get burned without feeling pain, and we can feel pain when no injury has occurred. Many of us have had the experience of cutting ourselves and not realizing that we are injured for some time. As a matter of fact, I (WJM) must admit that I did get burned when I firewalked. I got a dime-sized blister on my left foot, under my arch. Despite this evidence that I was burned, I remember feeling no pain, and I didn't discover the blister until the next morning.

The detection of pain caused by exposure to fire is not only a function of the temperature of our feet; it is also affected by the general sensitivity of our body and mind and by the presence of other, competing sensations. If we are in a quiet room and fully alert, we will be maximally sensitive to pain. If we are tired and surrounded by noisy, distracting events, we will be much less sensitive. Distraction can reduce the pain people experience, because they can attend to only a few things at once. Distraction is the basis for a number of techniques psychologists teach patients who suffer chronic pain. These techniques are quite effective.

In addition, the physiological responsiveness of our bodies is governed to a large extent by a circadian rhythm. When we stay awake well past our normal bedtime, our normal physiological functions are nevertheless somewhat depressed—as if the body expected to be asleep even though it wasn't. The people who walked on the coals at 1:00 A.M. were therefore much less likely to feel pain or heat than they would have been had they conducted the same walk at 1:00 P.M.

Furthermore, the instructions we had been given before the walk actually seemed calculated to distract our attention from the sensations of our feet. Concentrating on the "mantra," looking up at the sky, hearing the applause and shouts of elation, and breathing in an artificial and forced manner, all served to distract the walker.

Women and men who are familiar with the Lamaze technique for preparing women for the rigors of childbirth know that increasing one's breathing rate in a prescribed manner just before the moment of greatest pain helps to reduce the pain that the mother experiences. The controlled breathing taught in the seminar had the same effect of reducing the maximum pain the firewalkers experienced. The likelihood of their perceiving any pain even if they were burned was greatly reduced.

Scientific Assessment

Firewalking, as practiced in this country, is being used as the keystone of a self-improvement program. It is claimed that by using special techniques the student can walk on hot coals. It is further claimed that these same techniques can be applied to solve the problems of ordinary life. Firewalking can be so surprising to us that it can have a powerful effect on our beliefs. Students frequently speak of having their entire system of beliefs blown after succeeding. Thus there can be no doubt about its powerful persuasive effect.

Nevertheless, as we have explained, the training has nothing at all to do with whether or not a firewalker will be burned. It does have some effect on whether you will want to walk and on what you will experience as you walk, but whether you avoid a burn is determined by the ordinary behavior of heat on the soles of your feet.

It is probably safe to say that the seminar we witnessed at least temporarily increased the self-esteem and confidence of most of the participants. The training effectively used techniques like behavioral modeling, verbal persuasion, and group pressures, which are well known to psychologists. We did not assess how long the benefits last or how well they might translate into increased success at the more mundane tasks of life. Many of the formulas for success were no different from those available in conventional self-help and positive-thinking programs. Those who were burned or who ultimately lacked the courage to walk, on the other hand, were very likely to experience a decrease in self-esteem and confidence because they would be likely to believe that their minds were weak.

The firewalk is an unusual and very persuasive technique. The seminar students are led to believe that it represents the first of what will be many examples of the wonderful effects of the seminar training. As we have shown, however, the training has nothing to do with not getting burned, since anyone can walk on the embers without much chance of injury. The students, unfortunately, do not know this, and it is this deceptive but persuasive practice that is our greatest concern.

The students may be led to accept the correctness of all that is offered during the seminar. In fact about one-fifth of the firewalkers pay as much as $375 for a full weekend course involving neurolinguistic programming, and, we are told, new and exotic theories of nutrition. Now neurolinguistic programming may be a useful addition to mainstream psychology, but from the material presented in the seminar it is certainly impossible to make a sensible judgment. The extreme claims for psychic- or faith-healing-style cures certainly cast doubt upon its truth, as does the use of a trick to supposedly show its effectiveness.

We are, of course, unable to read the minds of those who teach and profit from firewalking. We cannot tell if they are themselves deceived, simply ignorant, or charlatans. In any case, some people are clearly being harmed. Because elementary physics is not known, some are being burned. Because success is attributed to mental strength, those who are burned or fail to walk are damaged. And those who succeed are likely to believe much of the rest of the teachings of the trainers. Some of these teachings are fine, but others are quite exotic and strange, if not actually dangerous. What will happen to the believers when, inevitably, they learn the truth? Because all of the beneficial aspects of these trainings are available from other, more conventional sources, such as college courses in psychology and nutrition (perhaps, even a physics course), which have few harmful effects, we cannot find any justification for the deception that is being practiced and we would advise everyone to stay away. The firewalk trainers are misleading us about the keystone of their program—that it is their training that makes it possible for us to walk on hot coals. Considering that this basic principle of their program is wrong, the rest of it cannot be trusted.

Postscript

We have written a more extensive report on firewalking for the European biomedical journal *Experientia*, 44, 310-315 (1988). The article contains information about firewalking in many cultures and a survey of earlier accounts and research. This particular issue of *Experientia* contains a series of review articles on investigations of the paranormal.

References

Doherty, Jim. 1982. Hot feat—firewalkers of the world. *Science Digest,* August, p. 66.
Krakauer, Jon. 1984. Get it while it's hot. *Rolling Stone,* August 30, p. 22.
Krier, Beth Ann. 1984. The curious hotfoot it to a new fad. *Los Angeles Times,* April 11, Part 1, p. 1.
Walker, Jearl. 1977. "The Amateur Scientist," *Scientific American,* August, p. 126.

JOE NICKELL and JOHN F. FISCHER

Incredible Cremations: Investigating Spontaneous Combustion Deaths

Having seemingly struck intermittently over the centuries, the specter of "spontaneous human combustion" appeared to have claimed yet another victim in St. Petersburg, Florida, one morning in mid-1951. On July 2, a Monday, at eight o'clock, the landlady of a four-unit apartment building on Cherry Street attempted to deliver a tenant's telegram, for which she had just signed. As she walked to the apartment of the tenant, Mrs. Mary Reeser, and attempted to open a hall door, she found the knob too hot to grasp. Her cries for help summoned two house painters who ran over from across the street.

Advancing through the smoke-filled hallway into the 67-year-old widow's efficiency apartment, the men came upon the evidence of a gruesome mystery. It was to become *the* case in the annals of the alleged phenomenon known as "spontaneous human combustion" (SHC)—a case that demanded answers to many questions: What was the nature of a fire that had no apparent cause, that could leave a room relatively undamaged yet so completely consume the body of a large woman that there remained little more than ashes, a slippered foot, and an eerie shrunken skull? Might proponents be correct in suggesting that SHC is related to "geomagnetic fluctuations" (Arnold 1981) or to man's "electrodynamic being" (Gaddis 1967)?

To answer such questions we launched a two-year investigation that focused on Mrs. Reeser's death but began with a historical overview of the alleged phenomenon. Our lengthy two-part report was published in the journal of the International Association of Arson Investigators (Nickell and Fischer 1984).

We found that a widely publicized mid-nineteenth-century debate over the supposed phenomenon is typical of the continuing controversy. That debate was sparked (so to speak) by Charles Dickens's novel *Bleak House,* wherein a sinister, drunken Mr. Krook perished by "spontaneous combustion." George Henry Lewes, the philosopher and critic, had publicly accused Dickens of perpetuating a vulgar superstition. Lewes (1861) insisted that such a death was a scientific impossibility, a view shared by the German chemist Justus von Liebig, who wrote: "The opinion that a man can burn of himself is not founded on a knowledge of the circumstances of the death, but on the reverse of knowledge—on complete ignorance of all the causes or conditions which preceded the accident and caused it" (Liebig 1851).

Thus rationalists like Lewes were seizing the scientific high ground with the question of *cause:* Dickens, on the other hand, was arguing primarily from *effect:* He cited several cases of the alleged phenomenon, some of which had been attested to by medical men of the time. To assess these contrary views we began by researching a number of seemingly representative cases that spanned more than two and a half centuries.

One of the earliest cases took place in February 1725 at Rheims. The burned remains of a Madame Millet were found on her kitchen floor, a portion of which had also burned. Although her husband was subsequently convicted of murdering her, a higher court reversed the decision, attributing the death to spontaneous combustion. Actually, the woman was one who "got intoxicated every day," had gone to the kitchen "to warm herself," and was discovered only "a foot and a half's distance" from the hearth. Therefore, Thomas Stevenson (1883), in his treatise on medical jurisprudence, suggested her clothes had "accidentally ignited."

In contrast to this was another early case, the 1731 death of the Countess Bandi of Cesena, Italy, aged 62, who was *not* given to intoxication. Although her body was supposedly reduced to "a heap of ashes," part of her head remained, and her legs and arms were not burned. The ashes contained "a greasy and stinking moisture," soot floated in the air, and from the window there "trickled down a greasy, loathsome, yellowish liquor with an unusual stink." However, this case—which served as one of Dickens's sources—seems quite explicable when further data are added: On the floor was an empty, ash-covered lamp on which the countess had apparently fallen, its burning oil no doubt aiding in the immolation.

At least three other eighteenth-century cases involved women who drank: Grace Pett of Ipswich, who perished in 1744, her burned remains attended by a fatty stain and lying near both a fireplace and a fallen candle; a French-woman, Madame De Boiseon, aged 80 in 1749, who supposedly "drank nothing but spirits for several years," whose body was still burning in a chair placed "before the fire"; and, sometime prior to 1774, 52-year-old Mary Clues of Coventry, who was "much given to drinking" and whose death a medical investigator attributed to her shift having caught fire, either from "the candle on the chair or a coal falling from the grate."

Sometime before 1835 (when Theodric and John Beck published the case in their *Elements of Medical Jurisprudence*), an intoxicated 30-year-old Hannah Bradshaw burned to death in New York. A four-foot hole had burned through the floor of her room, and her bones and a burned-off foot were found on the ground underneath. Significantly, a candlestick, with a portion of a candle in it, was found near the edge of the hole.

Other nineteenth-century cases include the 1852 death of John Anderson, a wood hauler and "notorious dram drinker," who was seen to get down from his cart, stumble, and burn to death. His body was only *charred,* which is consistent with his clothes having caught fire and with there being no additional fuel source. Anderson's lighted pipe was found under his body.

In 1870, in France, the body of a drunken woman was found on her bedroom floor, which was still smoldering. There was considerable damage to the torso, with a "greasy black soot adhering to the vertebrae." Although there was supposedly "no fire in the grate"—at least none remaining—the body nevertheless lay partially *across the hearth;* the drunken woman may have set her clothes ablaze while attempting to light the fire.

After the turn of the century, in 1908, a retired English schoolmarm named Wilhelmina Dewar was found dead. Her body was burned, but the bed on which it was lying remain unscorched. Under questioning at the inquest, her sister admitted that she had actually discovered Wilhelmina "burned, but still alive" and that she had "helped her walk to the bed, where she had died."

From the cases above (typical of the 30 we researched) some patterns emerged. For example, there did seem to be some correlation between drunkenness and supposed instances of SHC. Early theorists, including members of the temperance movement, had suggested that alcohol-impregnated tissues were rendered highly combustible, but scientists refuted the notion by experimentation and pointed out that a person would die of alcohol poisoning long before imbibing enough alcohol to have even a slight effect on the body's flammability. We determined instead that the correlation was most likely due to heavy drinkers' being more careless with fire and less able to properly respond to accidents.

We also found an even more significant correlation: In those instances where the destruction of the body was relatively minimal, the only significant fuel source seems to have been the individual's clothes, but where the destruction was considerable, additional fuel sources—chair stuffing, wooden flooring, floor coverings, and so on—augmented the combustion. Such materials under the body appear also to have helped retain melted fat that flowed from the body and then volatilized and burned, destroying more of the body and yielding still more liquefied fat to continue the process known as "the candle effect." (Stevenson explained that in one case a hempen mat had become so combustible because of "the melted human fat with which it was impregnated" that it "burnt like a link"—i.e., like a pitch torch.)

Such correlation of the amount of destruction with the utilization of available fuel sources makes a forceful argument against the notion of "preternatural combustibility." And the presence of plausible sources of the ignition—proximate candles, lamps, fireplaces—makes the postulation of "spontaneous human combustion" completely unwarranted.

Proponents of SHC argue that bodies are difficult to burn because of the great amount of water they contain, but the water is boiled off ahead of the advancing fire. Again, they argue from comparisons to the destructive force of crematories, asserting for instance that a temperature of 2,500° Fahrenheit or more is required to destroy a body in three hours (Allen 1951). Actually, an authoritative forensic source states that a period of only one and a half hours at 1,600° to 1,800° is required (Spitz and Fisher 1980). In any case, if a longer time is involved, a lower temperature would be sufficient. As D. J. X. Halliday of the Fire Investigation Unit of London's Metropolitan Police Forensic Science Laboratory explains, "Cremation is intended to destroy a body in the shortest possible time and is therefore carried out under extreme conditions, but a relatively small fire can consume flesh and calcine bone if it is allowed to burn for a long time" (Halliday 1986). And many hours were typically involved in the cases we researched wherein the destruction was extensive.

But what of a case in which there was no known cause for the ignition, the body was almost completely destroyed—except for a foot and a "shrunken skull"—and yet the surroundings were relatively undamaged? That is the way the celebrated "cinder woman mystery"—"probably the best-documented modern case" of SHC (Gadd 1981)—is sometimes portrayed. But our reinvestigation of that case—which involved our obtaining the police report, the death certificate, and contemporary news accounts—provides a lesson in the need for treating instances of alleged SHC on a case-by-case basis.

For example, one account (Gadd 1981) neglects to include some essential facts: When last seen, Mary Reeser was wearing a flammable nightdress and housecoat, sitting in the overstuffed chair in which she subsequently died, and smoking a cigarette. Also omitted was the fact that earlier that day she had told her son, a physician, that she had taken two sleeping pills and intended to take two more before retiring (Allen 1951).

Other accounts concede that Mrs. Reeser may have indeed died as a result of dropping her cigarette as she dozed off and that SHC may not have been the cause, but they postulate a related phenomenon termed "preternatural combustibility." For example, Vincent Gaddis scoffs: "That flames from a nightgown, housecoat, and a chair that doesn't flare up but smolders, could create sufficient heat to cremate a large human body is ridiculous. And the notion that fluid-saturated fatty tissues, ignited by an outside flame, will burn and produce enough heat to destroy the rest of the body is nonsense" (Gaddis 1967).

Gaddis was reacting to the conclusion stated in the official police report (the text of which is given in Blizin 1951) that "once the body became ignited, almost complete destruction occurred from the destruction of its own fatty tissues." In fact, we learned that Mrs. Reeser was a "plump" woman and that a quantity of "grease"—obviously residue from her body—was left at the spot where the chair had stood.

As to Gaddis's insistence that the chair would not burn, Thomas J. Ohlemiller—an expert in smoldering combustion at the Center for Fire Research, Department of Commerce—told us: "Fire deaths caused by cigarette ignition of bedding and upholstery are among the most common in the U.S. . . . The smoldering spreads slowly and can sometimes consume the entire piece of furniture with no flames." Ohlemiller added, "More commonly the smoldering process abruptly ignites the gases coming from the object; this may occur an hour or more after the smoldering process was initiated" (Ohlemiller 1982).

In the Reeser case, what probably happened was that the chair's stuffing burned slowly, fueled by the melted body fat and aided by partially open windows. From the time the widow was last seen sitting in the chair until her remains were discovered, almost 12 hours had elapsed.

Gaddis had further questioned why, if the fatty tissue had indeed burned, it did not spread the fire. The answer is that the fire did spread more than some accounts acknowledge: An adjacent end table and lamp were destroyed and a ceiling beam had to be extinguished when firemen arrived. Besides, the melted fat would have been slowly absorbed by the chair's stuffing, and in any event the floor was of concrete.

That one of the widow's feet remained intact may have been due to the fact that Mrs. Reeser had a stiff leg, which she extended when sitting. Or, as the burning chair collapsed and the body rolled out onto its right side, the foot reached beyond the radius of the fire.

One of the strangest and most frequently reported elements of the case—the alleged shrinking of the skull—probably never happened. The self-styled "bone detective" who is often quoted on the subject merely referred to second-hand news accounts and thus spoke of "a roundish object identified as the head" (Krogman 1953). Actually, as a forensic anthropologist theorized at our request, Mrs. Reeser's skull probably burst in the fire and was destroyed, and the "roundish object" could have been merely "a globular lump that can result from the musculature of the neck where it attaches to the base of the skull" (Wolf 1983).

In conclusion, what has been described as "probably the best-documented case" of alleged spontaneous human combustion is actually attributable to the deadly combination of a lit cigarette, flammable nightclothes, and sleeping pills. And the notion of preternatural combustibility must yield to the evidence supporting the "candle effect"—in which a body's fat liquefies and thus participates in its own destruction.

However, although even a lean body contains a significant amount of fat (present even in the bone marrow), other factors may be involved in a given instance. We therefore urge investigation of cases on their own evidence. The operative word is *investigation*, not merely debunking—although the former may surely result in the latter in instances of alleged spontaneous human combustion.

References

Allen, W. S. 1951. Weird cremation. *True Detective*, December, pp. 42-45, 93-94.
Arnold, Larry E. 1981. Human fireballs. *Science Digest*, October, pp. 88-91, 115.

Beck, Theodric R., and John B. 1835. *Elements of Medical Jurisprudence*, 5th ed., vol. 2. Albany, N.Y.: O. Steele et al., pp. 60-68).

Blizin, Terry. 1951. The Reeser case. *St. Petersburg Times* (Fla.), August 9.

Gadd, Laurence D., and the Editors of the World Almanac. 1981. *The Second Book of the Strange*, New York: World Almanac Publications, pp. 33-36.

Gaddis, Vincent H. 1967. *Mysterious Fires and Lights.* New York: David McKay.

Halliday, D. J. X. 1986. Letter to editor, *New Scientist,* May 29.

Krogman, Wilton M. 1953. The improbable case of the cinder woman. *General Magazine and Historical Chronicle,* Winter: 61-69.

Lewes, George Henry. 1861. Spontaneous combustion. *Blackwood's Edinburgh Magazine,* 89 (April):385-402.

Liebig, Justus von. 1851. Letter XXII in *Familiar Letters on Chemistry.* London: Taylor, Walton & Maberly.

Nickell, Joe, and John F. Fischer. 1984. Spontaneous human combustion. *Fire and Arson Investigator,* 34(3):4-11; 34(4):3-8.

Ohlemiller, Thomas J. 1982. Personal communication.

Spitz, Werner U., and Russell S. Fisher. 1980. *Medicolegal Investigation of Death.* Springfield, Ill.: Charles C. Thomas.

Stevenson, Thomas. 1883. *The Principles and Practice of Medical Jurisprudence,* 3rd ed. Philadelphia: Henry C. Lea's Son & Co., pp. 718-727.

Wolf, David. 1983. Personal communication.

ADRIAN FURNHAM

Write and Wrong:
The Validity of Graphological Analysis

It is one of those nice but sad ironies that, as popular interest and especially commercial application of handwriting analysis, or graphology, is on the increase, scientific scrutiny of its claims remains limited and may be on the decrease. Like many of the other "ologies" that claim to be useful in describing and predicting human behavior, it has a long past, with many notable figures like Goethe speculating that somehow one may expect that a person's character is projected in the way he or she writes. The term *graphology* in fact was first used in 1871 by the French cleric Michon, who spent 30 years studying handwriting.

Since the beginning of this century there has been more and more interest in the topic, and it is difficult to go into any large bookstore without finding among the self-help, occult, or even psychology/social-science books some texts on how to analyze handwriting. These tomes tell you what factors to look at (i.e., size, slant, zone, pressure) and what traits (temperament, mental, social, work, and moral) are revealed. In fact there are schools of graphology, each with a slightly different history, approach, and "theory." However what appears missing most from the area is not a method of analysis so much as a theory of how or why individual differences are manifest in handwriting. For instance, is one to assume that personality traits are the result of genetic biological differences that predispose all social behavior, including handwriting, or is writing style, like other social behaviors, a product of complex primary, secondary, and tertiary education?

Despite the lack of any sound, illuminating, or indeed falsifiable theoretical base, there has been a great deal of interest in graphology by hard-pressed managers and administrators anxious for a valid and nonfalsifiable way of measuring the desirable and less desirable traits of employees. Dispassionate and disinterested research, however, has severely questioned the usefulness of graphological analysis.

A review of the literature shows, as ever, equivocal results. Some, albeit few, studies show extra-chance results linking handwriting to such personality traits as neuroticism, but a large number of studies reveal no clear pattern between graphological analysis and psychological assessment. Consider, for instance, the following conclusions taken from various studies:

1. "It was concluded that the analyst could not accurately predict personality from handwriting." This was based on a study of Vestewig, Santee, and Moss (1976) from Wright State University, who got six handwriting experts to rate 48 specimens of handwriting on 15 personality variables.

2. "No evidence was found for the validity of the graphological signs." This is from Lester, McLaughlin, and Nosal (1977), who used 16 graphological signs of extroversion to try to predict from handwriting samples the extroversion of 109 subjects whose personality test scores were known.

3. "Thus the results did not support the claim that the three handwriting measures were valid indices of extroversion." This is based on the study by Rosenthal and Lines (1978), who attempted to correlate three graphological indices with the extroversion scores of 58 students.

4. "There is thus little support here for the validity of graphological analysis." This was based on a study by Eysenck and Gudjonsson (1986), who employed a professional graphologist to analyze handwriting from 99 subjects and then fill out personality questionnaires as she thought would have been done by the respondents.

5. "The graphologists did not perform significantly better than a chance model." This was the conclusion of Ben-Shaktar and colleagues (1986) at the Hebrew University, who asked graphologists to judge the profession, out of eight possibilities, of 40 successful professionals.

6. "Although the literature on the topic suffers from significant methodological negligence, the general trend of findings is to suggest that graphology is not a viable assessment method." This conclusion comes from Klimoski and Rafael (1983), based at Ohio State University, after a careful review of the literature.

It is apparent that these tests of the validity of graphological analysis were very different and perhaps not entirely adequate.

Hans Eysenck, whose research in the area spans a 40-year period, and his Icelandic collaborator Gisle Gudjonsson have made the point that because there appear to be two different basic approaches to the assessment of both handwriting and personality (holistic vs. analytic), this leaves four basic types of analysis:

Holistic analysis of handwriting. This is basically impressionistic. The graphologist, using his or her experience and insight, offers a general description of the kind of personality he or she believes the handwriting discloses.

Analytic analysis of handwriting. This uses measurement of the constituents of the handwriting, such as slant, pressure, etc. These specific, objective, and tabulated measures are then converted into personality assessment on the basis of a formula or code.

Holistic analysis of personality. This too is impressionistic and may be done after an interview, when a trained psychologist offers a personality description on the basis of his or her questions, observations, and intuitions.

Analytic analysis of personality. This involves the application of psychometrically assessed, reliable, and valid personality tests (questionnaires, physiological responses to a person, and the various grade scores obtained).

As a result of this fourfold classification there are quite different approaches to the evaluation of the validity of graphological analysis in the prediction of personality. These are:

1. Holistic matching, which is the impressionistic interpretation of writing matched with an impressionistic account of personality.

2. Holistic correlation, which is the impressionistic interpretation of writing correlated with a quantitative assessment of personality.

3. Analytic matching, which constitutes the measurement of the constituents of the handwriting matched with an impressionistic account of personality.

4. Analytic correlation, which is the measurement of the constituents of the handwriting correlated with a quantitative assessment of personality.

Clearly, of these four widely used methods the final analytic correlational method is the most empirically based. A colleague (Barrie Gunter) and I decided to do a study along the lines of the analytic correlational method advocated by Hans Eysenck.

We had 64 adults of highly diverse backgrounds, ages, and professions do two things. First they copied out a text (of about 100 words) on the topic of tea onto a sheet of white unlined paper. They were each given identical, recently sharpened pencils to do the job. They also filled out the Eysenck Personality Questionnaire, which purports to measure the three fundamental dimensions of personality: extroversion-introversion, neuroticism, and psychoticism. The literature on the reliability and validity of this measure is voluminous and the dimensions it measures have been shown to relate consistently and theoretically predictably to physiological, psychopathological, cognitive, and social variables.

The personality questionnaire yields three scores for each subject. In order to do an analytic appraisal of the handwriting a number of graphological books were consulted to ascertain which factors to look at. There was no apparent agreement on which factors were most important, or indeed on what particular styles indicated. Nevertheless it was decided to select a dozen or so of the factors most commonly referred to. They were: size of writing; percentage of the page used; slant of letters; width of words; connectedness of letters with words; pressure on the page; spacing of words; regularity of crossed t's; regularity of dotted i's; where the t's are crossed; where the i's are dotted; and finally whether the subject loops letters below the line or above the line.

Each factor was rated on a 3- or 5-point scale. Thus slant was rated as 5 points (1 = extreme left, 2 = moderate left, 3 = upright, 4 = moderate right, 5 = extreme right) and pressure as 3 points (1 = slight, 2 = moderate, 3 = great.)

The 64 writing specimens, all of the same passage, were then given to two independent judges, neither a graphologist, who rated each script according to the 13 factors. In order to check the reliability of their assessment a correlation coefficient was calculated. It turned out to be nearly 90 percent (5 = .89). A third judge then examined all scripts and resolved the dispute on the 10 percent of disputed items. This meant that each subject was left with 13 different objective measures of his or her handwriting.

The whole point of the analytic correlational method is that objective and quantitative measures of both personality and handwriting are correlated. More than 70 correlations were computed. Less than 6 percent proved significant—one with extroversion and three with neuroticism, indicating that neurotics tended to have small handwriting, with a left slant and consistently dotted i's.

Despite numerous other statistical analyses, including analysis of variance, multiple regression, and discriminant analysis, none of the results proved significant. Thus we were forced to conclude, as others had done before us, that graphological analysis was invalid. In fact we concluded thus: "Even if graphological analyses were valid. In fact we concluded thus: "Even if graphological analyses were valid, the *theoretical* basis of the method appears weak, nonexplicit and nonparsimonious. Furthermore, it is unclear why it should be used if clearly valid and reliable measures exist to measure the same thing (i.e., personality) more cheaply, accurately, and efficiently. Perhaps one should be forced to conclude, rather uncharacteristically for researchers, that no further work needs to be done in the field," (Furnham and Gunter 1987, p. 434).

It would be unwise not to anticipate criticisms of this relatively small study and not to consider possible responses:

—The 13 variables missed out on all or some of the critically important graphological variables. Indeed, I had correspondence with a graphological consultant who suggested both *speed* of handwriting (supposedly a determinant of naturalness, genuineness, and spontaneity in personality) and consistency of height and slant (supposedly an index of balance and control in life) were crucial. The consultant did not dispute the importance of other factors. This may well have been a valued criticism if there was an agreed-upon set of criteria. I happened to use a number of textbooks and to attempt some consensus.

—Experienced graphologists would have come to different conclusions. The point of this study was to derive reliable, objective, numeric indices of handwriting, not impressionistic accounts. Graphologists could have been used, but the crucial factor was the reliability of the judge. This was in fact achieved, and hence meant experienced graphological analysis was rendered redundant.

—Graphology does not relate to the three major variables predicted, namely, extroversion, neuroticism, and psychoticism. This is simply not true, as it most frequently purports to do just this.

—The personality test was at fault; whereas graphology does predict personality, psychometricized questionnaires do not. While the absolute validity of nearly all (and particularly some) questionnaires remains in doubt, there is more than sufficient empirical evidence for the validity of the test used here.

Many graphologists consider psychological evaluation of their "trade" a threat. Graphologists, it seems, tend to regard psychological research as cynical rather than—as I believe it actually is—skeptical. Research into the validity of graphology has, for all its faults, appeared to be disinterested. But even if graphology had merits and was valid, it would remain nothing less than a technique in search of a theory.

Ben-Shaktar and his Israeli colleagues (1986, p. 652) have thoughtfully concluded thus:

> 1. Although it would not be surprising if it were found that sloppy handwriting characterized sloppy writers, stylized calligraphy indicated some artistic flair, and bold, energetic people had bold, energetic handwriting, there is no reason to believe that traits such as honesty, insight, leadership, responsibility, warmth, and promiscuity find any kind of expression in graphological features. Some may have no somatic expression in graphological features. Some may have no somatic expression at all. Indeed, if a correspondence were to be empirically found between graphological features and such traits, it would be a major theoretical challenge to account for it.
> 2. There are not enough constraints in graphological analysis, and the very richness of handwriting can be its downfall. Unless the graphologist makes firm commitments to the nature of the correspondence between handwriting and personality, one can find ad hoc corroboration for any claim.
> 3. The a priori intuitions supporting graphology listed above operate on a much wider range of texts than those graphologists find acceptable. As graphologists practice their craft, it appears that from a graphological viewpoint, handwriting—rather than being a robust and stable form of expressive behavior—is actually extremely sensitive to extraneous influences that have nothing to do with personality (e.g., whether the script is copied or not, or the paper lined or not).
> 4. It is noteworthy that most graphologists decline to predict the sex of the writer from handwriting, although even lay people can diagnose a writer's sex from handwriting correctly about 70% of the time. They explain this by insisting that handwriting only reveals psychological, rather than biological, gender. Although common sense would agree that some women are masculine and some men are effeminate, it would be somewhat perverse to argue against the presumption that most women must be feminine and most men masculine. Could the graphologists simply be reluctant to predict so readily verifiable—or falsifiable—a variable?

Readers familiar with the techniques of cold reading will be able to understand why graphology appears to work and why so many (otherwise intelligent) laypeople believe in it. The growth of graphology may be due to the inability of empirical scientists to discover or invent a simple, single, robust, and predictive measure of personality themselves. But one cannot allow graphologists to fill this void, given that from any objective and dispassionate evaluation of their wares, graphology is quite simply invalid.

References

Ben-Shaktar, G., M. Bar-Hillel, F. Bilin, E. Ben-Abba, and A. Flug. 1986. Can graphology predict occupational success? Two empirical studies and some methodological ruminations. *Journal of Applied Psychology,* 71:645-653.

Eysenck, H. J. 1945. Graphological analysis and psychiatry: An experimental study. *British Journal of Psychology,* 35:70-81.

Eysenck, H. J., and G. Gudjonsson. 1986. An empircal study of the validity of handwriting analysis. *Personality and Individual Difference,* 7:263-264.

Furnham, Adrian, and Barrie Gunter. 1987. Graphology and personality: Another failure to validate graphological analysis. *Personality and Individual Differences,* 8:433-435.

Klimoski, R., and A. Rafael. 1983. Inferring personal qualities through handwriting analyses. *Journal of Occupational Psychology,* 56:191-202.

Lester, D., S. McLaughlin, and G. Nosal. 1977. Graphological signs for extroversion. *Perceptual and Motor Skills,* 44:137-138.

Rosenthal, D., and R. Lines. 1978. Handwriting as a correlate of extroversion. *Journal of Personality Assessment,* 42:45-48.

Vestewig, R., A. Santee, and M. Moss. 1976. Validity and student acceptance of a grapho-analytic approach to personality. *Journal of Personality Assessment,* 40:592-597.

ROBERT BASIL

Graphology and Personality:
"Let the Buyer Beware"

Nearly four hundred years ago Shakespeare told us, "There's no art/To find the mind's construction in the face." The bard's appraisal was shrewd, but it has not dissuaded others from seeking heretofore unseen physical keys to personality. The graphology panel at CSICOP's Chicago conference was a case in point. There the question was: "Is there art, or a science, to find the mind's construction in . . . *penmanship?*"

The answer was clear yet tentative: "No . . . at least not yet."

The panel made for an odd morning, with the skeptics providing better arguments for graphology—the system of determining personality traits via handwriting analysis—than did the graphologists themselves. While graphologists Rose Matousek, president of the American Association of Handwriting Analysts, and Felix Klein, vice president of the Council of Graphological Societies, relied on anecdotes, intuition, and bold, totally untested theories to validate their discipline, it was the rigorous statistical analysis of Professors Richard J. Klimoski and Edward Karnes that demonstrated graphology's limited, problematic accuracy. Said panel moderator Barry Beyerstein afterward, "The pro-graphology people presented as good a case as they could, but I was a little disappointed. We didn't want them to tell us about their satisfied customers or how their particular brand of graphology works, but about new evidence not in the literature. They ignored that."

Beyerstein, a psychologist and neurophysiologist, opened the discussion by outlining some key questions that must be asked of graphology: Are trained graphologists, given particular handwriting samples, capable of giving more or less identical diagnoses? Do their tests really measure what they say they do? Are they *predictive*—that is, when the personality trait being measured bears no obvious relationship to the thing being tested (say, the way one makes an *s*)? By what criteria are these samples analyzed—which aspects of the immensely complicated design of handwriting are especially meaningful? And how are these samples standardized?

Rose Matousek, the first panelist, did not address these questions. Quite eager to concede that "more work needs to be done," Matousek compared the status of contemporary graphology to that of psychology in its early days, before it had achieved professional, accredited standing. And that standing will come, Matousek asserted confidently. "Since handwriting analysis does not require mystic or paranormal explanations," she said, "I thought it would be easy to convince the CSICOP audience of the field's worth." She attempted to do so by declaring: "Handwriting is *brain-writing*. It's an expressive, spontaneous movement, a unique personal performance similar to the fingerprint." No human activity, she said, is less "conditioned by conscious process."

There was no question that Matousek had assembled an impressive taxonomy of handwriting styles. Less convincing, however, were her interpretations of these styles. According to graphology's "zonal theory," for example, penmanship's "upper," "middle," and "lower zones" are related to a person's "intellectual," "practical," and "instinctual" selves, respectively. And handwriting that sticks to the left-hand side of the page belongs to those who are attached to "the self, the past, and mother," while writing that zooms to the right comes from the pens of those more concerned with "others, the future, and father."

The problems with this model are both clear and typical of the field as a whole: Does the zonal theory assert that a person cannot be attached to the self, the past, and *father?* Matousek noted that these aren't hard and fast categories, put together as they were in an intuitive, empirical fashion.

Felix Klein's approach matched Matousek's, his presentation largely consisting of showing slides of handwriting to the audience. Mohandas Gandhi's writing, small and neat, showed that Gandhi loved peace. Napoleon's, wild and jagged, proved that the French general's temperament was not a whole lot like Gandhi's. And so on. While Klein claimed that competent graphologists could compose penetrating psychological profiles on the basis of handwriting samples, he admitted: "I don't believe that a scientific method has yet been devised to validate graphology."

Ed Karnes, a psychologist at Metropolitan State College in Denver, described a study he conducted on nine college administrators. The participants were given two kinds of personality profiles, one made through graphological analysis and the other through more standard "psychometric" tests. The administrators were then asked to choose their own from the assembled profiles and assign each of the rest to the other eight. Karnes's findings were illuminating: Graphology's success, he said, "is based on the P. T. Barnum effect, the tendency of people to ascribe great validity to general statements as long as they think the statements are made specifically about them." Example: While a high number of administrators identified with graphological profiles *not* written especially for them, few did so when presented with psychometric analyses (which tended to be much more detailed) not written for them.

Ohio State University psychology professor Richard Klimoski shared Karnes's conclusions, recommending that graphological analysis not be included in the hiring or promotion process. Indeed, the use of this utterly unvalidated technique in employment decisions became this panel's alarming subtheme. Klein claimed, for example, that 91 percent of Israel's corporations employ graphology in making personnel decisions—as does the Israeli government. Klimoski added that American corporations, such as Sears, U.S. Steel, and Bendix, have been known to use graphological consultants. These consultants, he said, "are usually brought in at the end" of the personnel-selection process "as validators"—that is, to assure bosses they've chosen the right guy or gal for the job. How sage is their advice? "Let the buyer beware," Klimoski said.

Each panelist agreed with Klimoski's assessment that graphology "is a fascinating area, amenable to scientific research." Douglas Hofstadter, who received CSICOP's In Praise of Reason Award the preceding evening, noted in the question-and-answer period that "it seems very plausible that all sorts of aspects in handwriting are revealing." Cracking the code will be a difficult project, he said. "We don't even have any system to describe faces yet."

In an interview following the panel, Beyerstein agreed. "Handwriting may indeed reveal some very helpful things. But all methods used so far have failed and failed dismally" to discern them. At bottom is the vexing question of "personality" itself. "Trying to define somebody's personality," said Beyerstein, "is a fool's errand. Many psychologists seriously doubt whether there is an 'inner core' of fixed and immutable characteristics in the human mind." Which leaves us with the obvious question: As the notion of "personality" as an inherent human trait becomes more difficult to sustain, will there be anything there for graphology to measure once the field gets its act together?

ARLEEN J. WATKINS and WILLIAM S. BICKEL

A Study of the Kirlian Effect

An interesting photographic phenomenon called Kirlian photography can be demonstrated by applying a high-voltage (15,000–60,000 volts) high-frequency discharge across a grounded object placed on a sheet of film lying on the high-voltage plane. A typical configuration and one used for this study is illustrated in Figure 1. When the object placed on the film plane is grounded to complete the current loop, a discharge occurs between the object and the high-voltage conducting plane creating an air-glow discharge, which appears to the eye as a purple-blue fuzzy light called an aura. The aura is a very real physical phenomenon and can be recorded directly on photographic paper, on film (black and white or color), or on photo plates. When the plates are developed, the aura appears as a fuzzy glow around the boundary of the image.

Beginning in the 1970s, the origin of this aura image and its relevance to the state or condition of the object producing it became a topic of great popular interest. Claims were made that the aura of human objects—fingers, toes, etc.—contain information about the physiological, psychological, and psychic state of the individual. For plant and animal parts—leaves, stems, legs, wings, tissue, bone cross-sections, etc.—the aura was claimed to carry information about the "life-force," "life-energy," or "bioplasma" of the object. If the aura were indeed a probe for such conditions and carried information about important parameters inaccessible by, or more accurate than, other techniques, it would be a powerful and important technique for such studies.

Literature on Kirlian photography reports many studies by various people and groups. One universal and puzzling point is that it is often discussed with an air of mystery. It has been referred to as a new phenomenon, an unknown phenomenon, and a mysterious phenomenon carrying important information about life. For most physicists, the first guess is that the effect (the aura) is a corona discharge in air. If this is the case, the phenomenon, although it may be complicated to explain in detail, is well known and will be governed by the laws of physics. Therefore, any scientist setting out to investigate it

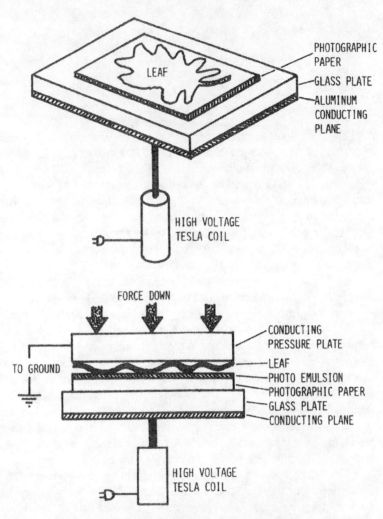

FIGURE 1. Arrangement of the high-voltage coil, glass plate, conducting plane, photographic paper, sample, and press for making Kirlian photographs.

will first document all observables relating to this phenomenon. Many serious studies have done exactly this. This was the motivation of this study, which we carried out with an apparatus we constructed to generate Kirlian photographs.

We took more than 500 Kirlian photographs to study the aura from three sets of objects with various configurations—animal, plant, and mechanical. Figure 2 displays some typical aura images. It is rather easy to recognize the objects used. The boundaries are quite distinct; the aura is rich in detail and shows much variation from object to object. We now discuss specific aura patterns to substantiate or refute certain claims and interpretations made by Kirlian investigators.

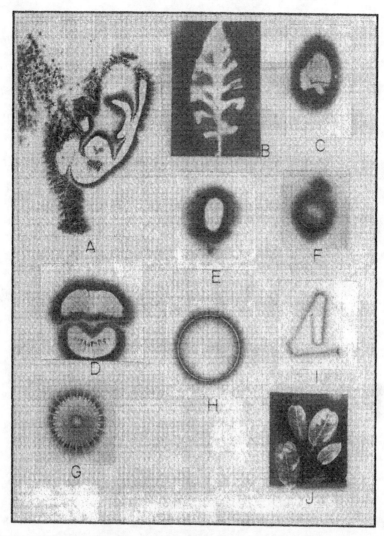

FIGURE 2. Kirlian photographs: (A) ear, (B) leaf, (C) thumb, (D) lips, (E) finger, (F) diamond ring, (G) brass gear, (H) metal ring, (I) paper clip, (J) leaves.

Claim 1

The aura is related to the "life-energy" or "bioplasma" of the animal or plant. The shape, size, intensity, and structure of the aura depend on the psychic energy, state of mind, emotion, well-being, illness, etc., of the object. Figure 3 shows a set of aura pictures of three different individuals. Figure 4 shows the aura of mechanical objects—coins, wire, water, gears, and sharp metal points.

Question: If the aura is due to the "bioplasma" or "photo energy" of

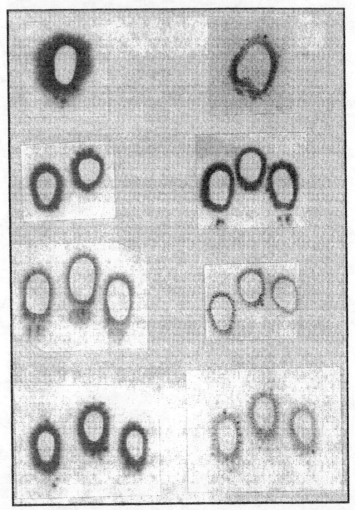

FIGURE 3. Assorted auras of thumb and finger prints of different people using different photographic paper, exposure, and development times.

the living object, then why does it appear from mechanical objects?

Answer: Since the aura appears from dead and organic objects as well as living or once living objects, the aura does not represent a "bioplasma."

Claim 2

The aura is supposed to represent the condition of the object via its size, shape, intensity, and structure. Figure 5, A, B, C show three sets of finger auras, from three different people. Each set was taken within a period of

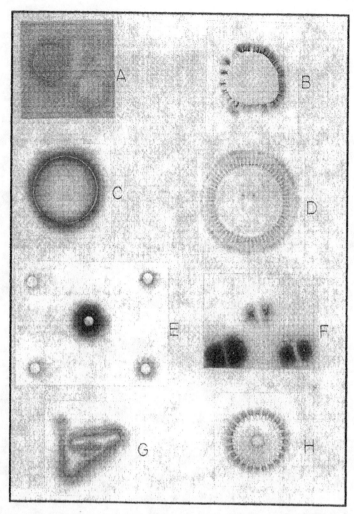

FIGURE 4. Assorted auras of mechanical objects: (A) penny ring, (B) water droplet, (C) ring, (D) brass spur gear, (E) small metal discs, (F) sharp point pairs, (G) paper clip, (H) brass gear.

15 seconds. Note that the aura varies from finger to finger in each set and very markedly in set 5C.

Question: If the aura represents the condition of the object, what interpretation do we give for the markedly different patterns?

Answer: For this set, none. We do not suspect at this point that the change in aura from one print to the next in any strip represents a change in mental or physical condition or personality of the individual. We suspect instead the cause of the differences is due to lack of experimental control, which will be discussed in detail in the last section.

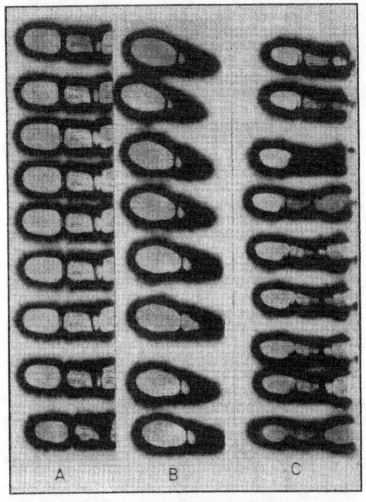

FIGURE 5. Aura of three sets of fingers from three different people.

Claim 3

When two fingers of two *different* people are placed side by side simultaneously, the aura pattern shown in Figure 6 results. Note the combined aura shows a sharp boundary between the two images. This is said to be due to the incompatibility of the two individuals; and the sharpness of the boundary, a measure of the degree of incompatibility. However, Figure 6B shows an aura created by two fingers of the *same person* simultaneously, 6D and 6F the aura of two pennies, 6E of a dime and a quarter, 6G of a metal bar and a dime, 6C of three fingers of the same hand simultaneously. In *all* cases

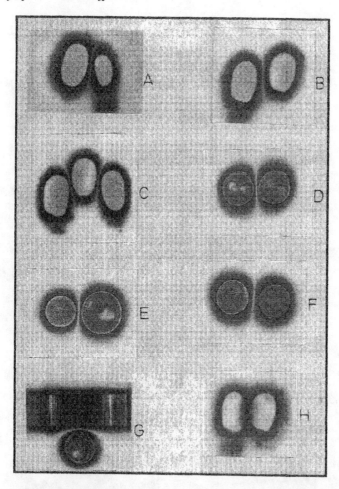

FIGURE 6. Aura pairs showing equipotential boundary between aura patterns: (A) fingers of two different people, (B) and (C) fingers of the same person, (D) and (F) two pennies, (E) penny and quarter, (G) dime and metal bar, (H) aura pattern of two fingers of same person not taken simultaneously.

a sharp boundary occurs between the auras.

Question: If the sharpness of the boundary is an indicator of incompatibility, what interpretation do we give for the sharp boundary between the auras created by two fingers on the *same hand* of the *same person,* as well as between metal objects?

Answer: The interpretation of incompatibility is wrong unless two or three fingers on the *same* hand are incompatible with one another, or a dime is incompatible with a quarter, etc. The sharp boundary and its shape is easily explained in all cases using well-known physics laws. The boundary where no aura occurs is caused by the lack of electron motion in the film plane. Since

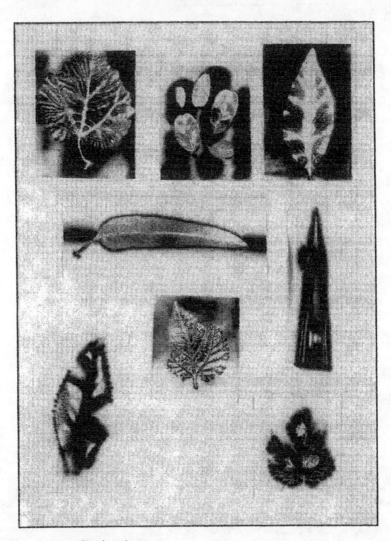

FIGURE 7. Aura of various leaves.

both objects are at equipotential and both seek to neutralize the surrounding
film plane, an area somewhere between the two objects will be at zero potential,
i.e., an electron there is attracted with equal force in both directions. Therefore,
it doesn't move. There is no electron flow, no current, no excitation of the
air molecules, and therefore no aura. These physics principles also exactly explain
the curved boundary of Figures 6E and 6G. Figure 6H shows the aura of two
"compatible" fingers. However, it was made by placing first one finger, *then*
the other, on the photographic paper. Since the images are not made simultaneously, the electrons can flow into the other image area, causing an aura there.

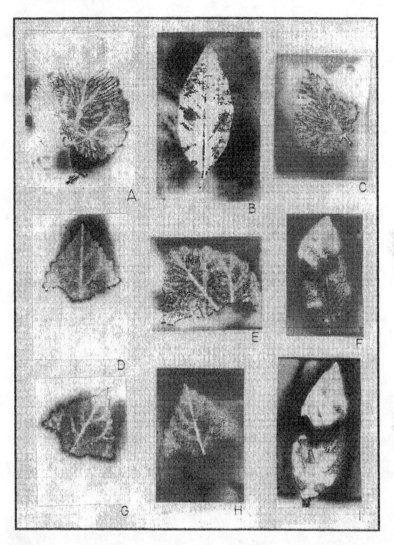

FIGURE 8. Aura of whole and broken leaves: (A), (B), and (C) whole leaves; (D), (E), (G), and (H) broken leaf with one piece missing; (F) and (I) broken leaf with broken sections separated.

Claim 4

When Kirlian photographs are taken with color film, in addition to the size, intensity, structure, and shape of the aura, we obtain the new parameters of color and color distribution. Color photographs of auras are very dramatic, showing a rich color distribution, which is claimed to contain information about the emotions of the subject—red = anger, strong emotion; blue = cool-

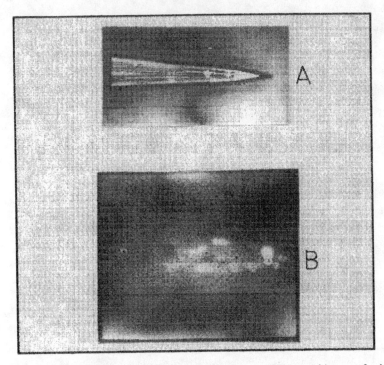

FIGURE 9. "Phantom Leaf Effect"; (A) entire leaf aura, (B) aura of image of missing leaf shown in (A).

ness and composure; etc. Although no color photography was done in this study, an examination of many color slides from a previous study brings up the same questions as the black and white pictures do *and more*.

Question: Are emotion and personality related to the color distribution of the aura?

Answer: Until proper controls show that the color photograph contains information *in addition to* what can be explained by laws of physics and the photographic process, this question cannot be answered. (Color emulsions contain three color dyes with different sensitivities to photon and electron stimulation.)

Claim 5

The aura of leaves and stems shown in Figure 7 is very rich in detail. As with finger auras, there is a large difference in aura patterns even though they are of the same leaf. There is a rather remarkable claim that one can get an aura pattern of an entire leaf even though part of the leaf has been torn or cut away or otherwise removed. This is called the "phantom-leaf effect."

It is evidently difficult to reproduce but has been reported by several investigators and recorded on movie film. (There is also a "phantom salamander-tail effect.") None of our photographs demonstrated the phantom-leaf effect. Figure 8 shows the aura of torn and cut leaves. In no case was an aura detected in the region of the missing leaf or around its boundary.

Question: Is the "phantom-leaf effect" due to the bioplasmic body of the leaf?

Answer: The several caes we investigated where a phantom-leaf effect seemed to occur were *artifacts* and quite easy to explain. When a whole leaf is pressed against the film plane with a metal plate, moisture, dust, and a minute amount of juices are squeezed from the leaf, leaving an image. Sometimes this image could even be seen with the eye. When the whole leaf was removed and the metal plate returned to its original position, the new photograph showed a weak, fuzzy, ill-defined, and "mysterious" shape of the missing leaf. Figure 9B. shows the aura of the entirely missing leaf shown in 9A. This image, however, is totally an artifact.

The Kirlian Technique: Parameters Involved and Controls Needed

Although Kirlian photographs are very easy to make—needing a minimum of talent, equipment, and money, the entire Kirlian process from sample preparation to photo interpretation involves very many parameters and a very complex interplay between parameters and conditions. The following list contains 22 of the many parameters that must be controlled. Some of the more crucial ones will be discussed in detail. The parameters can be grouped into the following areas: (A) electronic and mechanical configuration, (B) sample and environment, (C) film, plates, and photographic process, (D) photographic image interpretations.

A. *Electronic and Mechanical:* (1) Voltage discharge. (2) Current in discharge. (3) Current density through sample. (4) Frequency of the discharge (pulses per second). (5) Pulse shape. (6) Total resistance in circuit consisting of the air, emulsion, and sample. (7) Electric field configuration; point-plane, high-voltage plane, shape (square, rectangular, circular), grounding plane shape. (8) Sample holder and pressures used, size, shape and electrical characteristics of the press used to flatten sample against film plane.

B. *Sample and Environment:* (1) Size, shape, and surface regularity of sample. (2) Conductivity (moisture content), salt or other local chemicals. (3) Surrounding atmosphere: (*a*) atmospheric pressure, (*b*) humidity, (*c*) temperature, (*d*) chemical content. (4) Complete history of the sample. (5) Cleanliness—lack of dust, specks, stains, moisture.

C. *Films, Plates, and Photographic Process:* (1) Film and paper type, ASA, RMS granuality, H-D characteristics, and *all* other film properties. (2) Emulsion properties—dielectric constant and conductance sensitivity to electron excitation, contact pressures (these are not usually published data). (3) Exposure time—

continuous, pulsed, long, short, over-exposed. (4) Development—time, temperature, and chemicals used. (5) Reproduction of negatives—magnification, types of paper used (all paper characteristics). (6) For color photographs, peak wavelength sensitivity of dyes and sensitivity to pressure, electrons, and temperature.

    D. *Photographic Image Interpretation:* (1) Qualitative—comparisons, relative intensity, shape, size, and structure of aura usually made by the eye. (2) Qualitative—microdensitometer studies of intensities recorded on *calibrated* plates. (3) Color—spectral studies of radiation emitted and of images formed according to quantitative color theory.

    In the set of experiments reported here we found lack of control of the pressure on the sample, exposure time, development time, sample structure and preparation, and applied voltage caused the greatest variations in aura for the *same* sample. Indeed, even in cases where the aura was recorded under "identical" conditions, for the same object, the aura had significant variations in its properties. Of course a "significant variation" is a qualitative, subjective opinion.

Conclusions from These Experiments

In this rather short scientific investigation of the Kirlian technique, we were able to turn up a number of artifacts and puzzling signals that after a little thought and study were found to fit into the normal scheme of things. Moist fingers, varying pressures, different paper sensitivity, exposure and development times were responsible for most of the variations in the auras. *We conclude there is no need to evoke psychic phenomena to explain results and there is no evidence that psychic conditions affect the aura patterns.*

    There is no reason to relate the aura to a "bioplasma." The body of course does radiate in the infrared. (It is a black body at 98.6° F.) The Kirlian aura is a visual or photographic image of a corona discharge in a gas, in most cases the ambient air. Its color depends on the composition of the air, pressure, and impurities emanating from the sample as well as the voltage and current of the source. Other gases, such as nitrogen, helium, argon, and carbon dioxide, that we have used also produce auras, but with color differences and shapes that depend on the spectroscopic and electrical properties of the particular gas. (Caution! In no case should hydrogen be used. It is extremely explosive when mixed with air and ignites with a spark. Remember the Hindenburg!)

    The discharge ionizes and excites the molecules and atoms in the air, causing light to be emitted. The radiation emitted from excited gases in virtually all physical conditions has been extensively studied by spectroscopists since the early 1800s. The shape, size, intensity, and fine structure of the aura depends on exposure times, conductivities, pulse rates, voltages, and photographic properties of the plates and film used. When there is no applied voltage, there is no discharge. Then there is no aura because there is no light.

    There is no evidence as yet that any *feature character* or *property* of the aura pattern is related to the physiological, psychological, or psychic condition of the sample. Although the aura surely depends on some physical properties of the system—i.e., the conductivity of the sample (sweaty fingers, perspiring hands), force exerted on the sample—it also depends on many other complicated effects. There is no doubt that some psychological and physiological conditions do manifest themselves in external signals: lie-detectors do work, heat sensors can detect tumors, shaking hands represent nervousness or illness, and so on. However, the Kirlian technique has not yet been shown to be a direct or meaningful link to these conditions. In fact, while most Kirlian investigators acknowledge the effects of the physical parameters, they make no attempt to standardize their research by controlling the parameters, nor do they appear to be concerned with the significance of changing parameters. Indeed, for the most part, the parameters within their research are only vaguely reported if at all, making replication studies by other researchers impossible.

The difficult and pressing challenge then would be to control the parameters and demonstrate in several specific cases that the aura produces information inaccessible to, or better than, other techniques. Its usefulness then would not need to be advertised; it would be picked up immediately by all laboratories that can use it to extend their research into new directions. Within two months after Roentgen discovered X-rays, his device was used by doctors to examine bones.

The Kirlian aura will most likely remain a fascination to *nonscientific* people because of the ease in producing the aura and its "mysterious manifestations" of sparks, discharges, corona, and aura coupled with the words "life force," "photic energy," "bioplasma," "life energy," and so on. Most Kirlian claims will come from "experimenters" who will combine the complicated effects of gaseous discharges with samples having complicated structure and electrical properties, and film recordings involving complicated photographic processes and interpretations based on ignorance of the phenomena and the need for proper controls.

Bibliography

Aaronson, Steve. 1974. Pictures of an unknown aura. *The Sciences.* New York: New York Academy of Sciences.

Gardner, Martin. 1957. *Fads and Fallacies in the Name of Science.* New York: Dover.

Krippner, Stanley, and Daniel Rubin. 1974. *The Kirlian Aura: Photographing the Galaxies of Life.* New York: Doubleday.

Moss, Thelma. 1975. *The Probability of the Impossible.* New York: New American Library.

Ostrander, Sheila, and Lynn Schroeder. 1970. *Psychic Discoveries Behind the Iron Curtain.* Englewood Cliffs, N.J.: Prentice-Hall.

Pehek, J. O., H. J. Kyler, and D. L. Faust. 1976. Image modulation in corona discharge photography. *Science,* October 15, 263-270.

Singer, Barry. 1981. Kirlian photography. In *Science and the Paranormal,* ed. by George Abell and Barry Singer. New York: Scribner's.

I. W. KELLY, JAMES ROTTON, and ROGER CULVER

The Moon Was Full and Nothing Happened: A Review of Studies on the Moon and Human Behavior

It is commonly assumed that a full moon brings out the worst in people. Those who do research in this area invariably begin reports by reminding readers that "lunacy" and "lunatic" are derived from *luna,* the Latin word for moon. Although lunacy is an outdated concept, investigators have tried to link the phases of the moon to such behaviors as alcoholism, madness, epilepsy, somnambulism (sleepwalking), suicide, homicide, arson, and, of course, lycanthropy (werewolfism).

Arnold Lieber (1978), a Miami psychiatrist, used the term *lunar effect* when referring to supposed links between phases of the moon and behavior. Critics prefer the term *Transylvania effect* (Shapiro et al. 1970). As one might guess, those who defend the lunar hypothesis have objected to the latter "because it conjures up visions of werewolves and Draculas" (Garzino 1982, 399). In our view, however, neither term is appropriate, since the word *effect* implies that investigators can establish something more than a correlation in this area. Obviously, without having had a "control group" on a planet without a moon (perhaps a random sample of Venusians), researchers cannot show that a full moon exerts a causal influence on behavior.

In the first part of this article, we describe results from a meta-analysis of studies that examined relationships between phases of the moon and behavior (Rotton and Kelly 1985a). We also note several studies that appeared after we completed our meta-analysis. In the second part, we speculate about why lunar beliefs persist despite the absence of reliable linkages between phases of the moon and behavior.

Research on Lunar Cycles and Behavior

Rotton and Kelly (1985a) combined data from 37 published and unpublished studies in a meta-analysis that had examined relationships between the moon's synod (4-phase) cycle and abnormal, deviant, and criminal behavior. Meta-analysis is a statistical procedure that combines results from empirical investigations. It allows reviewers to do three things: (1) estimate the overall or combined probability of results from different studies; (2) assess the size of relationships when results are averaged; and (3) identify factors that might help to explain why some studies have obtained apparently reliable results while others have not. This meta-analysis differed in one important respect from those that have been undertaken to resolve controversies in other areas: It included a reanalysis of results from previously published studies.

Of the 23 studies we checked, nearly one-half contained one or more statistical errors. Some of these were serious enough to prompt us to publish interim reports (Kelly and Rotton 1983; Rotton, Kelly, and Frey 1983) to correct errors that had crept into the literature. For example, we found that Lieber and Sherin (1972) had employed inappropriate and misleading statistical procedures in their often-cited study of homicides in Dade County, Florida. On the basis of binominal tests of significance, they claimed that a disproportionate number of homicides occurred during the 24-hour period before and after full moons. We found that this claim was based upon 48 tests of significance, which are not reported in their article. To make matters worse, their tests were not independent. For example, in one set of analyses they looked "at the three days before and after, three days before, three days after, two days before and after, two days before, two days after, one day before and after, one day before, one to two days after, and one to three days after full moon" (Rotton, Kelly, and Frey 1983, 111; Rotton and Kelly 1985c). Applying more conventional test procedures, it was found that homicides were evenly distributed across phases of the moon.

In another study, Templer, Veleber, and Brooner (1982) claimed that a disproportionate number of traffic accidents occurred during the night hours of the three-day periods of the new moon and the full moon. However, as Kelly and Rotton (1983) noted, a larger number of the full- and new-moon nights cited in the study fell on weekends. They suggested that apparent relationships might stem from the fact that more accidents occur on weekends than on weekdays. This suggestion was later confirmed by reanalysis of their data. Templer, Corgiat, and Brooner (1983) found that relationships vanished when they included controls for holidays, weekends, and months of the year. To their credit, they were willing to revise their original hypothesis: "It is likely that some, perhaps all, of the significant phase-behavior findings in the literature are a function of day of week or holiday or season artifact" (Templer et al. 1983, 994).

As these examples illustrate, a meta-analysis is no better than the studies on which it is based. In our meta-analysis, we took several steps to locate relevant articles and papers, including a computer search of the literature. Correcting for errors in original reports, we found that there was no consistent relationship between phases of the moon and acts usually described as lunatic. Taken as a whole, our results confirm the generally negative conclusions reached in prior reviews (Abell 1981; Cooke and Coles 1978; Campbell and Beets 1978; Kelly 1981). For every study that had found that people behave more strangely than usual when the moon is full, another had found that people's behavior was not affected.

Indeed, phases of the moon accounted for no more than 3/100 of 1 percent of the variance in activities usually termed *lunacy*. Estimating the percentage of unusual episodes that occurred during the quarter (25 percent) of the time when the moon is full, we found that about 25.7 percent of the episodes had occurred during full-moon periods. Of course there may be some who will claim that a difference of 0.7 percent is theoretically interesting. However, we are not impressed by a difference that would require 74,477 cases to attain significance in a conventional (i.e., chi-square) analysis.

Some might also object that we averaged over important differences when we combined data from different studies. To deal with this objection, we considered factors thought to mediate relationships between phases of the moon and behavior: sex of subject, type of lunar cycle (synodic vs. anomalistic or apogee-perigee), geographical features, publication practices, and type of lunacy (namely, mental hospital admissions, disturbed behavior in psychiatric settings, calls to crisis centers, homicides, and other criminal offenses). In only one of these subsidiary analyses did a difference approach significance. There was a slight (but not statistically significant) tendency for stronger relationships to appear in "pay" journals than in refereed sources and unpublished theses.

Several additional studies have since come to our attention. In one of these, Russell and Dua (1983) examined relationships between phases of the moon and aggressive episodes during Western Hockey League games. They based their conclusions upon 426 aggressive infractions recorded during the 1978-79 hockey season. After looking at several types of aggression, they concluded that "the present investigation offers no support for a lunar-agression hypothesis" (p. 43). More recently, Russell and de Graaf (1985) replicated the earlier study on hockey infractions on a new season (1983-84) of the Western Hockey League. As in the earlier study they found no evidence of a relationship between hockey aggression and moon phase. In another study, MacMahon (1983) examined suicide data in the United States over a 7-year period. After plotting suicide rates by lunar phase (using a corrected 30-day cycle), she concluded that "deviations from the mean were small and present no obvious pattern" (p. 747). Likewise, Atlas (1984) uncovered no relationship between lunar phases and violent episodes in Florida prisons.

Finally, Sanduleak (1985) examined relationships between lunar cycle and homicides in Cleveland, Ohio. His study is noteworthy, because Lieber and Sherin (1972) previously claimed that they had uncovered a reliable relationship between lunar cycles and homicides in this city. They based this conclusion on data between 1958 and 1970, whereas Sanduleak covered the period from 1971 through 1981 in his follow-up study. Sanduleak's results are aptly summarized by the title of his article: "The Moon Is Acquitted of Murder in Cleveland."

On the other hand, Davenhill and Johnson (1979) claimed to have detected a relationship between various personality variables as measured by the Eysenck Personality Inventory (EPI) and Cattell's 16 Personality Factors (PF) and changes in the lunar cycle. However, Startup and Russell (1985) criticized their research, pointing out that the Davenhill and Johnson study employed only a very small sample (12 males and 12 females) and only covered a short period of time (two lunar cycles). In addition, using 881 subjects over a two-year period and a more powerful statistical technique, Startup and Russell could replicate none of the findings obtained in the earlier study on the Eysenck Personality Quotient (EPQ, a revised form of the EPI) and only one with the 16 PF. However, the minuscule size of the relationship precludes any practical use, and the authors caution that it would be unwise to attach theoretical significance to this finding until it can be replicated by others.

Belief in Lunar Effects

Rotton and Kelly (1985b) found that one-half (49.7%) of the students in a Florida university agreed that some people behave strangely when the moon is full. Similar levels of beliefs have been recorded for students at a Canadian university (Russell and Dua 1983) and in Singapore (Otis and Kou 1984). What accounts for belief in lunar effects? Although we have only begun to pursue this question, we suspect that belief in lunar effects can be traced to three factors. One of these can be termed *media effects*. Another is misconceptions about physical factors. The third, and in some ways the most interesting, is cognitive biases that lead individuals to look to the moon when they witness unusual and apparently senseless types of behavior.

Media Effects

Newspapers, television programs, and radio shows favor individuals who claim that a full moon influences behavior. Arnold Lieber, one of those favoring the lunar hypothesis, has appeared on several talk-shows, including the nationally syndicated "In Search of . . ." On November 8, 1984, his research was highlighted on ABC's "20/20." This supposedly objective report began with its host, Hugh Downs, suggesting that lunar effects provided evidence for astrology: "The moon's effects are legendary and, according to some, the most obvious example of astrology—that ancient belief that has in the past twenty years become big business."[1] Likewise, Mirabile's (1984) presentation at the Institute of Child Development was widely disseminated by newspapers throughout the United States. Finally, on August 27, 1984, Ann Landers answered a reader's question by telling him, "It's true . . . some people get loonier than others when the moon is full."

Newspapers, of course, are in the business of telling people what happened. "The moon was full, and nothing happened" may be accurate, but it is not a very interesting headline. In research on curiosity and information-seeking, it is something of a truism that "good news is no news" (Rotton, Heslin, and Blake 1983, 49). When reporters call us on the phone, they would probably be happier if we assure them by saying, "The streets are full of loonies when the moon is full." Unfortunately, when one scientist doesn't give them a quotation that can be turned into an interesting headline, they can always find an "expert" who will provide the quotation they need.

For a reporter interested in writing a story, it is not hard to find somebody who will talk about an uncle, say, who acted peculiarly when the moon was full. (Who doesn't have a peculiar uncle?) Those who defend the lunar hypothesis are not above resorting to case histories and personal anecdotes. For example, after failing to uncover a statistical relationship between the moon's apogee-perigee (far-near) cycle and behavior, Lieber and Sherin (1972) indicated that a "perusal of official narratives on individual incidents of homicides indicates that homicides occurring during these periods are often of a particularly bizarre or ruthless nature" (p. 105). As Meyers (1983, 120) has observed, "anecdotes are often more persuasive than factual data." To dramatize the supposed effects of the full moon, for example, "20/20" showed pictures of Miami police being called out to keep a young man from killing himself. The announcer's voiceover:

> Even before the moon has risen and the sun still commands the sky, it starts: A confused young man has a cocked pistol to his head. The special response team is in place. If the subject points the gun at anyone else, he will be shot. . . . There are scenes like this somewhere every day, but in Dade County, Florida, at least, the special response team call-outs to incidents like this peaked at the time of the full moon—month after month.[2]

Misconceptions

Given the moon's obvious effects upon ocean tides, it is not surprising that scholars as well as students have jumped to the conclusion that it might also affect people's behavior. "If the moon can do that to oceans," our students say, "imagine what it can do to us!" In a similar vein, Lieber (1978) advocates a biological tide hypothesis. He contends (p. 115): "Because the [human] body [like the earth] is composed of 80 percent water and 20 percent 'land' or solids, it is reasonable to assume that gravity exerts a direct effect on the water mass of the body, just as it does on the water mass of the planet."

Lieber's analogy fails because it is too weak to warrant the inference he wants to draw. As Campbell (1982, 421) points out: "Only the *surface* of the earth has this 80:20 ratio . . . yet gravity involves attraction between three-dimensional structures (and their total masses, not just surface composition). Hence, the argument based on a similar water-solid ratio between earth and the human body is 'untenable.' " In addition, the moon causes tides only in unbounded bodies of water like the world's oceans (Abell 1981; Campbell 1982; Culver, Kelly, and Rotton 1986). Bounded bodies of water, such as land-locked lakes, unless they are very large (like the Great Lakes), are negligibly influenced. Clearly the water contained in the human body falls into the "bounded waters" category.

Even if we surmount these problems—for example, by assuming an idealized human who is uniformly covered by a layer of unbounded perspiration—gravitational mechanics still offers no support for the idea of biological tides. The expression for the tidal force F_{TIDE} to which an object of radius R will be subjected can be readily derived from the principles of classical mechanics: $F_{TIDE} = 2GRMm/d^3$ where G is the universal gravitation constant, M is the mass of the tide-raising object, m is the mass of the object upon which the tidal force is exerted, and d is the distance between the center of mass of two objects involved. A comparison of the tide-raising capabilities F_1 and F_2 of two separate objects on a given person can then be written as

$$\frac{F_1}{F_2} = \left(\frac{M_1}{M_2}\right)\left(\frac{d_2}{d_1}\right)^3 ,$$

where M_1 and M_2 are respective masses and d_1 and d_2 the respective distances of the tide-raising objects. As an example, suppose we wish to compare the tidal forces of a mother, the attending doctor, and the building on a newborn child with that of the moon. If the hospital is located on the side of the earth's surface nearest the moon, then the moon's center of mass will be about 378,000 km distant. Assuming the mother's distance from the child while she holds it is 15 cm or so, then a 55 kg mother will exert

$$\left(\frac{55kg}{7.35 \times 10^{22}kg}\right) \left(\frac{3.78 \times 10^8 m}{0.15m}\right)^3 = 1.2 \times 10^7$$

or 12 million times as much tidal force on her child as the moon. Calculations for the doctor and the fractional mass of the building contained within a radius equal to the child–building center of mass distance will yield similar results. In fact, it can be easily shown that we would have far more tidal concerns from a downtown area with lots of large-mass buildings and crowded streets than from the sun or the moon.

The biological-tide hypothesis fails on a number of other counts. In our review (Rotton and Kelly 1985a) we found six studies that have looked at the distance of the moon from the earth and various types of behavior. Only one obtained significant results, and these were contrary to the biological-tide theory: More undesirable behavior occurred when the moon was *farthest* from the earth. In addition, Lieber argued that we would expect lunar-related behaviors to be more pronounced at the Equator than at more distant latitudes and to have an amplitude variation in keeping with the times of lunar perigee and apogee. We found no evidence for this contention in our review. Sanduleak (1985) did not obtain significant results when he examined relationships between homicidal assaults and a tidal index that was proportional to the magnitude of the combined lunar and solar tide action. Finally, Russell and de Graaf (1986) found no relationship between the distance of the moon from the earth and aggression in hockey games.

Although Lieber and Sherin (1972) originally attributed supposed correlations between phases of the moon and behavior to water imbalances, Lieber (1978), Katzeff (1981), and others have proposed competing hypotheses.

Garzino, for example, has speculated about ion effects:

Because the moon modulates the earth's magnetic field, the entering ions follow a lunar cycle. During the full-moon phase, positive ions come down to earth *in great abundance*. But positive charged ions are now suspected by some scientists to create depression and irritability by increasing levels of serotonin in the nervous system. Serotonin is a mood-modifying chemical, a "downer." (Garzino 1982, 408, italics added. See also, Abel 1976; Katzeff 1981; Ossenkopp and Ossenkopp 1973.)

Although early research on air ions could be criticized on a number of grounds—for example, use of shoddy equipment that produced ozone as well as air ions—more recent studies have demonstrated that people's moods are altered by very high levels of ionized air (Baron, Russell, and Arms 1985; Charry and Hawkinshire 1981). There is fairly compelling evidence that the effects of negative ions are beneficial (e.g., improved mood and better performance on simple tasks), whereas the effects of positive ions appear to be less benign (Fisher, Bell, and Baum 1984). However, these effects depend upon personality factors, such as excitability, and are only found when individuals are exposed to very high concentrations of ions in a controlled (i.e., laboratory) setting.

Although positive ions are more prevalent when the moon is full, positive ion concentrations related to lunar variations are small when compared with those related to air-conditioning and air pollution (Campbell 1982). One is much more likely to feel the effects of positive air ions while working in an enclosed building. However, the question is not "Is there an ion effect?" It is, instead, "Are ion levels high enough when the moon is full to produce effects attributed to them?" The answer to this question appears to be no. Gilbert (1980) measured ion levels in a school for mentally retarded children.

He found no evidence for the ion hypothesis. Indeed, in his study, he observed more disturbed behavior when the moon was new than when it was full.

Although ion effects appear to be mediated by serotonin, we have not been able to locate any study that has examined correlations between lunar cycles and serotonin levels. The absence of research on physiological processes is, in many ways, surprising. Some of those who favor the lunar hypothesis are physicians, such as Lieber and Mirabile, who often speculate about physiological processes. Why have they not obtained blood or urine samples to determine if there is, in fact, a lunar component in hormone levels? As Asimov (1985:8) has observed, such evidence would be much more convincing than statistical analyses of homicide and crime rates: ". . . If these rhythms affect such things as our response to drugs or our tendency to violence or depression, then the rhythms must affect our internal workings. There must be a 14-day rise and fall in hormone production; or such a rise and fall in the activity of our immune system, or our cerebral drug receptors, or various aspects of our neurochemistry."

For several years now, investigators have been monitoring individuals' biochemical levels in hospitals and physiological laboratories in research aimed at answering other questions (e.g., Reinberg and Smolensky 1983). In most cases, they use spectral analysis to detect day-to-day and hour-to-hour changes in biological assays and electrodermal activities.[3] Given the large number of scientists involved in this research, it is hard to believe that a 14-day cycle could go undetected. Those who favor the lunar hypothesis often cite Brown's work on the activity patterns of oysters (Brown 1954) and hamsters (Brown and Park 1967). Some of these same authors have published books on biological rhythms (e.g., Garzino 1982). Strangely enough, they do not report anything resembling a 29-day cycle in human activity levels. Given the large number of studies done (in both the United States and West Germany) on the effects of social isolation (e.g., Luce 1971; Minors and Waterhouse 1981), it is surprising that none of them have reported that subjects act restlessly, talk to themselves, or eat or drink more when the moon is full.

Cognitive Biases

A number of cognitive biases contribute to belief in lunar effects. One is selective perception: Individuals are more likely to notice events that support their beliefs than those that do not. Further, individuals are more likely to look for a cause when they notice unusual behavior. Because the moon is conspicuous and its absence is not, it will be an object commonly invoked to explain odd events and behavior. When something odd happens, what other object is so impressively in view as a full moon? However, in research that is now being done at Florida International University, we have found that students do no better than chance when they are asked to guess the moon's phase. As Sanduleak (1985:6) observed, it does not seem likely that "even the most ardent proponent of a lunar effect could specify the phase of the moon. . . . I have tested audiences and found that only a very small percentage could." Social psychologists have found that most of us look to others when we have to make decisions (i.e., what they call "social reality"), and we often act like "cognitive misers"—that is, we look for simple solutions and base our decisions upon the first piece of information we receive (Fiske and Taylor 1983; Hansen 1980). Thus, we have to wonder how many individuals check to see if the moon is full when an unusual event occurs and somebody says, "Must be a full moon tonight."

Selective recall is another bias that contributes to belief in lunar effects. We often recall positive instances and forget negative ones (Nisbett and Ross 1980). Individuals may recall all the full-moon nights when something untoward happened while forgetting the uneventful full-moon nights and the many more *non*-full-moon nights when they witnessed unusual behavior.

Selective attention and recall contribute to illusory correlations (Rotton 1985a). Individuals find it hard to believe that events are random and unrelated, especially when they vary over time. For some, "Everything is related to everything else" is not just an ecological slogan; it is, instead, a principle that guides their thinking and leads them to interpret randomly distributed events as confirming their beliefs. As Meyers (1983, 129) has observed: "When we believe a correlation exists between two things, we are more likely to notice and recall confirming than disconfirming instances."

Illusory correlation is a special instance of a more general and confirmatory bias (Mahoney and DeMonbreun 1978; Snyder and Swann 1978; Watson and Johnson-Laird, 1972). Most of us seek data that support our beliefs, preconceptions, and hypotheses. It is commonly assumed that scientists are mainly interested in obtaining data that will support their theories and hypotheses. Unfortunately, as philosophers (e.g., Hempel 1966; Salmon 1984) have suggested, thinking rarely advances when one adopts a confirmatory strategy. We learn a great deal more when our hypotheses are shown to be inadequate.

Yet another bias is selective exposure, which leads believers to watch TV shows and read books that confirm their beliefs. Although research on the selective-exposure hypothesis has produced mixed results, Otis (1979) found belief in one paranormal phenomenon (UFOs) predicts movie preferences. In her study, individuals standing in line to see *Close Encounters of the Third Kind* were more willing to endorse pro-UFO items than were individuals waiting to see other movies (specifically, *The Gauntlet* and *Saturday Night Fever*). There is evidence that beliefs in lunar effects comprise part of a constellation of belief in paranormal phenomena. Rotton and Kelly (1985b) found that students who scored lower on tests of logical ability, and those who believed in reincarnation, ESP, and astrology were more likely to endorse beliefs in lunar effects.

Any of these biases may act as a self-fulfilling prophecy, leading to actions that confirm people's beliefs (Russell and de Graaf 1985). For example, if police officers believe that a full moon causes criminal behavior, they might become more vigilant and make more arrests on full-moon than other nights (Frey, Rotton, and Barry 1979). In this regard, it is interesting to note that Rotton, Kelly, and Elortegui (1985) found that police officers were more likely to endorse items indicative of belief in lunar effects than a haphazard sample of pedestrians (the proverbial "man and woman on the street").

Conclusion

This article outlines the results of a meta-analysis of 37 studies and several more recent studies that examined lunar variables and mental behavior. Our review supports the view that there is no causal relationship between lunar phenomena and human behavior. We also speculate on why belief in such relationships is prevalent in our society. A lack of understanding of physics, psychological biases, and slanted media reporting are suggested as some possible reasons.

It is important to note that there are two hurdles to overcome before any findings on lunar variables and human behavior are deserving of public attention. The first hurdle is that *reliable* (i.e., replicable) findings need to be reported by independent investigators. The second hurdle is that the relationship should not be a trivial one. The lunar hypothesis fails on both counts.

Postscript

Several reviews of areas of research involving human behavior and lunar cycles have been published since this article appeared. Culver, Rotton, and Kelly (1988) have reported on new studies and have examined in detail explanations (e.g., gravity, tidal effects, light, geomagnetism, etc.) that have been put forward by advocates of lunar effects on human behavior. Martens, Kelly, and Saklofske (1988) critically reviewed 21 studies that have investigated the relationship between lunar cycles and human births. They conclude that insufficient evidence exists for such an association. Kelly, Laverty, and Saklofske (1990) describe studies that have examined disasters of various types and moon phase and contend that no good reason exists for belief in such a relationship. Finally, Martin (1990) reports on a study of suicides in Saskatchewan, Canada, over a six-year period, finding no association between suicide frequency and lunar cycles.

Notes

1. These quotes were transcribed from a cassette recording of the November 8, 1984, broadcast.

2. See note 1.

3. Any complex curve can be described in terms of a number of pure sinewaves that differ in amplitude, frequency (cycles per unit time), and initial phase or starting time. Spectral analysis is simply a mathematical procedure that allows an investigator to describe a wave in terms of pure waves. As Rotton (1985b) has noted, it is ideally suited for uncovering "hidden periodicities" in behavior.

References

Abel, E. L. 1976. *Moon Madness.* Greenwich, Conn.: Fawcett.

Abell, G. O. 1979. Review of "The Lunar Effect" by Arnold Lieber. *Skeptical Inquirer,* 3 (3):68-73.

———. 1981. Moon madness. In Abell and Singer (eds.), *Science and the Paranormal,* 95-104. New York: Scribner's.

Asimov, I. 1985. Moonshine. *Isaac Asimov's Science Fiction Magazine,* 9 (5):4-6, 8.

Atlas, R. 1984. Violence in prison: Environmental influences. *Environment and Behavior,* 16 (3): 275-306.

Baron, R. A., G. W. Russell, and R. L. Arms. 1985. Negative ions and behavior: Impact on mood, memory, and aggression among Type A and Type B persons. *Journal of Personality and Social Psychology,* 48: 112-119.

Brown, F. A. 1954. Persistent activity rhythms in oysters. *American Journal of Physiology,* 178: 510-514.

Brown, F. A., and Y. H. Park. 1967. Synodic monthly modulation of the diurnal rhythm of hamsters. *Proceedings of the Society for Experimental Biology and Medicines,* 125: 712-713.

Campbell, D. E. 1982. Lunar-lunacy research: When enough is enough. *Environment and Behavior,* 14 (4): 418-424.

Campbell, D. E., and J. L. Beets. 1978. Lunacy and the moon. *Psychological Bulletin,* 85: 1123-1129.

Charry, J. M., and F. B. W. Hawkinshire. 1981. Effects of atmospheric electricity on some substrates of disordered social behavior. *Journal of Personality and Social Psychology,* 41: 185-197.

Cohen, J. 1977. *Statistical Power Analysis for the Behavioral Sciences,* rev. ed. New York: Academic Press.

Cooke, D. J., and E. M. Coles. 1978. The concept of lunacy: A review. *Psychological Reports,* 42: 891-897.

Culver, R., and P. Ianna. 1984. *The Gemini Syndrome: A Scientific Evaluation of Astrology.* Buffalo, N.Y.: Prometheus Books.

Culver, R., I. W. Kelly, and J. Rotton. 1985. The moon and human behavior: A critique of the biological tide hypothesis (manuscript in preparation).

Culver, R., J. Rotton, and I .W. Kelly. 1988. Moon mechanisms and myths. *Psychological Reports,* 62:683-710.

Davenhill, R., and F. N. Johnson. 1979. Scores on personality tests correlated with phase of the moon. *IRCS Medical Sciences,* 7: 124.

Fisher, J. D., P. A. Bell, and A. Baum. 1984. *Environmental Psychology,* 2nd ed. New York: Holt, Rinehart & Winston.

Fiske, S. T., and S. E. Taylor. 1983. *Social Cognition.* Reading, Mass.: Addison-Wesley.

Frey, J., J. Rotton, and T. Barry. 1979. The effects of the full moon upon behavior: Yet another failure to replicate. *Journal of Psychology,* 52: 111-116.

Garzino, S. J. 1982. Lunar effects on mental behavior: A defense of the empirical research. *Environment and Behavior,* 14 (4):395-417.

Gilbert, G. O. 1980. Relationship of behavior of institutionalized mentally retarded persons to changes in meteorological conditions. (Doctoral dissertation, Union Graduate School-West, British Columbia.)

Hansen, R. D. 1980. Commonsense attribution. *Journal of Personality and Social Psychology,* 39: 996-1009.

Hempel, C. G. 1966. *Philosophy of Science.* Englewood Cliffs, N.J.: Prentice-Hall.

Katzeff, P. 1981. *Full Moons.* Secaucus, N.J.: Citadel Press.

Kelly, I. W. 1981. Cosmobiology and moon madness. *Mercury,* 10: 13-17.

Kelly, I. W., and J. Rotton. 1983. Comment on lunar phase and accident injuries: The dark side of the moon and lunar research. *Perceptual and Motor Skills,* 57: 919-921.

Kelly, I. W., Laverty, and D. Saklofske. 1990. An empirical investigation of the relationship between worldwide automobile traffic disasters and lunar cycles: No relationship. *Psychological Reports,* 67.

Lieber, A. L. 1978. *The Lunar Effect: Biological Tides and Human Emotions.* Garden City, N.Y.: Anchor Press/Doubleday.

Lieber, A. L., and C. R. Sherin. 1972. Homicides and the lunar cycle: Toward a theory of lunar influences on human emotional disturbance. *American Journal of Psychiatry,* 129: 101-106.

Luce, G. G. 1971. *Body Time: Physiological Rhythms and Social Stress.* New York: Bantam.

MacMahon, K. 1983. Short-term temporal cycles in the frequency of suicide, United States, 1972-1978. *American Journal of Epidemiology,* 117 (6): 744-750.

Mahoney, M. J., and B. G. DeMonbreun. 1978. Problem-solving bias in scientists. *Cognitive Therapy and Research,* 1: 229-238.

Marks, D., and R. Kammann. 1980. *The Psychology of the Psychic.* Buffalo, N.Y.: Prometheus.

Martens, R., I. W. Kelly, and D. Saklofske. 1988. Lunar phase and birthrate: A 50-year critical review. *Psychological Reports,* 63:924-934.

Martin, Sandra. 1990. Distal variables in a completed suicide population. Unpublished master's thesis, University of Saskatchewan, Saskatoon, Canada.

Meyers, D. G. 1983. *Social Psychology.* New York: McGraw-Hill.

Minors, D. S., and J. M. Waterhouse. 1981. *Circadian Rhythms and the Human.* Bristol: Wright.

Mirabile, C. S. 1981. Everything under the sun and moon: Orientation theory: A new view of man's brain function and behavior. Paper presented at the fifth annual IEEE Symposium on Computer Applications in Medical Care, Washington, D.C., November. (Published in Proceedings.)

———. 1984. Solar and lunar cycles in disturbed behavior. Paper presented at the fourth annual conference of the Institute of Child Development Research, New York.

Nisbett, R., and L. Ross. 1980. *Human Inference.* Englewood Cliffs, N.J.: Prentice-Hall.

Ossenkopp, K. P., and M. D. Ossenkopp. 1973. Self-inflicted injuries and the lunar cycle: A preliminary report. *Journal of Interdisciplinary Cycle Research,* 4: 337-348.

Otis, L. P. 1979. Selective exposure to the film "Close Encounters." *Journal of Psychology,* 101: 293-295.

Otis, L. P., and E. C. Y. Kou. 1984. Extraordinary beliefs among students in Singapore and Canada. *Journal of Psychology,* 116: 215-226.

Popper, K. 1978. *Conjectures and Refutations.* London: Routledge & Kegan Paul.

Reinberg, A., and M. H. Smolensky. 1983. *Biological Rhythms and Medicine: Cellular, Metabolic, Physiopathologic and Pharmacologic.* New York, N.Y.: Springer-Verlag.

Rotton, J. 1982. Review of "Full Moons" by Paul Katzeff. *Skeptical Inquirer,* 7 (1): 62-64.

———. 1985a. Astrological forecasts and the commodity market: Random walks as a source of illusory correlation. *Skeptical Inquirer,* 9 (4):339-346.

———. 1985b. Time-series analysis: A tool for environmental psychologists. *Population and Environmental Psychology Newsletter,* 12 (1): 15-19.

Rotton, J., and I. W. Kelly. 1985a. Much ado about the full moon: A meta-analysis of lunar-lunacy research. *Psychological Bulletin,* 97: 286-306.

———. 1985b. A scale for assessing belief in lunar effects: Reliability and concurrent validity. *Psychological Reports,* 57:239-245.

———. 1985c. The lunacy of it all. Unpublished manuscript, Florida International University, North Miami, Fla.

Rotton, J., I. W. Kelly, and J. Frey. 1983. Detecting lunar periodicities: Something old, new, borrowed, and true. *Psychological Reports,* 52: 111-116.

Rotton, J., R. Heslin, and B. F. Blake. 1983. Good news is no news: Some determinants of preattribution information search. *Representative Research in Social Psychology*, 13: 49-56.

Rotton, J., I. W. Kelly, and P. Elortegui. 1985. *Belief in lunar effects: Known group validity.* Manuscript in preparation.

Russell, G. W., and M. Dua. 1983. Lunar influences on human aggression. *Social Behavior and Personality*, 11: 2, 41-44.

Russell, G. W., and J. P. de Graaf. 1986. Lunar cycles and human aggression: A replication. *Social Behavior and Personality* (in press).

Salmon, W. C. 1984. *Logic*, 3rd ed. Englewood Cliffs, N.J.: Prentice-Hall.

Sanduleak, N. 1985. The moon is acquitted of murder in Cleveland. *Skeptical Inquirer*, 9 (3): 236-242.

Shapiro, J. L., D. L. Streiner, A. L. Gray, N. L. Williams, and C. Soble. 1970. The moon and mental illness: A failure to confirm the Transylvania effect. *Perceptual and Motor Skills*, 30: 827-830.

Snyder, M., and W. B. Swann, Jr. 1978. Hypothesis-testing processes in social interaction. *Journal of Personality and Social Psychology*, 35: 656-666.

Soyka, F. 1977. *The Ion Effect*. New York: Bantam.

Startup, M. J., and R. J. H. Russell. 1985. Lunar effects on personality test scores: A failure to replicate. *Personality and Individual Differences*, 6 (2): 267-269.

Templer, D. I., M. Corgiat, and R. K. Brooner. 1983. Lunar phase and crime: Fact or artifact. *Perceptual and Motor Skills*, 57: 993-994.

Templer, D. I., D. M. Veleber. 1980. The moon and madness: A comprehensive perspective. *Journal of Clinical Psychology*, 36: 865-868.

Templer, D. I., D. M. Veleber, and R. K. Brooner. 1982. Lunar phase and accident injuries: A difference between night and day. *Perceptual and Motor Skills*, 55: 280-282.

Watson, P., and P. N. Johnson-Laird. 1972. *Psychology of Reasoning: Structure and Content.* Cambridge: Harvard University Press.

PAUL KURTZ, JAMES ALCOCK, KENDRICK FRAZIER,
BARRY KARR, PHILIP J. KLASS, and JAMES RANDI

Testing Psi in China:
Visit by a CSICOP Delegation

Introduction, by Paul Kurtz

Five members of the Executive Council of the Committee for the Scientific Investigation of Claims of the Paranormal and one member of the CSICOP staff spent two weeks in China (March 21 to April 3, 1988). We were invited by China's leading scientific newspaper to appraise the state of psychic research and the extent of paranormal belief in China and to offer critical scientific evaluations where feasible. Our hosts were Mr. Lin Zixin, editor-in-chief, and other members of the editorial staff of *Science and Technology Daily.* Our group visited Beijing, Xian, and Shanghai, where we lectured at large public meetings and seminars and conferred with scientists, scholars, and journalists, including influential scientific critics as well as defenders of paranormal claims. While we were there, we carried out a number of tests of various subjects and claimants.

We are grateful for the gracious hospitality of our hosts, the openness and candor with which the meetings were conducted, and the many fine banquets and tours that were arranged.

Belief in the paranormal in mainland China has been growing rapidly in recent years. Spiritualism and psychical research had been pursued in China, as in the West, in the 1910s and 1920s. There had even been a Chinese psychical research society in Shanghai. But from 1949, when Marxism was officially installed by Mao, through the cultural revolution of the 1970s, China was cut off from the outside world. During this period, the ideological competition between idealism (which spiritualism and psychic research were viewed to be) and Marxist materialism led to the suppression of paranormal beliefs. It has been only since 1979, when greater freedom was permitted, that parapsychological influences again began to be felt in China, though the forms these beliefs have taken are in many ways unique to Chinese culture. Interestingly, belief in the paranormal has had a field day since then, and there has been very little public criticism.

Paranormal beliefs have taken two main forms: First, many people claim to have special "psychic" powers. Reports have filtered out to the West about so-called paranormal children who proponents claim are able to read Chinese characters written on bits of paper and placed in their ears, under their armpits, or even under their rumps—presumably a demonstration of ESP. These children are also said to be capable of psychokinesis (PK), in that they allegedly break matchsticks or repair broken ones by the power of their minds. Other "supermen," as they are called, are claimed to have a wide range of "psychic" abilities. They can supposedly extract vitamin C pills from a bottle while the sealed cap and bottle remain unbroken. Others are said to be able to open locks hidden in a box, move objects, make clocks run faster or slower, and/ or bend forks (à la Uri Geller). Many public demonstrations of these powers have been presented.

A book entitled *Wojiao*, or *Chinese Supermen*, edited by Zhu Yiyi and Zhu Kunlong, published in 1987, promotes paranormal powers and has sold 356,000 copies. Even one of China's most distinguished scientists and its leading rocket expert, Qian Xue Seng, formerly a professor at Caltech and chairman of the China Association of Science and Technology Societies, has been impressed by these demonstrations. And *Nature*, one of China's science journals, published in Shanghai, has carried articles supporting the reality of paranormal phenomena. The incompatibility of such claims with Marxist ideology has been circumvented by categorizing such research as "physiological." Extraordinary Functions of the Human Body (EFHB) societies have sprung up all over China, and they generally have supported psychic claims.

The second area of belief that has enjoyed considerable popularity of late and seems to be growing is the use of *Qigong* (pronounced "chi-gung") to treat certain illnesses. Qigong is a form of traditional Chinese medicine going back more than 2,500 years and is based on the theory of "meridians," undefined channels in the human body through which flows the fluid or gas known as *Qi*. There are two forms of Qigong: internal Qigong, in which a person practices deep breathing, concentration, and relaxation techniques; and external Qigong, in which the Qigong master is said to be able to affect and cure others. It is claimed that with external Qigong a kind of energy or radiation is emitted from the fingertips that can cure and/or prevent illnesses. Among the various maladies Qigong masters can allegedly heal are hypertension, neurasthenia, circulatory problems, glaucoma, asthma, peptic ulcers, tumors, and cancers. Qigong is practiced throughout China in many traditional hospitals and institutes of medicine. A marriage of psychic powers and Qigong occurs in such places, as masters use alleged psychics to diagnose illnesses by seeing into a person's body without the use of expensive X-ray machines. During the cultural revolution, the Gang of Four attacked Qigong, but a movement is now under way to restore respectability to this "treasure" of Chinese culture.

China is now making massive efforts to catch up with the rest of the world. Thus the key word is *modernization,* and high on the list of priorities is expansion in science and technology. Some of China's most distinguished scientists suffered repression during the cultural revolution, and many were sent out into the countryside to work. Scientific research languished during that period.

Are parapsychological and paranormal studies part of the new frontiers of science as some proponents in China maintain? Have there been significant breakthroughs in this area, or is this research, conducted in the name of science, simply pseudoscience? We were asked repeatedly if it was true that the CIA and the KGB are studying psychic phenomena for use by the military. If so, some wonder, will China be left behind in the psychic arms race?

How does traditional Chinese medicine—which includes herbal remedies and acupuncture as well as Qigong—compare with Western medicine? Can the claims of Qigong be validated? Many scientists in China are skeptical about these practices and see the need for scientific evaluation and criticism, but many others resent any Western involvement at all in traditional Chinese culture.

In any case, a number of Chinese scientists deplore the growth of irrational belief and welcome critical scientific investigation and skepticism—not on ideological grounds, but purely in terms of the quality of the research and the evidence. The only book critical of paranormal claims that has been published in China in recent years is *Psi and Its Variant—Extraordinary Functions of the Human Body* (1982), by Yu Guangyuan, and it had a very small circulation. Recently, however, excerpts from books and articles by skeptics, principally from the *Skeptical Inquirer,* have been translated into Chinese, though they are read by a limited and mostly scientific audience.

Fortunately we were given the opportunity to conduct a number of preliminary tests of various subjects who claimed to have special powers. Other Western scientists and reporters have visited China in recent years and some have been impressed with the demonstrations of "psychic" abilities they observed. Ours is one of the very few efforts of scientific testing by Western scientists in collaboration with Chinese scientists.

The following report is an account of the highlights of these tests. It is not an official report of CSICOP, but only of the individuals who took part in the tests.

As we left for home, the Chinese scientists expressed gratitude for our visit. They are in the process of translating other articles and books by skeptics and said that they would continue to do so. They also indicated that they hoped to form a Chinese Society for the Scientific Investigation of the Paranormal. CSICOP looks forward to working with them.

Preliminary Testing

While in the Chinese capital of Beijing, members of the CSICOP team lectured at the Institute of Scientific and Technical Information of China. We held several seminars and two public meetings. Paul Kurtz spoke on the history of paranormal and parapsychological research; Ken Frazier, on the recent National Research Council report on parapsychology; and James Alcock, on "The Psychology of Extraordinary Belief." James Randi demonstrated psychic surgery and "psychokinetic feats," and Phil Klass spoke on UFOs.

Informal polls taken by James Alcock of the public audiences of 300 to 350 indicated that approximately 50 percent of those present believed in psychic phenomena—about the same as in North America. The audiences showed little interest in UFOs. We asked how many believed that UFOs were extraterrestrial, and the response was fairly low.

James Alcock and Ken Frazier also spoke at an informal seminar at the Institute of Psychology of the Chinese Academy of Sciences, and Phil Klass lectured at the Beijing University of Aeronautics and Astronautics.

The CSICOP team was given two special demonstrations under the auspices of the Beijing College of Traditional Chinese Medicine. Approximately 75 people crowded into a classroom. We were given an informal demonstration of the powers of a young "psychic" woman. It was claimed that she was able to diagnose illnesses by seeing into a subject's body. She was asked to demonstrate her abilities with two members of our group. She announced that Phil Klass had an irregular heartbeat and that James Alcock had gallbladder trouble. According to Klass and Alcock, these diagnoses were incorrect. She was mildly embarrassed when Klass said that his health at age 68 was fine and that he frequently takes skiing vacations that would be impossible if he had such a heart problem. At one point, the psychic said she had been able to see into Jim Alcock's jacket and saw three pens in his pocket. When told he had only two pens, she shifted and said she must have been looking at Phil Klass. Klass *did* have three pens in his inside pocket. Since only minutes earlier she had asked Klass to move to a seat closer to her, she might have seen the pens inside his jacket as he did so. Next she asked Barry Karr, Paul Kurtz, and three others to stand on the stage in a relaxed position for ten minutes with their eyes half shut. As they did this, she stood in front of them and explained that she would direct her energies into them by a series of hand motions. She said this would make them "feel better." None of the subjects noticed a difference.

The most vivid demonstration was by a Qigong master, Dr. Lu. He placed one of his patients on the table in the front of the room. She lay on her stomach, facing away from Dr. Lu. He said that she had been suffering for 11 years from a lump on her lumbar vertebrae disk. He claimed that after he treated her the lump was reduced in size and her pain had lessened. Dr. Lu stood about eight feet behind the patient and began a rhythmic motion of his arms, which he continued for several minutes. The patient meanwhile began to move on the table. Sometimes her movements were slow and measured; at other times they were violent and convulsed. Dr. Lu maintained that Qi was emanating from his fingertips and that the patient's movements were in response to his efforts. He told us that he did not have to be in the same room with her for his power to work, that it would operate between walls and over some distance. He also informed us that 15 seconds was an adequate amount of time to transmit his Qi.

He agreed to let us conduct a test. Our design was simple: Lu would be in one room, the subject in another; the windows of the room would be shielded; and the Qigong master would begin his rhythmic motion upon our command. The test included ten 3-minute trials. For each trial, a coin was thrown into the air by a Chinese observer. Heads meant the master would attempt to transmit his Qi for the first 15 seconds of the trial; if it turned up tails, he would remain seated and not attempt to influence the subject.

James Randi, James Alcock, Barry Karr, and three Chinese witnesses were in the room with the Qigong master. In an adjacent room were Paul Kurtz, Ken Frazier, Phil Klass, and the subject, also accompanied by witnesses. Our watches were synchronized. During each three-minute interval we observed the subject's movements and kept meticulous notes. When the patient did move, we would ask her to stop after 15 seconds. After the test, we correlated the toss of the coin and the behavior of the patient. During one stretch of time, the coin came up tails four times in a row; this meant that the Qigong master did not transmit his Qi for 14 minutes and 45 seconds. However, the subject writhed during the entire session quite independent of what the Qigong master did. The only two trials during which the subject did not move were trials in which the coin had turned up heads and Dr. Lu was attempting to influence her. The results of these preliminary tests thus showed no significant correlation between the subject's movement and the Qigong master's efforts.

Although attempts have been made to detect infrared radiation emanating from the Qigong master's fingertips, no one had ever conducted the simple experiment that we devised to determine if there was a correlation between the movements of the Qigong master and those of the patient. How then can we account for the movement of the woman in the demonstration that took place when she and the Qigong master were in the same room? We reasoned that in the context of their roles of master and patient, both knew what was expected of them. They both believed in the power of Qi to make the woman move, hence she moved. It was clear to us that Dr. Lu's movements *followed* those of the patient when they were tested in the same room.

At a special meeting with philosophers and scientists at the Institute of Scientific and Technical Information, Paul Kurtz, James Randi, and Barry Karr were able to view videotapes—not widely distributed in China or in the West— of 16 children being tested for paranormal abilities. Two of these tests were conducted at the Medical Institute of Sichuan Province on January 14, 1982, by Professor Xu Ming Ding. Chinese characters were randomly selected from a dictionary as targets. Each character was drawn on a piece of paper and then folded and put into an envelope (any tampering with the envelope could be easily detected) or put into a matchbox (with a thread inside that would break if the box was opened). In the first test, using envelopes, 16 children were tested. Eleven of the children were unable to provide the correct answer; their envelopes were found not to have been tampered with. Four children were able to identify the correct target; in each of those four cases, there was clear evidence that the envelopes and targets had been tampered with. One child was able to draw half of the target character. However, the piece of paper had been pulled partway out of the envelope, uncovering that same half of the target. This was also clear evidence of cheating.

In a second test, using the matchboxes, the four children who had scored correctly in the first test were reexamined, but this time under the direct scrutiny of two video cameras. Two of the children were able to make the correct guesses, but the threads on their matchboxes had been broken and on the videotape they could be clearly seen peeking inside the box. The other two children, under close scrutiny, did not cheat and got no positive results.

The Chinese investigators concluded that there was no evidence that the children tested had powers of clairvoyance, for in every case where the child guessed correctly (only half-correct in one case) the sample had been tampered with in such a way that the target characters could have been seen by the subject.

Summary

Test I

11 children	negative results	targets not tampered with
4 children	positive results	unambiguous evidence of cheating
1 child	one-half positive result	unambiguous evidence of cheating

Test II

| 2 children | negative results | targets not tampered with |
| 2 children | positive results | unambiguous evidence of cheating |

While in Beijing, the CSICOP team was approached by several psychics and Qigong masters, or their emissaries, who said that they would like to be tested by our group. We specified that the tests would be conducted under rigorous conditions. In three cases the subjects and masters failed to appear. For example, a representative of an Extraordinary Functions of the Human Body group in Beijing said she would bring two gifted children to be tested the next day, but they never came. Another time a Qigong master, who allegedly could pull pills from an unopened bottle, never materialized. The latter had been highly recommended to us as someone with genuine psychic ability.

Sometimes, though, the subjects did keep their appointments. One evening a medical doctor who was also a Qigong master came to our hotel room. He displayed a large news clipping from a German newspaper reporting an extensive interview with him. He also said that he had been tested by an Italian doctor. The main focus, however, was his younger sister. He claimed she was a psychic who could diagnose illnesses by looking into the subject's body. Again clairvoyance was being claimed, though Qigong was invoked as an explanation. He further stated that she could correctly diagnose the physical conditions of our relatives back in the United States and Canada if we just supplied their names and relationship. She stressed that she was not reading minds, but was actually able to "see" the person and what they were doing at that particular moment. A test was agreed upon. The names, relationship, and state of health of persons known to us were written down on separate pieces of paper, which were then folded and put on the floor in the center of the room. These names and illnesses could be confirmed, if necessary, by a telephone call.

The diagnoses then began. (1) James Randi wrote the name of his sister, who had been suffering from breast cancer but had been effectively treated for the condition. In our opinion, the woman appeared to be using a typical cold-reading technique. She said there was no major illness. She mentioned everything from anemia to insomnia, but not breast cancer. (2) The second subject was James Randi's mother. The psychic talked about liver trouble and arthritis in her legs. She said that she could at that moment see his mother talking to someone. The results were again negative. James Randi's mother had been dead for two years. It should be noted that at one point during the test the psychic was asked by an observer (who should not have asked) if Randi's mother were alive. The woman said "Yes." (3) The psychic was then asked by Randi to diagnose him on the spot. She said he had trouble with his neck, which was incorrect. She did not mention that Randi had suffered a lower back injury last year. (4) The last subject was Paul Kurtz's daughter, who has diabetes. The psychic said she was pregnant, which was not true.

At that point the test was concluded. We asked the Qigong master's permission to publish the preliminary results of the test. He became very upset and insisted we not use his name or that of his sister. He explained that, because his sister's powers were so great, they were supposed to be kept secret. Though not convinced of her powers, we agreed not to use their names.

*　　*　　*

Xian has become a tourist mecca ever since the recent remarkable discovery of thousands of terra-cotta warriors and horses entombed more than 2,000 years ago by an emperor of the Qin Dynasty. In Xian, the CSICOP team again met with officials of the local Qigong institute. They too maintained that Qigong was able to cure people of a variety of illnesses. China, they said, had given the world five great discoveries: gunpowder, the compass, printing, paper, and Qigong. They referred to the theory of "meridians" to explain Qigong. Some people are especially meridian sensitive, they said. They also related Qigong to the four phases of the moon.

They presented us with two young women, both students at the police academy, who they said were able to see into a subject's body by psychic means to diagnose illnesses. James Randi proposed a test. The girls agreed to demonstrate their powers on the six members of our team by telling us whether our tonsils and appendix were missing or intact. All had had their tonsils removed except Randi, and none had his appendix removed. We wrote down our actual conditions in advance of the diagnoses. The girls thus had to make 12 decisions. In two instances they declined to make a choice. Of the remaining 10 cases, they called 5 right and 5 wrong. When we pointed out that their score was the same as could be expected by chance, they became upset. Earlier they had said that they were invariably correct.

By far the most interesting tests we conducted while in China were in association with Mr. Ding Wei Xin, secretary-general of the Xian Paranormal Function Application Association and editor of the magazine *Paranormal Function Probe*. Mr. Ding emphasized that his organization is unique in China, since it is concerned with the *practical* applications of the paranormal. A former journalist with an academic background, he said he was the first person to make public the tests of children with clairvoyant and PK powers, and claimed to have trained more than 100 people who possessed paranormal powers. The most practical use of these abilities, he said, was in diagnosing illnesses, which could save a considerable amount of money. He boasted that psychic diagnoses were far more accurate than X-rays and could "match CAT scans." The people who work with him, he said, had diagnosed several thousand patients by psychic means. He told us that psychics are also helpful in locating natural resources and in archaeology. Paranormal powers had been used, he said, in searching for criminals, and psychokinesis (PK) could help remove kidney stones and gallstones. Mr. Ding's claims were astonishing to us, but he proposed providing various subjects for testing.

Professor Fan Yu Lin, president of the group, said in a formal talk that he believed the paranormal phenomena produced by the children were genuine and he tried to explain this as a product of the evolutionary process.

We were in a crowded room with about 80 people. In the rear were several of the "psychic" children, all about 11 years old. Mr. Ding brought forth two young girls for us to test. He first claimed that they could read characters in sealed envelopes. James Randi produced a number of envelopes that had been prepared beforehand, each containing one randomly chosen Chinese character. The children were permitted to hold the envelope in only one hand and for some 20 minutes they were under constant scrutiny by those present. The results were negative and the children became very distressed. Mr. Ding was evidently surprised. The next test, he said, would be of PK.

He brought up two other young girls and provided two matchboxes. Inside one box was placed a green match, which we insisted be marked on all sides with two red stripes. Mr. Ding said one of the girls would break the match using her psychic power. A second match, similarly identified, was broken into several pieces and inserted into the second box. He asserted it would be restored to its original condition by the other girl. Again our controls were stringent. The boxes were marked on the inside and outside with the initials J.R. They were then sealed with tape. The girls could only touch the box with one hand or lay it on the floor in front of them. They could not remove it from sight. In evident distress at these strict conditions, they were not able to perform, and these tests were also negative. Mr. Ding admitted that he had never before marked the matches.

Mr. Ding was himself dismayed at the negative results. He implied that they might be due to the fact that the children were nervous and under the scrutiny of too many people. He agreed that we should continue testing the next day in our hotel under quieter conditions. That evening Mr. Ding held a huge banquet at which we were able to meet and dine informally with six of his prime subjects, including two young boys. We had a wonderful time and ended the night by disco dancing with the youngsters. The atmosphere was friendly and the attitude was positive, something that Mr. Ding had insisted was necessary for proper testing the next day.

The following morning the testing resumed in the People's Hotel in Xian. Present were Mr. Ding, three young girls (and later two boys), four members of Mr. Ding's group, the CSICOP team, Mr. Lin (our host in China), a reporter for the *Science and Technology Daily,* and a translator. We designed the protocol for the first test for PK. There were three empty matchboxes. An unbroken toothpick was placed into box "B1." The box was closed and sealed with white medical tape. Another toothpick was broken into three pieces and placed into box "B2," which was also closed and sealed with medical tape. An unbroken toothpick was placed into box "B3," and it too was sealed with medical tape.

We conducted the tests under the most stringent conditions: The tooth-picks had all been indelibly marked for identification and the boxes were tightly wrapped with medical tape, numbered, and photographed. The children were not permitted to remove the boxes from our sight. They could hold a box with only one hand or put it on the floor in front of them. Most of the session was videotaped, as was the previous day's session, and we noted that Mr. Ding was gazing absently out of the window and took little interest in our procedures. Under such conditions, the girls were unable to produce positive results—when the boxes were opened, the toothpicks were in the same condition as before.

Mr. Ding again voiced surprise. The children were usually successful, he said. We then told him he could conduct the next experiment as he wished and we would simply observe his method of testing. So we began anew. This time there were four subjects—three girls and a boy. For this experiment

Mr. Ding used cellophane tape and matches with *green* heads. He also wrapped the boxes in paper. The procedure was the same as before, though Mr. Ding told us he would tape the boxes even more tightly. The first box ("A") contained broken matchsticks; the second box ("B"), a whole match; the third box ("C"), a broken matchstick; and the fourth box ("D"), three matches. He said the children would restore the broken matches or, if whole, break them by the power of their minds. Mr. Ding kept no records of any kind; nor did he mark any of the matches.

Under Mr. Ding's supervision, bedlam broke out. After a few moments, the children darted from the room with the matchboxes in their posssession. They ran up and down the stairs, in and out of the elevator, inside and outside the building. Mr. Ding saw nothing wrong with this. He did not bother to count the remaining matches, nor did he take notes as we had done of all the previous tests. The children went in and out of the room several times. At about noon, after an hour and a half of running around, the children sat quietly in their chairs for 15 minutes in an attitude of concentration. They then said they were tired, and Mr. Ding was not confident that they had been successful. The children asked if they could leave the hotel grounds with the boxes. Mr. Ding said yes, and proposed that we meet again at eight o'clock that evening. We were shocked at this loose protocol; but it was Mr. Ding's test, so we agreed.

That evening they all returned. We were told that one of the boxes, box "D," had been accidentally destroyed; it was not returned. We then proceeded to examine the other three.

The outer wrapping paper and tape on boxes A and B did not appear to have been tampered with or unsealed. When we opened them, the matches were exactly as they had been before. They had not changed. Box "C" was a different matter. Although somewhat the worse for wear, the box at first appeared not to have been tampered with. *But on closer inspection it was clear that the tape had been unwrapped and removed;* vegetative matter (most likely grass) and a strand of hair were found *under* the cellophane tape. We opened the box. It had previously contained five broken pieces of a match with a green head. Now we found an entirely intact match, but it had a *red* head! Moreover, we discovered that the girl in the experiment had given the matchbox to the two boys who returned it, but who had not even been part of the experiment. Mr. Ding apparently saw nothing wrong with this.

We ruled that there was obvious evidence of tampering and that cheating had taken place. Although Mr. Ding now admitted to us that some of his children had cheated in the past, he maintained that many such cases were genuine. Unwilling to admit that a child had cheated in this case, he argued that there may have been vegetative matter on the table when the matchbox was wrapped. This was simply not so. Moreover, he rationalized that the green matchstick had been miraculously changed to a red one. He reached into his brief case and produced a match, carefully wrapped in paper, which appeared to have both a red head and a green head. This, he said, had been produced by one of the young boys who had returned it; and he affirmed that this indicated an even more surprising power of psychokinesis! Later, when we confronted the two boys individually about what had happened, we got contradictory stories. One even blamed his father, who he said had told him if he could restore a broken green match he could just as easily change it into a red one!

What may we conclude from this fiasco? It was apparent that Mr. Ding was extremely naive and that he was unable to design a simple controlled experiment to detect fraud. Obviously the children were playing games and doing so with impunity. Yet Mr. Ding attributed their feats to a psychokinetic effect.

*   *   *

The last two days of our sojourn in China were spent in Shanghai, a city of faded elegance with very little new construction other than some tourist hotels. Here we met with the faculty and staff of the Shanghai College of Traditional Chinese Medicine. This group seemed most receptive to our suggestion that rigorous tests be made of Qigong. To our surprise, they were completely unaware of the importance of double-blind trials and expressed enthusiasm about doing such tests in the future. They focused primarily on internal Qigong, a form of relaxation that they claimed could reduce hypertension and have a beneficial effect on other illnesses. They were somewhat skeptical of external Qigong, where a master seeks to induce changes in a patient's health.

We also met at the Dong Hu Guest House with some skeptical scientists and philosophers who said that they had done tests with psychics and Qigong masters with invariably negative results. Unfortunately, they confided, the press was more interested in reporting the fantastic claims of paranormalists than in the more mundane, skeptical critiques. This, we noted, is similar to what occurs in other countries.

Conclusion

Our preliminary testing of various children, psychics, and Qigong masters produced negative results. The tests were recorded by still cameras, audiotape, and videotape.

Editor's Postscript

The Tiananmen Square massacre of June 1989 and subsequent government repressions changed everything in China. The new freedoms and openness that had led to interest in democracy and Western ideas were officially squelched.

So too was the accompanying rising public fascination with the paranormal, undoubtedly now merely driven underground, still further from scientific scrutiny. The political events also affected the hosts of our visit. Lin Zixin, editor of *Science and Technology Daily*, editorially supported the student demonstrations of May 1989 and was subsequently relieved of his duties. He had been scheduled to come to the United States for a series of talks, visits, and conferences but was prevented from leaving China. Another of our hosts soon after our visit came to the U.S. for graduate study. He returned to China to take part in the May 1989 pro-democracy demonstrations and was subsequently detained by authorities for three weeks before being released and allowed to return to the United States.—*K.F.*

Part 6: Medical Controversies

KARL SABBAGH

The Psychopathology of Fringe Medicine

Fringe medicine shares much common ground with the paranormal. First, there is often a *desire to believe* in the therapies and the practitioners of fringe medicine, just as there is a *desire to believe* in the paranormal. Second, it is often difficult to tell the difference between a fringe therapy and a paranormal claim. Astrology, palmistry, phrenology, and psychic surgery all crop up in the guise of alternative therapies and display themselves side by side with osteopathy, acupuncture, and homeopathy, all asking to be believed in for reasons that have nothing to do with the results of any investigations.

But why *do* so many people believe in the effectiveness of fringe medicine? In discussing this question, we will be exploring the scientific disciplines of psychology, physiology, and anthropology—and gullibility.

The question can be broken down into two further elements: How does fringe medicine work when it does work? And why do so many people believe it does when it doesn't?

I have used the word *psychopathology* in the title because in *my* dictionary, one of the meanings of the word is "a study of mental functions under conditions brought about by disorder or disease." Some of my analysis deals with people's mental functions when they are seeking help for disorder or disease; and I believe that they think pathologically, as I will suggest later.

When I say fringe medicine sometimes works, I mean that every day people are feeling better after some fringe treatment or other. The question is, Why do they feel better? Is it a genuine improvement in their physical condition due to the specific effect of the treatment, as practitioners claim? Or is it for much more complex reasons that have to do with human psychology, perceptions of probability theory, the physiological links between mind and body, and the natural variability of disease? In looking at fringe medicine I will be suggesting that, when it works, it works for none of the reasons given by fringe practitioners themselves.

Some of the most dramatic and convincing examples of "cures" by fringe medicine come about when people with undeniably serious physical illness get better while undergoing some fringe therapy. What other explanation could there be, particularly if conventional medicine appears to be only of limited success?

A few years ago, an American cancer-researcher presented a convincing analysis of this by no means rare event. Emil J. Freireich, of the M. D. Anderson Hospital in Houston, Texas, offers what he calls the "Freireich Experimental Plan," which enables anyone to set himself or herself up as a therapist and is "guaranteed to produce beneficial results." Although it's a tongue-in-cheek analysis, I think Freireich's plan tells us a lot about how an impression of effectiveness can come about in fringe medicine.

There are two essential requirements. The first is a *treatment* of some sort; it doesn't matter what. It could be some new form of psychotherapy or a really impressive physical procedure, some type of rubbing or hand-waving or a mechanical device of some sort, or the administration of some substance, a drug, a plant, or a chemical. The second requirement is that whatever treatment you choose must be absolutely harmless.

Starting from these two conditions, Freireich shows how almost any fringe technique can lead to a situation where any outcome can be interpreted as confirming its success. The key factor in his analysis is the natural variability of all disease. Figure 1 shows the typical course of a serious illness. Freireich points out that every disease, acute or chronic, has important periods of remission, the ups on the graph. At those times, a patient feels better than he has done for some time, and in fact he actually *is* physically better, by all objective measurements. This is true even if there is an inexorable trend downward, and it is even truer, of course, in the case of diseases that are not potentially fatal. "There is no disease I know," says Freireich, "where inevitable and continuous progression is the universal characteristic." On the basis of these observations, Freireich has devised a schedule for the budding fringe-therapist. The first rule is that treatment should only be applied to a patient after a period when he has been getting progressively worse. If you apply the treatment during one of the *ups* and the patient continues to improve, he can always say he would have got better anyway. If treatment is applied when the patient is getting *worse,* four possible things could happen. First, the patient could start to improve. Natural variability will ensure that this possibility is always present. If it happens, it immediately "proves" that the treatment is effective. The second thing that could happen is that the disease remains stable. This *also* "proves" that the treatment is working, because it has arrested the disease. What is needed now, says Freireich, is the application of the treatment at a higher dose, or for a longer period. This will of course cause no harm, because the treatment is harmless. A third possibility is that the patient continues to get worse. However, the practitioner need not be at all put out by this, even if the patient is, because this can be taken to mean that the dosage was inadequate and must be stepped up, or that the treatment hasn't been taken long enough. The fourth, and saddest, outcome is that the patient dies. Even in this case, the good fringe-practitioner need

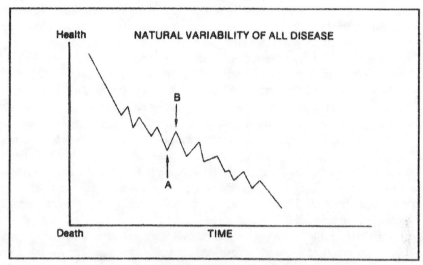

FIGURE 1. When to treat (A) and when not to treat (B).

not accept defeat. The death is an indication that the treatment was delayed too long and applied too late—"if only he'd come to me sooner." This also means that the case need not be included in the overall analysis of the results.

So far, so good. If you are a therapist following the Freireich Experimental Plan and have applied it to a number of patients, you have already had *some success*—the ones who started to get better. But you have still got a number of patients who stayed the same or continued to get worse, whom you are continuing to treat.

It is now time to move on to phase 2. Here you have to treat the patients who remain in the same way as you treated the patients in phase 1—applying your therapy *only when they have got worse*. There will be some more successes, and there will be some patients who need to continue with more intensive treatment. As you apply these procedures stage by stage, you will end up with two groups of patients—the ones who have shown objective improvement (apparently thanks to your treatment) and those who are dead, who can be excluded from the study because, in spite of all your efforts, their late arrival meant that they were beyond saving.

There's a third and even more ingenious stage that helps to confirm the effectiveness of the treatment among those patients who are showing improvement. Once the patient improves you *must reduce* the dose. This then produces two possible outcomes: (1) The patient continues to show improvement, which proves how effective the treatment was. (2) The patient stabilizes and stops improving or even starts to get worse again. This is *also* proof of the effectiveness of the treatment, because reducing the treatment has obviously made the disease active again, "proving" that the treatment was responsible for the earlier improvement. In fact, it is a well-known principle of homeopathy that when you apply a remedy the patient will often get worse and that this is a proof that the remedy is just right for the condition and the patient you are treating.

There is actually a serious purpose behind Freireich's analysis. First, a lot of it rings true as an account of relationships between patients, doctors, and their diseases. But it also highlights a point that is an important component in the way we perceive treatment of illness—the tendency in all of us to ascribe cause and effect where none exists. There is no doubt that much disease *does* behave in the way Freireich describes, and there's no doubt in my mind that natural variability leads to fringe medicine getting credit for outcomes that would have occurred anyway. To anyone who has read or observed fringe medicine at work, the Freireich Experimental Plan seems to ring very true. Freireich's analysis applies to conditions where there's an *impression* of cause and effect between treatment and disease but where in fact the treatment is having no effect at all.

There's another group of conditions that seem to benefit from fringe medicine; and this is not really surprising, because they probably do not exist at all as clinical entities. These are the half-perceived discomforts or anxieties that come and go in all of us for all sorts of reasons and that fringe practitioners elevate to the status of a disease by labeling them. Here's a list of illnesses taken from a recent compendium of fringe remedies: diffuse aches, dandruff, loss of appetite, insomnia, apoplexy, dislike of being alone, "doormat tendency," pins and needles, numbness sensations, lassitude, neuromuscular difficulties, a tendency to lymphatic stagnation and weakness in the lymphatic cleansing system, congestion, sluggish liver, constipation, hiccups; and, there is one other group of patients who can be cured by fringe medicine, "those who feel they have set blocks on their own free emotional expression and want to go back . and evaluate their reaction to those blocks," whatever that means.

If we have one of these "illnesses" and if we take a homeopathic pill one day and feel less sluggish or apoplectic the next, we are ready to see cause and effect where none exists. It seems to be the case that, if B follows A, it is difficult for most people not to believe that A and B are linked in some way. ("Every time I wash the car, it rains shortly afterwards," or "The new Bishop of Durham was installed in York Minster and the following day lightning struck.") These beliefs are strengthened when we have no objective standard to assess likelihood. In the case of a particular disease, we do not know how likely it is that someone undergoing treatment would have got better anyway, for example. And, indeed, it is often difficult for anyone to make that calculation. The course of an individual illness depends on too many factors.

All we can really say, like Emil Freireich, is that people often get better or go into remission for no discernible reason when they have *not* been given acupuncture or homeopathic remedies. Most of us are just not familiar enough with probability figures or the natural history of disease to make the sort of informed judgments that apply in the assessment of therapeutic effectiveness. We are often in the same state of ignorance about physiology as my young daughter was about meteorology when she looked out of the window one windy day and said: "Those trees are certainly pushing the wind about."

We all have a tendency to look for patterns in the world and make links where none exist. If your constipation disappears shortly after someone has inserted a needle with great ceremony in a very specific part of your earlobe, few of us would doubt that one causes the other, even though constipation around the world gets better every hour with little or no insertion of needles into ears.

In fact, in addition to nondiseases, there are other situations where a range of conditions does genuinely benefit from contact with some form of fringe medicine. These are conditions that have a known psychosomatic component. It is interesting how the links between mind and body have been monopolized by fringe practitioners as if the subject of psychosomatic medicine had never existed. In fact, of course, the past 30 years of medical research have seen a very respectable body of research work into the links between psychological events and physical illness. But, although many fringe therapies place great emphasis on the indissoluble link between mind and body, they fail to draw the obvious conclusion—that many of the beneficial results of specific fringe therapies come about because of the belief in the practitioner, rather than because of any particular manipulation, preparation, or device.

Many fringe treatments include the relief of headache, backache, and general pain among the indications of their success, and they undoubtedly "work" in the sense mentioned earlier. In other words, people feel better; the pain is diminished or entirely absent after treatment. The important question is: Why? If your headache goes away after your toes have been manipulated by a warm friendly therapist who exhibits all the signs of caring and love that we all need so much these days, is it the manipulation that removes the headache or some other factor, such as belief in the therapist? If an iridologist looks deeply into your eyes and gives you a firm diagnosis of "lymphatic stagnation and weaknesses in the lymphatic cleansing system," it would be churlish of you and your body not to respond by feeling better afterward. But Occam's Razor leads us toward parsimony in the search for explanations in the world around us.

We have to choose an explanation for the following situation. A remarkably similar list of conditions all improve after diagnosis and treatment with a wide range of fringe therapies. There are two possible explanations. One is that *each therapy* is specifically responsible for the improvement in the patient treated. Mrs. A's headache improved because her ear was pricked; Mr. B's backache improved because he took a homeopathic remedy; Miss C's digestion improved because the energy zone in her navel was manipulated; and so on.

But there is a second possible explanation. *All* of these therapies could be effective because they all act in the same way. They create the same psychosomatic improvement through belief in the therapy or the therapist, enhanced by the full panoply of unusual devices and charts, pseudoscientific terminology, and single-minded concern shown by the therapist for the patient. To take one of the most common medical problems, headaches, there is strong evidence that 90 percent of the headaches presented to doctors are tension headaches, probably linked to some mental state of anxiety or depression.

This does not, of course, make them any the less painful or any the less physical. But if they *arise* for psychological reasons, is it not likely that they go away for psychological reasons? To summarize this particular point, I am suggesting that in this situation illness is cured *either* by foot manipulation *or* by belief in foot manipulation, *either* by homeopathy *or* by belief in homeopathy, but not by both. The belief in the therapy and the therapist, enhanced by the time and the close attention given by the therapist, leads to an effect on the mind that in turn produces an effect on the body, in a range of conditions that have been shown in the past to be related quite closely to emotional states.

All of this means that we must be very careful to assess the nature of the evidence offered for the effectiveness of fringe medicine. Any "proof" offered must include a rigorous definition of the condition in the first place, so that we know a real illness is being treated. It must also exclude effects based on the randomness of the human physiology and on the effect of the relationship between the practitioner and the patient. Experiments to establish such proof are not difficult to organize, in spite of the protests of the fringe practitioners. The trouble is that, when attempts are made, the strength of the proof is in inverse proportion to the rigor of the experimental controls. This is certainly the case with one of the fringe techniques that has been tested: iridology (*Skeptical Inquirer,* Spring 1983). This is a method of diagnosing disease using the patterns in the iris. It was developed and promoted by Ignatz von Peczely of Hungary.

Dr. von Peczely had a pet owl, and one day he accidentally broke its leg. Peczely then noticed something odd in the owl's eye—a black stripe had appeared on the iris where no black stripe existed before. This incident led Peczely to the belief that the organs and systems of the body are represented on the iris and that if anything is wrong inside the body a corresponding change will be seen in the iris.

Each part of the body is represented by a segment of the iris, with the right eye dealing with the right side of the body, and the left with the left side. One fringe-medicine text endorses iridology in the following terms: "As a diagnostic tool, iridology has several obvious advantages over other systems. Merely by studying the two irises, the therapist can obtain information about all parts of your body simultaneously. And he or she can not only deduce your present state of health but also predict potential trouble spots. In addition, iridology is completely safe"—remember Freireich.

By looking at the iris, the iridologist comes up with such diagnoses as "severe gastrointestinal toxicity and congestion, resulting in irritation in the bladder and lower back; underactivity in the ovaries and uterus; spasticity throughout the large intestine; sinus headaches and," not surprisingly after that list of problems, "depression." This is, of course, a good example of the category I mentioned earlier and includes several "nondiseases." But iridology also deals with well-recognized physical conditions, and its use with these can be, and has been, put to the test.

In one trial, published in the *Journal of the American Medical Asso-*

ciation, there were two very interesting results. One iridologist in the study correctly identified 88 percent of patients who were actually suffering from kidney disease, merely by spotting the relevant blotch on their irises. That is good news. The bad news is that he also diagnosed 88 percent of his normal controls as suffering from kidney disease, when they weren't. The more important part of the study presented six iridologists with iridology photographs of the irises of 143 subjects, 48 of whom actually had kidney disease. The overall results were no better than chance.

Another controlled clinical trial carried out in Australia presented iri-dologists with iris photographs before and after the subjects developed an acute disease and asked iridologists to determine whether a change in the iris had occurred and, if so, which organ was affected. The only photograph identified as showing changes was actually one of two photographs taken two minutes apart, as a control.

The significance of these results is not so much their negative quality but the fact that the studies were done at all. With a fringe technique like this, the studies are not difficult to design or control; nor are they expensive to carry out. They reproduced as closely as possible the diagnostic situation for the iridologists, who do use detailed photographs, and so it was difficult for someone who claimed to be a practitioner of the technique to refuse to take part.

These studies are an example of the kinds of tests the practitioners them-selves should be organizing if they are truly confident of the basis of their techniques. What have they got to lose? If they are right, they will prove it to a lot of skeptical doctors and patients; and, if they are wrong, they will save themselves a lot of wasted effort and disappointed patients in the future.

Unfortunately, negative evidence rarely has any impact on fringe prac-titioners, who sometimes seem more like priests than doctors, basing their convictions on faith rather than evidence. How otherwise can we explain re-marks like the following, made by seemingly sane and intelligent analysts of the fringe-medicine scene: Rogerian therapy has "no set techniques and no set procedures to follow," or "All teachers of t'ai chi emphasize the impossibility of describing it in print," "The Feldenkrais method is difficult to explain quickly and easily." As Feldenkrais himself said: "The first principle of my work is that there isn't any principle."

These are statements of such monumental vagueness that it can only be blind faith that leads anyone to take these therapies seriously. In fact, fringe practitioners seem generally unworried about the shortage of convincing expla-nations for the phenomena they so firmly believe in. If you look at the various theoretical bases offered for the different fringe therapies, you will find a bewildering variety of hypotheses about how the body really works, some of them contradicting one another. On the whole, this is not true of orthodox medicine. When a surgeon operates on the brain, he is dealing with the same organ that a physician might treat with drugs or a psychiatrist with ECT. They would all agree on roughly how many cells it has, how large it is, and the important physiological and biochemical factors involved.

When a reflexologist treats a body, he has an entirely different perception of how that body works from, say, an iridologist or an acupuncturist. This is clear from the impressive but meaningless charts that purport to show the roots of human disease according to various different fringe theories. Is kidney disease rooted in the bottom sector of the iris or the center of the sole of the foot, or several different points in the ear or perhaps at the lower end of this complex set of energy zones? Or could it even be, although this suggestion may be too revolutionary to consider, that kidney problems make themselves known in the kidneys?

Any scientific explanation of the occasional effectiveness of fringe medicine is more likely to lie in the realms of orthodox psychology and physiology than in the more exciting worlds of forces, energy fields, meridians, and vibrations. And this is where we come to "psychopathology." Because, in spite of the lack of evidence and even in spite of evidence against the validity of most fringe-medicine techniques, people still believe in them—their thought patterns seem to run along irrational lines and lead to totally invalid conclusions. This may be what I can only describe as the "ratchet effect."

Some of us are old enough to remember watches that you had to wind up by hand, with a little knurled knob on the side. Indeed they even had little moving bars, quaintly called "hands," instead of glowing rectangular numbers. In those far-off days of the 1970s, we wound our watches by turning the knob backward and forward between our fingers. When you wound it one way, it caused a spring to tighten—but, and here's the cunning bit, when you wound it the other way *it didn't cause the spring to loosen!* At work inside the watch to cause this miracle of a decidedly unreligious nature was a small mechanism called a "ratchet." Rotation forward increased the tightness of the spring but rotation backward *left* it tight, so that the next rotation could add a further increment to the tightness.

I believe that the human brain has a ratchet, and it is one that swings into place whenever people are confronted with something they really want to believe in. Whenever they come across an example of a phenomenon that reinforces the belief they are interested in preserving, the mainspring of their belief tightens a little bit. But, if a little later they come across something that doesn't reinforce the belief, something that even contradicts the hypothesis they are fondly nurturing, the wheel rotates in the opposite direction but the spring doesn't loosen—it's still as tight as it was, and their faith is unshaken.

A fringe treatment that coincides with improvement on one occasion out of ten will nevertheless leave a patient or even an observer gasping at its effectiveness. It is like the medium who says, "I'm getting your mother or possibly your aunt or it may be your grandmother, and her name is something like Edna or Betty or a name beginning with M or D, and she says she hopes that you had a good holiday by the sea or do I see mountains?" and the ratchet swings into action to produce a satisfied customer who is convinced that the medium conveyed an accurate message from her dead Aunt Mary about the family holiday in Majorca.

This tendency operates throughout the rich variety of human mental activity. In fact, I believe we have several ratchets. We have a "bad news ratchet," for example. "Oh, isn't the world a depressing place," we say as planes crash, bank-robbers rob, and earthquakes quake, each of these sad events tightening the mainspring of our depression. Meanwhile, Boy Scouts help little old ladies across the road, jumbo jets take off by the minute and fail to crash and burst into flames, and parents and children frolic joyfully in the park on family picnics that don't end in tragedy, and it has no effect on the depression. We certainly have a "there are more things in heaven and earth, Horatio" ratchet that clicks inexorably tighter with every rumor of spoons bent, distances viewed, or telephones that fly across rooms with only the gentlest nudge from a disturbed teenager. Meanwhile, as fakes are unmasked or confess, and conjurers duplicate their feats and perform even more amazing ones, our little ratchets protect the spring from slipping back and plunging us into a boring real world where, on the whole, usual things happen usually and impossible things don't happen at all.

I began by suggesting how fringe medicine might work. I hope that, at the very least, I have shown how difficult a question that is to answer, because fringe medicine itself is such a ragbag of separate practices with little in common other than the fact that people sometimes feel better after indulging in one or the other of them. In that sense, they come under the same category as winning a lottery or having an enjoyable night out. It seems clear that there are few, if any, situations where, if someone does feel better, it is for the reasons offered by the therapists. The final question is: Does all this matter? As Freireich pointed out, most fringe remedies are harmless—that's one of the secrets of their success. Does it matter if gullible practitioners continue to treat gullible patients? One reason I think it matters is that it seems to be a form of semi-institutionalized dishonesty, and on principle that seems to be wrong. But there is a second, more important reason: Real harm can occur. And there are two pieces of evidence, picked at random over the past few months, that suggest how. One relates to a child with leukemia whose parents refused chemotherapy and took her to a homeopath. The diet of fresh fruit juice and homeopathic pills unaccountably failed to do the child any good, and she died. The second involves a woman with miliary tuberculosis, a highly infectious form of the disease, who refused hospital treatment and went to a fringe practitioner instead. He diagnosed constipation and gave her a mixture of Epsom salts and herbs. The woman died, after a period at large during which she distributed TB-laden sputum among those around her.

You may believe that people should be allowed to kill themselves with the treatment of their choice. But, even if you take that excessively libertarian view, we are faced here with the death of a child, because of her *parents'* views and the presence in the community of someone who was a threat to the environment because of her own mistaken beliefs. Even these two incidents alone are too high a price to pay for the freedom to offer unproved and unscientific remedies to the general public.

FRANK REUTER

Folk Remedies and Human Belief-Systems

> My oldest brother had a toothache, and a man told us: You know them places that be on the inside of a mule's leg looks like it been a sore or something. He told us to get a pocket knife and trim some of that off and put it in a pipe and let my brother smoke it. And we did that and it stopped it.
>
> Folk informant, Dermott, Arkansas, 1976

The use of the "chestnuts" on the inside of a mule's leg to cure toothache may not be a very widespread practice, but it is a small part of what is historically the world's most extensive system of medicine: the folk remedy. Medical science as we know it is rather new. The knowledge that microorganisms exist and play a role in disease is little more than a century old. For most of history, folk remedies were humankind's defense against disease and pain. In large parts of the world, folk medical practices still predominate, and remnants of them can readily be found in the midst of our technological culture.

If the toothache remedy seems silly on first reading, it is probably because most of us seek out dentists to treat our cavities. You might ask yourself what you would do were you equipped with a pipe, a knife, and the remedy in the opening quotation, but no pain relievers or other known remedies, and you experienced an excruciatingly painful toothache while traveling in a remote area by mule. The efficacy of the remedy is guaranteed by the folk practitioner, who bluntly proclaimed that the aching "stopped."

If you would not turn to the remedy under those or any other circumstances, your mind set suggests a question: Given the fact that so many folk remedies seem preposterous to a public accustomed to visiting medical doctors, why did such a large body of folk medicine develop? And, more important, how could it have been deemed to be successful? the answer to the first question is simple: The anguish of suffering from injury and disease is so great that remedies had to be sought. If you spared the mule's calluses on your remote expedition and your tooth still ached, it is fairly certain that you would begin experimenting with other simple expedients—perhaps placing cool or warm water on the tooth or massaging the gums. We all reinvent folk medicine, however crudely, when we are in pain and without help or useful knowledge.

The answer to the second question is far more complex. Two theories, one physical, one psychological, are commonly used to explain the success of folk medical systems. Studies show that perhaps a third of folk remedies, even if they cannot cure the targeted ailment, offer some benefits to the recipients. Herbal teas, for example, may be high in vitamins that serve as catalysts in cell reactions and therefore promote the body's healing processes. People have occasionally stumbled upon natural antiseptics and antibiotics. As to psychological benefits: since much disease is psychosomatic—a person's attitude has much to do with the severity of a disease and with the ability to recuperate—any attention and care given to a patient, even though the physical methods used are improper, may be beneficial. Mothers who kiss a child's minor scrapes and wounds intuitively understand this.

These two theories have a measure of truth to them, but their combined force is inadequate to explain both the extent of, and the belief in, folk medical practice.

Folk medical practitioners, at least those who are isolated from modern medicine and do not have to fear being chastised for their failures, are certain of the efficacy of their remedies. It is, in fact, ironic that medical doctors with all their formal training are far more likely to admit that their practices can fail than are folk practitioners. An allergist, for example, in explaining the procedures for desensitizing a patient to an allergen, has to admit that the process is not always successful, that there are nonresponders for whom the expensive shots simply may not work. Such doubt is seldom found in a shaman or an herb doctor. Yet the pervasiveness of folk medicine and the confidence of the folk practitioners cannot be fully accounted for merely by vitamin-laced concoctions or by engaging bedside manners.

To unravel this mystery—how can "medical" practices that have a low probability of success become rigidified into a system that is considered successful—I shall allude to the practices of Dr. Benjamin Rush, a physician, social activist, signer of the Declaration of Independence, and, in spite of his Edinburgh medical degree, a proponent of folk medical-practice.

In 1793, Philadelphia, the already crowded capital of the newly formed American nation, absorbed a large number of immigrants from a slave revolution in Saint Domingue. These French colonials brought with them an unwanted guest, the infectious viral disease yellow fever. In the hot days of August, the disease spread quickly and panic ensued, causing much of the city to be evacuated. A number of courageous physicians, among them Benjamin Rush, elected to stay behind to minister to those who could not afford to flee. These physicians were severely hampered in their attempts to control the disease. There was as yet no knowledge of infectious microorganisms, and theories of the mechanism by which the disease spread, though hotly debated, were inaccurate.

Dr. Rush, a deeply religious man, suffered considerably because of his inability to help his patients, who, tortured by frightful symptoms, which included the vomiting of black fluids, were dying without relief. Rush frantically searched the printed literature for some promising cure. One manuscript caught his attention. John Mitchell, who had treated yellow-fever patients in Virginia a half-century earlier, recommended using extensive purges to clear the abdominal viscera of bilious humours. Most physicians, Rush included, had been using mild tonics, mild bleedings, and cool baths to try to cure the fever. But nothing seemed to work. Now Rush was confronted with the idea that the cautious and conservative practices of most physicians might be responsible for the low rate of recovery. No matter how weak the patient seemed, the "putrid miasmata" in the body had to be expelled. A physician could not afford to be faint-hearted: the weaker the patient, the stronger must the purgative be.

Though intrigued by the theory, Rush was understandably reluctant to try such a radical procedure. An opportunity arose for him to test it, however, when he came upon a man who was almost dead and for whom no hope of survival remained. Rush administered a radical purgative, and the patient miraculously revived. When several more patients treated in the same manner recovered, Rush was convinced he had found a cure for the disease. He turned to the heaviest purgative that he knew of, a concoction of ten grains of calomel and fifteen of jalap. In addition to the purgative, he prescribed heavy bleedings. Some of his patients began vocally to testify to the success of his method and, in the ensuing months, he became a folk hero sought out by ever increasing numbers of the afflicted. In spite of working exhausting hours, he could not reach everyone who sought his aid, and so he published his remedy in the *Federal Advertiser* in hopes that those he could not reach could administer the cure to themselves.

Before the plague was over, it had claimed more than 4,000 victims. Rush, in *An Account of the Bilious Yellow Fever of 1793*, contended that the heavy loss of life had to be blamed on the physicians who failed to utilize his cure. Because he was convinced of the efficacy of his procedure and of the incompetence of his colleagues, he argued that "it is time to take the cure of pestilential epidemics out of the hands of physicians, and to place it in the hands of the people."

What Benjamin Rush did and what he wrote is illustrative of what happens in folk medicine. A hopeless situation arises in which human health is endangered. Some expedient is devised to combat the disease. It works, and then it becomes circulated and widely recommended, eventually rigidifying into a practice. In Rush's case, though he had a formal medical degree, he went so far as to disavow the formal practice of medicine when he became convinced that people could become medical practitioners, provided they were armed with his cure.

If Rush's remedy sounds a little too good to be true, it is. His procedures could have had no positive effect in curing yellow fever. Indeed, it is surprising that anyone who underwent the regimen survived it, much less the combination of it and yellow fever. Calomel contains mercury, an extremely toxic substance when ingested. Though yellow fever has a corrosive effect on the intestines, postmortem examinations of several of Rush's patients demonstrated that the damage to the intestines was greatly magnified by his purgative. The bleeding that Rush recommended was based on the faulty assumption that the body contains twice as much blood as it actually does; in several days, were his recommendations carefully followed, more blood would have been drained from a patient than the human body contains.

What then happened? I think an answer to the question helps explain the belief in folk medicine, especially since, in a sense, Rush's methods were folk methods.

Success in folk medicine is a result of a combination of the way the human body and the human mind work. An understanding of what is happening in one without the other is insufficient.

Modern science has taught us that the human body, insofar as it is cured, tends to cure itself. The body is its own greatest protector: the immunological system, which produces antibodies to fight antigens, accounts for almost all recovery from disease. Even the expensive trappings of a modern hospital are merely meant to be aides to this functioning. If a diseased human organ, an infected appendix or a kidney, is removed surgically, it is still the patient's body that must heal the wounds of surgery and repair damaged cells. The surgical procedure, in other words, does not cure the patient, it merely increases the body's chance of not being overwhelmed by infection. Nothing can save the patient if the internal system breaks down. This accounts for the terror associated with a disease like AIDS, which is so threatening because it is an immunological deficiency that renders the body defenseless. This is not to belittle medicine; its discoveries are prodigious and its contributions to health salutary, but the success of modern medicine depends on an understanding of how the healthy body protects itself.

In Benjamin Rush's case, what probably happened was that he administered his remedy to a number of patients just at a moment when they were about to recover; certainly this must have been the case with the first miraculous recovery. It seems to have been a fairly common pattern in the disease for an individual to seem to be on the verge of death only to be up and about within the course of the day. Probably the virus had been defeated by the immunological system, but the exhaustion and fatigue of fighting the disease left the victim in a seemingly hopeless condition. When one sees such a recovery take place and has a theory in his mind that a particular medical procedure should produce such a cure, it is hard not to assume a cause-effect relationship. Certainly a man of good will, who had seen much death and suffering and who was, by his own accounts, deeply attached to his patients, must be excused for failing to see that the miracles he observed were not of his making.

There is, of course, a problem with such an explanation. Simple mathematical odds would preclude a large number of miraculous recoveries taking
place in the manner just described. One would assume, then, that the physician
would see just how poor the rate of recovery for his patients was and that
the system would collapse.

It is here that the second part of the formula must be called upon: the
part that looks at the role of the human mind in belief systems. The human
mind is the universe's gullible machine. When there is a strong need to believe
in a system, the human mind can become extremely uncritical. The need for
using a double blind in scientific research attests to this. In his book *Flim-
Flam!* James Randi, a professional magician who understands well what tricking
an audience means, offers a detailed explanation of how people who uncritically
believe in paranormal phenomena delude themselves. One of the points he
makes is that, once someone is convinced of the truth of an idea, it becomes
very easy to either ignore or rationalize away negative evidence. Rush's own
publication on the yellow-fever epidemic demonstrates that he fell into precisely
this trap. Although he lost his own sister and a number of close associates
to this plague, in spite of administering his cure to them, he found ready
excuses for all his losses. It must be remembered that his initial thrill over
the success of his method resulted from his believing that a radical procedure
could undo radical damage and bring patients back from the edge of the
grave. And yet, when he had to account for his sister's and his pupils' deaths,
he lamented that the large number of patients who came to visit him caused
his house to be filled with such a "concentrated miasmata" that remedies simply
could not be expected to save its inhabitants. The patients he did save, and
he counted himself among them, were proof of the success of his method.

I would suggest that the example of Benjamin Rush is a paradigm for
the way in which folk remedies develop. Someone is sick; in desperation,
something is tried. The person recovers, as most humans do from most ailments, and the cure is ascribed not to the body's ability to cure itself but
to the remedy. The remedy then is taken into the general system of folk
beliefs, whence it becomes hard to dislodge. When an individual fails to be
cured by the remedy, the practitioner resorts to one of a number of explanations:
the patient was already too sick; a careless underling did not properly administer the remedy; the patient was so immoral the gods did not allow a cure.
Rush used the first two of those excuses even though he came to use his
system because the purgatives seemed to save the most hopeless patients.

If this explanation of the pervasiveness of folk remedies in light of their
seeming foolishness is not convincing, remember this: The human body can
recover from virtually any infectious disease. The efficacy of a remedy is
constantly being reinforced by samples of "success," even when the remedy
makes no contribution to the recovery at all; or, as in Rush's case, probably
reduces the chance of recovery.

If we return to the original toothache remedy, this theory might seem to be confounded. The rotting of a tooth is degenerative, and relief would not be brought until the tooth were repaired or extracted. Therefore, it might be argued, success could not result from the remedy, and the remedy's existence could not depend upon an accidental cure. But, since the actual ability to cure a targeted ailment is not a necessary condition for the existence of a folk remedy, one of several things might have happened. the throbbing associated with a toothache is often intermittent, especially in the early stages of decay. The remedy might have seemed to bring temporary relief. There are also such things as "false" toothaches, which result from pressure in the sinus cavities. Were such a condition to clear up within a reasonable time after the smoking of a mule's "chestnuts," then the remedy might survive several failures. It takes only some positive evidence, not an actual cure, to keep a remedy in circulation. This is certain: the remedy is recollected because someone *thought* it worked.

In folk medicine, then, a pattern something like this works: Because the human body repairs and heals itself, anything that is temporally interposed between sickness and recovery can come to be interpreted as the cause of healing. Once it is assumed that a procedure is efficacious, a small amount of "positive evidence" is enough to overcome failure in a majority of cases, because the human mind, protecting its belief-systems, reasons away the failures.

Note

The remedy that appears in the introductory quotation comes from Freddie Vaughn and Frank Reuter, "Negro Folk Remedies Collected in Southeast Arkansas, 1976," *Mid-South Folklore.* 4 (Summer 1976):71. Much of the information about the 1793 yellow-fever epidemic comes directly from Benjamin Rush's *An Account of the Bilious Yellow Fever of 1793.* An excellent historical analysis of the Philadelphia plague can be found in J. H. Powell's *Bring Out Your Dead* (Philadelphia: University of Pennsylvania Press, 1949).

WILLIAM JARVIS

Chiropractic: A Skeptical View

Chiropractic is the most significant nonscientific health-care delivery system in the United States. As a result of their high level of organization and aggressiveness, chiropractors are licensed to practice in all 50 states and several foreign countries. Chiropractic is so well entrenched that it must be viewed as a societal problem, not simply as a competitor of regular health-care.

"Chiropractic" literally means "done by hand" (*chiros* = hand; *praktos* = practice), referring to manipulation of the spine. Manipulation (i.e., "the forceful, passive movement of a joint beyond its active limit of motion," according to *Dorland's Illustrated Medical Dictionary*) is not the exclusive domain of chiropractors. Folk practitioners sometimes called "bonesetters" have long used the notion of bones "going out of place" to explain maladies, and they employ manipulation as a panacea. Andrew Taylor Still invented "osteopathy" based upon the theory that luxated bones interfere with blood circulation, producing all manner of diseases. (Osteopathy officially abandoned Still's theory in 1948.)

Today, physiotherapists, athletic trainers, and several medical specialists sometimes employ manipulation for neuromusculoskeletal conditions. There is sufficient evidence that manipulation can at least temporarily improve the range of motion of impaired joints and relieve pain—sometimes dramatically—to make it a worthwhile, albeit limited, medical procedure. Manipulation requires a good deal of individual skill, which many chiropractors apparently possess.

History

Chiropractic's uniqueness is not in its use of manipulation but, rather, in its theoretical basis for doing so—which also explains why chiropractors overutilize spinal manipulation therapy (SMT), often applying it without justification.

Chiropractic is the brainchild of Daniel D. Palmer, a late-nineteenth-century dabbler in metaphysical approaches to health care. Palmer had practiced phrenology and magnetic healing, and had some osteopathic training. He reported that a spiritualist medium inspired him in his search for "the single cause of all disease." He puzzled over the fact that pathogenic germs were found in both healthy and sick people and searched for an explanation. (Today, we know that the immune system makes the difference.) He claimed that one day in 1895 he restored the hearing of janitor Harvey Lillard and experienced an illumination that the spine was the key to health and disease.

Unique Theory

Palmer contrived the notion that "subluxations" of the spine impinge nerves, interfering with nerve flow, which he dubbed the Innate Life Force, and that all a practitioner had to do was to adjust the spine—the healing powers of nature would do the rest. Neither Palmer nor any other chiropractor has ever been able to reliably demonstrate the existence of "subluxations," much less validate their importance to health and disease. Nevertheless, chiropractic has grown and thrived and boasts of 24,000 practitioners nationally.

When chiropractors are challenged to explain precisely what effect nerve impingement is supposed to have upon a nerve impulse (i.e., frequency of propagation, amplitude, etc.), they either fall back upon metaphysical notions of the Innate Life Force or evoke one of many common ploys: (1) make a virtue of their ignorance by retorting that they don't know how it works but that it does; (2) claim that studies to determine the mechanism are now under way or just completed but unpublished (the "Oh, haven't you heard? You're behind the times!" ploy); or (3) do as the American Chiropractic Association has done—evade the issue by officially changing the rhetoric and adding uncertainties: "Disease *may* be caused by disturbances of the nervous system. . . . Disturbances of the nervous system *may* be caused by derangements of the musculoskeletal structure. . . . Disturbances of the nervous system *may* cause or aggravate disease in various parts or functions of the body. . . ." (ACA 1984, pp. 8-9). They do this while continuing to practice as if subluxations were an established reality.

Lack of Validity of Chiropractic Theory

A comprehensive critique of chiropractic's lack of scientific validity was written by the College of Physicians and Surgeons of the Province of Quebec (1966) in 1963 and remains relevant today. In 1973, Yale University anatomist Edmund Crelin demonstrated that subluxations severe enough to impinge upon the nerves exiting the spinal foramina were impossible to produce without total disablement. Crelin (1985) states in *Examining Holistic Medicine* that instead of the scientific response of attempting to replicate his research, the ACA wrote a tirade of verbiage, concluding that his work was invalid because it was done on cadavers. In fact, Crelin states, the absence of a reflex response in a dead body should make subluxations easier to produce. Faced with this evidence, a true-believing chiropractor once remarked to me that the reason Crelin had failed to demonstrate the chiropractic hypothesis was that he worked with cadavers in which Innate Life Force was no longer present!

An excuse chiropractors employed for years for the lack of scientific evidence for their theory was that the government wouldn't provide the necessary research funding. The falsity of this claim was exposed in 1972, when the International Chiropractic Association reported (ICA 1972) that the federal government had approved a grant for Dr. Suh, at the University of Colorado, to develop a method for measuring spinal configurations to determine the existence of chiropractic subluxations and that the grant *application* was the *first* in chiropractic's history.

Scott Haldeman, D.C., M.Sc., Ph.D., M.D., a third-generation chiropractor whose personal commitment to validating chiropractic led him to obtain a medical degree and advanced degrees in neurophysiology, has criticized Suh's and others' attempts "to find more accurate ways of measuring a subluxation in the absence of any solid data that the subluxation is worth measuring" (Haldeman 1977). Although providing chiropractic public relations personnel with fodder for a decade, Suh's work on the illusive subluxation never got anywhere and now seems fruitless at best.

Lack of Clinical Reliability

Apologists have some room for defensive debate because of the difficulty of establishing mechanisms of cause and effect in many other fields of science. An easier and more practical test of chiropractic is that of the reliability (i.e., consistency) of chiropractors' clinical ability to identify subluxations. They have not fared well in these tests.

Smith (1969) visited the Palmer Clinic in Davenport, Iowa, and the National College Clinic at Lombard, Illinois, on two successive days. At Palmer he was told that he had subluxations at the ninth dorsal and the fifth lumbar; at National a subluxation was diagnosed at the fifth dorsal only. Neither found other subluxations alleged by another chiropractor earlier.

Stephen Barrett (1980) "sent a healthy four-year-old girl to five chiropractors for a 'check up.' The first said the child's shoulder blades were 'out of place' and found 'pinched nerves to her stomach and gall bladder.' The second said the child's pelvis was 'twisted.' The third said one hip was 'elevated' and that spinal misalignments could cause 'headaches, nervousness, equilibrium or digestive problems' in the future. The fourth predicted 'bad periods and rough childbirth' if her 'shorter left leg' were not treated. The fifth not only found hip and neck problems, but also 'adjusted them' without bothering to ask permission." Completely inconsistent findings were also diagnosed in two adult women.

Mark L. Brown, a reporter for the *Quad City Times,* serving the Davenport, Iowa, area, conducted his own five-month investigation of the practice of chiropractic. He produced an insightful 36-page Sunday newspaper supplement (available from the National Council Against Health Fraud, Inc.[1]) revealing some bizarre as well as some useful practices. Brown also found numerous inconsistencies, including diagnoses that his left leg was shorter than his right leg and vice versa!

Isolation

Chiropractors commonly blame others for their lack of science, claiming to have been isolated by organized medicine. In reality, chiropractic's isolation is self-imposed. Chiropractors substitute chiropractic philosophy for science and commonly boast of intellectual superiority. True-believers say that someday the world will acknowledge the greatness of chiropractic. It would be impossible for chiropractors who adhere to chiropractic theory and philosophy to work with scientific health-care providers. Chiropractors' concepts of the causes and treatments of disease differ radically; and, although health scientists worldwide can bridge barriers of language and culture via the common ground of basic science, they cannot work with pseudoscientists.

There is considerable concern about the wisdom of permitting chiropractors to serve as entry-level health-care providers. Practitioners devoted to a pseudoscientific approach to disease are apt to miss serious diseases when hearing patient complaints. Reformist chiropractor Peter Modde states that malpractice is an inevitable result of chiropractic training and philosophy.

How widely the subluxation theory is believed among chiropractors is uncertain. Quigley attempted to measure this in 1981. Eighty-eight percent of 268 responding chiropractors gave subluxations in musculoskeletal problems an importance of 70 percent or more. Sixty percent of 265 respondents gave subluxations in visceral disorders a rating of 70 percent or more (Quigley 1981). Even less is known about the proportion of chiropractors who believe in the metaphysical Innate Life Force.

Factionalism

Chiropractic has become a conglomeration of factions in conflict, bound together only by opposition to outside critics. At least a dozen different notions about how the spine should be corrected divide practitioners. Some say only the Atlas needs adjusting; others go to the other end of the spine and say only the sacral area is important. Still others use both ends (sacro-occipital). Several adhere to specific vertebral levels for specific organs or diseases. Some measure leg lengths or test muscles—called "applied kinesiology" (A-K)—for weakness or strength in association with foods, colors, music, and just about anything else. (Even A-K's originator expressed skepticism about the technique's being used to determine one's personal state.)

The most obvious rift among chiropractors is between "straights" and "mixers." Straights adhere more to chiropractic's original theory and practice, while "mixers" (a term applied by the straights and abhorred by the mixers) may incorporate almost any modality into their practices. The ICA is the straights' national organization, and the ACA represents mixers.

The utter confusion within chiropractic over subluxations, scope of practice, and other important health-care issues was rediscovered by the U.S. Department of Health and Human Services' Office of Inspector General (OIG) in 1986. According to its report: "Heated controversy regarding chiropractic theory and practice continues to exist. . . . On-site and telephone discussions with chiropractors and their schools and associations, coupled with a review of background materials . . . result in a picture of a profession in transition and containing a number of contradictions. . . . There continues to be some disagreement within the profession regarding which conditions are appropriate for chiropractic care and regarding appropriate parameters for treatment."

The OIG investigators also made note of "the problem side of chiropractic." The report states: "Despite the evidence which was presented during the study regarding the increased emphasis on science and professionalism . . . there also exist patterns of activity and practice which at best appear as overly aggressive marketing and, in some cases, seem deliberately aimed at misleading patients and the public regarding the efficacy of chiropractic care."

There are also chiropractors who practice rational, conservative spinal manipulative therapy for neuromusculoskeletal disorders. They do not claim to be alternative practitioners but offer their skills as manipulation therapists when such treatment is medically justifiable. What proportion of chiropractors fit this description is unknown. The National Association of Chiropractic Medicine (NACM) was formed in 1984 as an attempt to organize reform-minded chiropractors. There may be a substantial number of rational chiropractors "in the closet," in which case our view of chiropractic may be distorted by the hucksters and zealots among the guild.

Evaluating Individual Chiropractors

Because of the great diversity among chiropractors, it is impossible to evaluate individual practitioners according to the commonly advanced straights/mixers dichotomy. There is a common misconception that straights are old-fashioned (i.e., unscientific) and mixers are modern (i.e., scientific). Although straights may be labeled "cultists" for adhering to Palmer's dogma, the additional modalities mixers employ are apt to be just as nonscientific. Mixers utilize colonics, iridology, unproved devices, applied kinesiology (muscle-testing), megavitamins, herbology, crystals, variations of acupuncture, glandular therapy, craniopathy, and a seemingly endless array of dubious diagnostic, prescription, and therapy procedures.

Of course those who use questionable methods can be rejected as untrustworthy on that basis alone, but because chiropractors are very philosophical, individuals can be challenged on the basis of *why* they manipulate. True-believer chiropractors would be loath to admit that their ministrations are so mundane as to deal with effects rather than basic causes. This topples them from their imagined dominance as "the true physicians" to merely therapists. Rational chiropractors will readily admit that SMT (spinal manipulation therapy) relieves *effects* not *causes*. The fact that they mainly provide temporary somatic relief, rather than dealing with the causal factors of disorders, is evidenced by the large amount of repeat business they generate. Too frequent treating is a factor noted by nearly every commission that has evaluated chiropractic. Third-party payers commonly control chiropractic costs by limiting the number of treatments or the annual amount for which devotees may receive reimbursement.

Individual chiropractors may also be judged according to their opposition or endorsement of scientifically established public-health practices, such as fluoridation, immunization, pasteurization, and modern food technology. Chiropractors are often in the forefront of anti-scientific social movements opposing such practices.

Patient Loyalty

Despite its appalling lack of science, chiropractic has a loyal following of clients. I believe this can be credited to the somatic relief of SMT and the psychological aspects of chiropractic care. SMT involves the laying on of hands, which reportedly has the effect of relaxing the patient. The laying on of hands also is said to increase suggestibility, which enhances the placebo effects of SMT. Reformer Samuel Homola, D.C. (1963) says:

> The majority of the "subluxations" commonly found by many chiropractors are likely to be painless and imaginary. In replacing these imaginary subluxations, the practitioner places his hands upon the patient's back and applies a sudden thrust, causing the bones to come together, making a noise like the "crushing of an old basket." This thrust, with "popping" of the vertebrae, has a tremendous psychological influence over the mind of the healthy patient as well as over the mind of the sick patient. While the popping sound itself is quite meaningless . . . (such as "cracking" the knuckles) . . . this influence might be used to advantage in curing psychosomatic conditions—provided the patient is informed that the bone is "back-in-place" and will stay there. By the same token, however, such treatment used on the mentally unstable and nervous person can cause a great deal of harm;

that is, by perpetuating a psychosomatic condition or even creating a new psychological illness.

Thus Homola points out that the placebo effect is not without potential for harm. By experiencing relief, the patient is *taught* illness through operant conditioning. In addition to the direct physical effects Homola describes, chiropractors engage in a great deal of verbal conditioning. Manipulation itself can feel good (although it can also be painful) and can be addictive, according to reformer Charles Duvall, D.C. (1984).

Chiropractors have been shown to be better at satisfying patients than medical doctors are. This is because they validate patients' problems medical doctors tend to minimize, offer simplistic explanations about health and disease, and work at being friendly. Chiropractic also offers both a mechanistic and a metaphysical explanation for its effects, appealing to both needs. Chiropractic rhetoric has adapted itself to contemporary favors of "holistic" health-care. In fact, chiropractic is not holistic but spine-centered, but this seems to escape notice.

Selling Chiropractic

Success promoters who give seminars to train chiropractors in psychological patient manipulation are a notorious problem within chiropractic. I study a wide variety of health pseudosciences, but I know of no other guild that has formalized the education of practitioners in patient deception. While many of the procedures taught are simply good office practice, many others are fraudulent. (See Barrett 1980.)

Former ACA public-relations consultant Eric Baizer says the ACA conducts an aggressive public-relations program aimed at selling chiropractors as family doctors and primary-care providers. Baizer (1983) describes how as a PR expert he defended chiropractic publicly by responding to negative press reports. He says he employed stock answers and reusable clichés—what one writer termed "factoids" (i.e., statements designed to *resemble* facts). "For example," Baizer wrote, "if someone attacked the quality of chiropractic education, we would point out that chiropractors attended colleges accredited by an agency recognized by the U.S. Office of Education—*implying* that the schools must be of high quality. How good is the chiropractic curriculum? How qualified are the instructors? Are inspections of the colleges thorough? These are the kinds of issues best left unexplored."

Survival of a Pseudoscience

The illusionary "subluxation" not only is the theoretical basis for chiropractic but constitutes its legal basis as well. State acts describe chiropractic as the finding and removal of subluxations. It is enigmatic that such a system thrives late in the twentieth century, which has seen such progress in the biological sciences and health care. While biological scientists have unraveled the genetic double-helix, chiropractors have failed to scientifically define their theory or scope of practice, or to justify their very existence as primary health-care providers.

Chiropractic's survival and success is undoubtedly due to the reality that there is much more involved in health-care delivery than science. Politics, business considerations, and the clinical art often take precedence. Although it is the scientific validity of the methods employed that justifies modern health-care, in practical terms of survival in the marketplace, chiropractic demonstrates daily that the scientific aspects of health care are of least importance and in *greatest need of protection.*

The chiropractic guild is adept at having its way with politicians. This appears to be primarily due to its applications of business law to the health marketplace. Chiropractors pose themselves as competitors of "allopathic medicine" (a misnomer, since allopathic medicine, which employed bleeding, purging, and so forth, to balance the four Greek humors, was replaced long ago with the emergence of medical science). Unfortunately, regular medicine is regarded by many legislators as simply holding another opinion among differing viewpoints. Chiropractors encourage the concept that they are a separate but equal health-care delivery system. They find allies among those who present science as merely "Western thought" and find the Innate Life Force notion compatible with Eastern metaphysical world-views.

Politicians seem to have trouble differentiating between religious beliefs and beliefs in various forms of health care. Chiropractic's clientele is loyal, and its political clout is greater than that of its critics. Patients willingly cooperate with chiropractors when asked to send letters to lawmakers. Many subscribe to the myth of a vindictive medical profession out to crush its opposition.

In the final analysis, the validity of chiropractic is not a medical controversy as much as one of the basic biological sciences. Medicare reimburses chiropractors with millions of taxpayers' dollars each year for removing subluxations allegedly demonstrated by X-rays. Basic biological scientists have a public duty to objectively test a theory as radical as chiropractic's to determine if it is valid. The failure to require scientific validation of an entire health-care delivery system sets a disturbing precedent for other nonscientific systems to lay claim to the public purse. Transcendental meditators are trying to qualify for reimbursement in New Zealand, and Christian Science practitioners are paid for faith-healing in this country. Chiropractic and other nonscientific forms of health care will survive until the public demands that scientific justification become a primary qualification for legalization and reimbursement.

Note

1. The National Council Against Health Fraud, Inc., has published a position paper on chiropractic. A single.copy may be obtained free by sending a stamped, addressed business-sized envelope to NCAHF, P.O. Box 1276, Loma Linda, CA 92354.

References

American Chiropractic Association. 1984. *Chiropractic: State of the Art.* pp. 8-9.

Baizer, Eric. 1983. Inside the American Chiropractic Association: Selling the chiropractor as a family doctor." *CCAHF Newsletter,* 6(1).

Barrett, Stephen. 1980. *The Health Robbers.* Philadephia, Pa.: George Stickley.

College of Physicians and Surgeons of the Province of Quebec. 1966. *The New Physician,* September.

Crelin, Edmund. 1985. Chiropractic. In *Examining Holistic Medicine,* edited by Douglas Stalker and Clark Glymour, 197-220. Buffalo, N.Y.: Prometheus Books.

Duvall, Charles. 1984. *Chiropractic Claims Manual.*

Haldeman, Scott. 1977. The importance of research in the principles and practice of chiropractic. *Worldwide Report,* January.

Homola, Samuel. 1963. *Bonesetting, Chiropractic, and Cultism.* Critique Books.

International Chiropractic Association. 1972. *International Review of Chiropractic,* April.

Quigley. 1981. Chiropractic's monocausal theory of disease. *ACA J. of Chiro.,* June.

Smith, Ralph L. 1984. *At Your Own Risk: The Case Against Chiropractors.* New York: Simon and Schuster.

STEPHEN BARRETT, M.D.

Homeopathy: Is It Medicine?

During the past several years, increasing numbers of homeopathic remedies have been offered for sale in health-food stores and elsewhere. Their promoters suggest that they are safe, effective, natural remedies that have no side effects. This report summarizes the yearlong investigation of homeopathy I conducted on behalf of *Consumer Reports* magazine.

Homeopathy's Roots

Homeopathy dates back to the late 1700s, when Samuel Hahnemann (1755-1843), a German physician, began formulating its basic principles. Hahnemann was justifiably distressed about bloodletting, leeching, purging, and other medical procedures of his day that did far more harm than good. He was also critical of medications like calomel (mercurous chloride), which was given in doses that caused mercury poisoning. He then developed his "law of similars"—that the symptoms of disease can be cured by substances that produce similar symptoms in healthy people. The word *homeopathy* is derived from the Greek words *homeo* (similar) and *pathos* (suffering or disease).

Although ideas like this had been espoused by Hippocrates in the fourth century B.C., and by Paracelsus, a fifteenth-century physician, Hahnemann was the first to use them in a systematic way. He and his early followers conducted "provings," in which they administered herbs, minerals, and other substances to healthy people, including themselves, and kept detailed records of what they observed. Later these records were compiled into lengthy reference books called *materia medica,* which are used to match a patient's symptoms with a "corresponding" drug.

Hahnemann believed that diseases represent a disturbance in the body's ability to heal itself and that only a small stimulus is needed to begin the healing process. In line with this—and to avoid toxic side-effects—he experimented to see how little medication could be given and still cause a healing response. At first he used small doses of accepted medications. But later he used enormous dilutions and concluded that the smaller the dose, the more powerful the effect—a principle he called the "law of infinitesimals."

That, of course, is just the opposite of what pharmacologists believe today. As summarized in the 1977 report of an Australian Parliament committee of inquiry: "For each [drug] property, there is a clearly defined dose-response relationship in which increasing the dose increases the effect. . . . There is not one example in the whole area of pharmacology in which simple dilution of a drug enhances the response it produces any more than diluting a dye can produce a deeper hue, or adding less sugar can make food sweeter."

Homeopathy's Remedies

Homeopathic drugs are prepared as follows: If the medicinal substance is soluble, one part is diluted in either 9 or 99 parts of a water and/or alcohol solution and shaken vigorously; if insoluble, it is finely ground and pulverized in similar proportions with powdered lactose (milk sugar). One part of the diluted medicine is diluted, and the process is repeated until the desired concentration is reached. Dilutions of 1 to 10 are designated by the Roman numeral X (1X = 1/10, 2X = 1/100, 3X = 1/1,000, 6X = 1/1,000,000). Similarly, dilutions of 1 to 100 are designated by the Roman numeral C (1C = 1/100, 2C = 1/10,000, 3C = 1/1,000,000, and so on). Most remedies today range from 6X to 30X.

According to the laws of chemistry, there is a limit to the dilution that can be made without losing the original substance altogether. This limit, called "Avogadro's number" (6.023×10^{23}), corresponds to homeopathic potencies of 12C or 24X (1 part in 10^{24}). Hahnemann himself realized there is virtually no chance that even one molecule of original substance would remain after extreme dilutions. But he believed that the vigorous shaking or pulverizing with each step of dilution leaves behind a spiritlike essence that cures by reviving the body's "vital force." Hahnemann's theories have never been accepted by scientifically oriented physicians, who charge that homeopathic remedies are placebos (inert substances).

Because homeopathic remedies were actually less dangerous than those of nineteenth-century medical orthodoxy, many medical practitioners began using them. At the turn of the century, homeopathy had some 14,000 practitioners and 22 schools in the United States alone. But as medical science and medical education advanced, homeopathy declined sharply, particularly in America, where its schools either closed or converted to modern methods. The last pure U.S. homeopathic school closed during the 1920s, but Hahnemann Medical College in Philadelphia continued to offer homeopathic courses on an elective basis until the late 1940s. A few graduates from other modern medical and osteopathic schools later became homeopaths by taking courses here or abroad or by training with a practicing homeopath.

Homeopathic remedies were given legal status by the 1938 Federal Food, Drug, and Cosmetic Act, which was shepherded through Congress by Senator Royal Copeland (D-N.Y.), a prominent homeopathic physician. One provision of this law recognized as drugs all substances included in the *Homeopathic Pharmacopeia.* Now in its ninth edition, this book lists more than 1,000 substances and the historical basis for their inclusion: not modern scientific testing, but homeopathic "provings" conducted as long as 150 years ago.

Today's Marketplace

The 1985 directory of the National Center for Homeopathy, in Washington, D.C., lists some 300 licensed practitioners, half of them physicians and the rest mostly chiropractors, naturopaths, dentists, veterinarians, and nurses. But Jay P. Borneman, of Swarthmore, Pennsylvania, whose family has been marketing homeopathic remedies since 1910, believes that several hundred more consider themselves homeopaths and that many conventional physicians utilize one or a few homeopathic remedies for specific conditions. Larger numbers of homeopaths practice in England, France, India, Germany, the Soviet Union, and several other countries where homeopathy is more popular.

Laypersons are also involved in practicing homeopathy. Some operate offices, which may not be legal. A few unaccredited schools have offered correspondence courses leading to certificates or "degrees" in homeopathy. Consumers interested in homeopathic self-treatment can obtain guidance through lay study groups, books, and courses sponsored by the National Center for Homeopathy.

Most homeopathic practitioners still rely on *materia medica* in choosing among the thousands of remedies available. But a few utilize computerized electrical devices that they claim can help match the remedies to the patient's diseased organs. "Classical" homeopaths—who follow Hahnemann's methods closely—take an elaborate history of the patient to fit the remedy to the individual. The history typically includes standard medical questions plus many more about such things as emotions, moods, food preferences, and reactions to the weather. The remedy for symptoms on one side of the body may differ from that for identical symptoms on the other side. Classical homeopaths prescribe one substance at a time, while nonclassical homeopaths may prescribe several.

Homeopathic remedies are available from practitioners, health-food stores, and drugstores, as well as manufacturers who sell directly to the public. A few products are sold person-to-person through multilevel marketing companies. Home-remedy kits are available from several companies. Jay Borneman believes that U.S. sales of homeopathic products probably total no more than $15 million a year, with half of these made by five companies that have been in business for 75 to 150 years.

According to FDA officials, homeopathic remedies used to be marketed on a small scale by these five companies, mainly to serve the needs of licensed practitioners. "These drugs bore little or no labeling for consumers because they were intended for use by homeopathic physicians who would make a diagnosis and either compound a prescription, dispense the product, or write a prescription to be filled at a homeopathic pharmacy," says William G. Nychis, the FDA's expert on homeopathy. "The pharmacies also sold a limited number of nonprescription homeopathic products. During the past decade, however, the homeopathic marketplace has changed drastically. New firms have entered the field and sold all sorts of products through health food stores and directly to consumers."

Jay Borneman readily admits that "there is a lot of insanity operating under the name of homeopathy in today's marketplace. Companies not committed to homeopathy's principles have been marketing products that are unproven, untested, not included in the *Homeopathic Pharmacopeia,* and combination products that have no rational or legal basis. Some are simply quack products called homeopathic for marketing purposes."

Perhaps the most blatant promotion was that of Biological Homeopathic Industries, Albuquerque, New Mexico, which in 1983 sent a 123-page catalogue to almost 200,000 physicians nationwide. Among its products were BHI Anticancer Stimulating, BHI Antivirus, BHI Stroke, and 50 other types of tablets claimed to be effective against serious diseases. In 1984, the FDA forced the company to stop distributing several of the products and to tone down its claims for the rest.

In September 1985, agents of the FDA and the Pennsylvania Health Department seized some $125,000 worth of drugs sold person-to-person by Probiotic, Inc., of Reading, and Homerica, Inc., a subsidiary. The products were labeled "Skin Relief," "Human Power Recharger," and "Pain Control" and did not state what they were for, what was in them, or how to use them.

At least ten other companies offer questionable homeopathic remedies for over-the-counter sale. Some product examples are: Arthritis Formula, Bleeding, Kidney Disorders, Flu, Herpes, Exhaustion, Whooping Cough, Gonorrhea, Heart Tonic, Gall-Stones, Cardio Forte, Thyro Forte, and Worms.

Homeopathy's Legal Status

In most states, homeopathy can be practiced by any physician or other practitioner whose license includes the ability to prescribe drugs. Three states—Arizona, Nevada, and Connecticut—have separate homeopathic licensing boards. The Nevada situation is notable because some of its practitioners acquired licenses as homeopaths after other states revoked their medical licenses for cancer quackery.

Arizona's licensing boards are subject to "sunset" review, which means they will be abolished unless reauthorized by the Arizona legislature. Last year, as the expiration date for the homeopathy board drew near, the state's homeopaths joined forces with health-food stores to lobby vigorously. To counter the idea that a board might not be needed because there were only a handful of homeopaths in the state, the American Institute of Homeopathy (a group of about 100 classical homeopathic physicians) urged its members to apply for licensure in Arizona "to show that there are doctors interested in practicing homeopathy today." According to the National Health Federation, a health-food-industry group that helped with the campaign, close to 2,000 supporters attended hearings and state legislators got hundreds of handwritten letters supporting homeopathic licensing. The reauthorization bill passed unanimously.

Public protection regarding drugs is based on a framework of federal laws and regulations that require drugs to be safe, effective, and properly labeled. But the FDA has not applied this framework to homeopathic remedies. Since most homeopathic remedies contain no detectable amount of active ingredient, it is impossible to test whether they contain what their labels say. They have been presumed safe, but unlike most other drugs, they have not been proved effective against disease by scientific means, such as by double-blind testing. If the FDA were to require such proof for homeopathic drugs to remain on the market, the industry would not survive unless it could persuade Congress to change the law.

The American Association of Homeopathic Pharmacists, a group of leading homeopathic manufacturers, has proposed that homeopathic remedies remain marketable without a prescription for minor ailments that do not require complex medical diagnosis or medical monitoring. Traditional homeopathic remedies used for the treatment of serious diseases would be available by prescription only from physicians and others authorized by state laws to prescribe drugs. The FDA is considering this proposal and hopes to issue a policy guide for homeopathic products in the near future.

In January 1986, the North Carolina Board of Medical Examiners revoked the license of George A. Guess, M.D., the state's only licensed homeopathic physician, after concluding that he was "failing to conform to the standards of acceptable and prevailing medical practice." (The North Carolina Supreme Court has upheld this decision, but Dr. Guess is appealing to a federal court.) Dr. Guess is a 1973 graduate of the Medical College of Virginia and was board-certified in family practice from 1976 through 1983. But in 1978 he began practicing homeopathy. During hearings held by the Board, another family practitioner testified that although Dr. Guess is intelligent and well trained in orthodox medicine, "homeopathy is not medicine. It's something else."

Most pharmacy-school educators seem to feel the same way. Last year I sent a questionnaire to the deans of all 72 U.S. pharmacy schools. Faculty members from 49 schools responded. Most said their school either doesn't mention homeopathy at all or considers it of historical interest only. Hahnemann's "law of similars" did not find a single supporter, and all but one respondent said his "law of infinitesimals" was wrong also. Almost all said that homeopathic remedies were neither potent nor effective, except possibly as placebos for mild, self-limited ailments. About half felt that homeopathic remedies should be completely removed from the marketplace.

Homeopathic Research

Probably the best review of homeopathic research is the two-part article by A. M. Scofield, Ph.D., a British biochemistry professor. In the *British Homeopathic Journal* (73:161-180 and 73:211-226, 1984), he concludes: "Despite a great deal of experimental and clinical work there is only a little scientific evidence to suggest that homeopathy is effective. This is because of bad design, execution, reporting or failure to repeat promising experimental work and not necessarily because of the inefficacy of the system which has yet to be properly tested on a large enough scale. . . . It is hardly surprising, in view of the quality of much of the experimental work as well as its philosophical framework, that this system of medicine is not accepted by the medical and scientific community at large."

Scofield cautions against dismissing homeopathy simply because its underlying philosophy does not fit accepted scientific premises. Feeling that "some of the experimental work already done suggests that homeopathy may be of value," he recommends that carefully controlled experiments be done to test homeopathy further.

One apparently well-designed study was published in the British journal *Lancet* on October 18, 1986. In this study 56 hay fever patients who received a homeopathic preparation of mixed grass pollens had fewer symptoms than a comparable group of 52 patients who received a placebo. Whether this type of finding can be consistently reproduced remains to be seen.

Overview

During my lengthy investigation, I was impressed by the warmth and sincerity of the homeopathic leaders I met. But the key question is whether homeopathy is effective.

Consumer Reports concluded in its January 1987 issue: "Unless the laws of chemistry have gone awry, most homeopathic remedies are too diluted to have any physiological effect. . . . CU's medical consultants believe that any system of medicine embracing the use of such remedies involves a potential danger to patients whether the prescribers are M.D.'s, other licensed practitioners, or outright quacks. Ineffective drugs are dangerous drugs when used to treat serious or life-threatening disease. Moreover, even though homeopathic drugs are essentially nontoxic, self-medication can still be hazardous. Using them for a serious illness or undiagnosed pain instead of obtaining proper medical attention could prove harmful or even fatal."

Homeopathic leaders insist that their remedies are effective and that studies *do* support this viewpoint. They also suggest that homeopathy's popularity and long survival are evidence that it works. But the only way for homeopathy to gain acceptance by the scientific community would be to demonstrate positive results through repeated experiments designed with the help of critics and carried out with strict safeguards against experimenter bias and fraud.

If the FDA required homeopathic remedies to be proved effective in order to remain on the market, homeopathy would face extinction in the United States. But no indication exists that the agency is considering this. FDA officials regard homeopathy as relatively benign and believe that other problems should get enforcement priority. Moreover, if the FDA attacks homeopathy too vigorously, its proponents might even persuade Congress to rescue them. On the other hand, some level of enforcement is needed to prevent the homeopathic marketplace from getting completely out of hand.

For a look at a particularly dramatic homeopathic claim, see Martin Gardner, "Water With Memory? The Dilution Affair," in this volume.

Part 7: Astrology

GEOFFREY DEAN

Does Astrology Need to Be True?
Part 1: A Look at the Real Thing

Given the extraordinary ability of the human mind to make sense out of things, it is natural occasionally to make sense out of things that have no sense at all.

Richard Furnald Smith, *Prelude to Science*

The most popular arguments against astrology are (1) astrological signs bear no relation to the astronomical constellations, (2) astrology is earth-centered, whereas the solar system is sun-centered, (3) astrology is founded in magic and superstition, (4) there is no known way it could work, and (5) why moment of birth and not conception? The astrologer sees these arguments as no arguments at all, because if astrology works then it works, period.

Another popular argument is (6): If astrology can predict the future, why don't astrologers rule the world? Answer: Astrological prediction is far too tedious. To examine the birth charts (horoscopes) of every likely person or city or country in the hope of finding indications to your advantage is simply not practical.

Another more recent argument is (7): Research has shown that newspaper horoscopes and sun signs don't work. Here the astrologer replies that there is more to astrology than sun signs, and he then retires back to his charts convinced that such arguments reveal only ignorance and closed minds.

Unfortunately, these arguments are all too popular and tend to reoccur like old jokes. Thus, in their comprehensive review of the evidence for and against astrology, Eysenck and Nias (1982, p. 10) could say: "Much that passes for scientific criticism in the books and articles we have read is in fact little better than defamation and prejudice." Truzzi (1979) writes: "Attacking simple sun-sign astrology is largely a waste of time. . . . A manifesto denouncing newspaper astrology columns could as easily be signed by leading astrologers as by a group of respected scientists."

Recently astrologers in the United States have reacted to such arguments by a Media Watch project (AFAN 1985). This is designed to counter biased media reports on astrology, for example, those due to the "demagogical media grandstanding of self-appointed guardians of public morals and rationality like CSICOP." They point out that critics persistently mistake popular astrology for the real thing and that "just the last couple of decades has produced a psychologically and intellectually more mature astrology, of which the general population and the media remain totally unaware. How, then, are they to discover this, if we don't let them know?" A good point.

Obviously if we are to rise above the present shouting match we have to address astrology (the real thing, not popular nonsense) on the astrologers' terms. We have to go beyond the *popular* astrology of fairground tents and newspaper columns and seek out the *serious* astrology of consulting rooms and learned journals. It is not hard to find. In Western countries roughly 1 person in 10,000 is practicing or studying serious astrology (Dean and Mather 1977, p. 7), which is about the same as for psychology. In Western languages, serious astrology is currently the subject of more than 100 periodicals and about 1,000 books *in print* (1 in 2,000 of all books, or about the same as for astronomy), of which about half are in English. Since 1960 the annual output of new titles has doubled every ten years, at which rate the year 2000 will enjoy ten new astrology books every week, excluding almanacs and sun-sign books. Something this popular is clearly entitled to impartial investigation, especially since astrology has a solid core of testable ideas. We do not have to accept astrology on faith.

Unfortunately investigators of the real thing face daunting problems: (1) It takes at least a year to become even tolerably familiar with astrological theory and practice. (2) Competent criticism requires skill in astronomy, psychology, and statistics. (3) Relevant material in an astrological literature totaling some 200 shelf-meters of serious books and periodicals is highly scattered and usually inaccessible via normal library channels. (4) Relevant material in the orthodox literature is equally highly scattered over books, journals, and theses in psychology, education, sociology, and other disciplines, and is often accessible only with difficulty. Given these problems it is not surprising that debunkers with a living to make have taken the soft option of defamation and prejudice. Astrologers of course face the same problems, but they too have a living to make, so it is not surprising that U.S. astrologer Zipporah Dobyns can say that "astrology is almost as confused as the earthly chaos it is supposed to clarify" (Dobyns and Roof 1983). However, in the past ten years critical surveys have appeared that grapple with these problems, notably Dean and Mather (1977), Eysenck and Nias (1982), Kelly (1982), Culver and Ianna (1984), and Startup (1984). Unfortunately, these surveys do not properly address the real thing because relevant studies did not exist at the time they were written.

This situation has now changed. Relevant studies have now been made, and a consistent picture is emerging, most of it bad news for astrologers. In what follows we take a close look at the real thing, the reasons astrologers believe in it, and the very latest evidence. To be fair to a topic that has been so persistently misrepresented, and to allow adequate citation from a literature that is so difficult to access, this article will be a long one. But the findings have implications that extend to any character-reading technique, such as palmistry and numerology, so even if astrology is not your pet project you should find much of interest. We start with a look at the real thing and why astrologers believe in it.

The Real Thing

In broad terms the real thing boils down to a consultation between astrologer and client where something like this happens:

Birth chart of client, part-ner, com-pany, event, or whatever → Numerous (e.g., 40) interacting chart factors, each with its own meaning → Astrologer's interpretation → Client's situation, with Feedback from Client's situation to Astrologer's interpretation

The content of the consultation depends on where you are. Eastern astrologers concentrate almost exclusively on fate and destiny, i.e., prediction, for example, see Perinbanayagam (1981). But for every Western astrologer who concentrates on prediction there are probably another two who concentrate on psychology and counseling. Thus the popular view of Western astrology as consisting of prediction and nothing else is incorrect. Indeed, many astrologers eschew prediction; for example, the late Dane Rudhyar (1979), recognized as the leading U.S. proponent of humanistic (person-centered) astrology, says: "I am only interested in astrology as a means to help human beings to give a fuller, richer meaning to their lives. . . . I see no value in the prediction of exact events or even of precise character analysis." Rudhyar's approach has been critically examined by Kelly and Krutzen (1983), and their conclusions are cited in Part 2.

The chart interpretation itself is governed by the cardinal rule that no factor shall be judged in isolation. As noted above, a typical chart contains about 40 interacting factors, each with its own individual meaning, all of which will be relevant to the interpretation. (This total represents only the most basic factors, namely, planets, signs, houses, and aspects; many other factors, such as midpoints and dynamic contacts, can be included, which can increase the total to hundreds or thousands.)

However, as first shown by Miller (1956), our short-term memory cannot juggle more than about 7 ± 2 chunks of information at a time, as will be apparent when you try to dial a 10-digit telephone number. As a result the information content of the chart *always* exceeds our capacity to handle it.

In theory the astrologer overcomes this problem by a process called chart synthesis, whereby the relevant factors are balanced against one another on paper. In practice there is little agreement on how this balancing should be done, or even on what factors are relevant in the first place. As we shall see in Part 2, this has predictable consequences.

The Real Thing and Prediction

After any notable event, such as a major earthquake or an assassination, articles appear in astrology journals showing a clear correspondence between the event and its astrological chart. However the correspondence means nothing unless it can predict the event in advance. This was put to the test at the U.S. Geological Survey by Hunter and Derr (1978), who as part of a general evaluation of earthquake prediction systems invited the public to send in their predictions. The biggest response was from astrologers, with psychics and amateur scientists next. Hunter and Derr analyzed a total of 240 earthquake predictions by 27 astrologers and found their accuracy to be worse than guessing. The same was found for the predictions by psychics and amateur scientists.

Culver and Ianna (1984) surveyed 3,011 specific predictions made from 1974 to 1979 in U.S. astrology magazines, such as *American Astrology,* and found that only 338 (11 percent) were correct. Many of these could be attributed to shrewd guesses ("East-West tension will continue"), vagueness ("A tragedy will hit the eastern U.S. this spring"), or inside information ("Starlet A will marry director B"). After allowing for chance there seemed to be nothing left for astrology to explain. The same was found by Châtillon (1985), who surveyed 30 specific predictions for North America in 1984 made by Huguette Hirsig, one of Montreal's most famous astrologers. Only two (7 percent) were correct.

Reverchon (1971) surveyed a series of predictions made from 1958 to 1961 in the French astrological journal *Les Cahiers Astrologiques.* They were made by the renowned French astrologer André Barbault, a specialist in such predictions, and concerned the end of the French-Algerian war. As each prediction failed (the end was very protracted), Barbault was able to find further indications. No less than 11 successive predictions were made before the inevitable hit was achieved, thus reducing everything to a "childish game." Reverchon then compared Barbault's predictions of world crises for 1965 (published in 1963) against an independent list of 105 major world events for that year. There were 5 hits vs. 8 expected by chance. Specific predictions involving a dozen world leaders included many "high quality blunders"; for example, Kennedy would be reelected in 1964 (he was assassinated in 1963), Krushchev would remain in power until 1966 (he was deposed in 1964), de Gaulle would resign in 1965 (he was reelected), and both Erhard (Germany) and Wilson (UK) would enter a decline (both were reelected). Reverchon concluded: "What most surely appears . . . is the perfect inanity of the astrological undertaking. . . . What was announced did not happen, what happened was not announced."

Of course it may be that the specific predictions involved in these four surveys are more difficult than those made in a personal consultation, which may be concerned only with general trends. But, as pointed out by Culver and Ianna, astrologers who make such predictions presumably feel competent to do so; hence there is no reason to suppose that the results are not typical of astrologers generally. For a rare description of a typical technique at work see Stearn's (1972) account of a U.S. astrology class held in 1970; the astrologer makes numerous predictions for Richard Nixon, including "unprecedented popularity . . . peaking in 1975" (Watergate occurred in 1973; Nixon resigned in 1974).

Dean (1983) analyzed 18 years of Nelson's daily forecasts of shortwave radio quality and found no support for Nelson's quasi-astrological claim that planetary positions correlate with radio quality. Now radio quality is quantified into numbers to avoid guesswork, and Nelson's technique (which is based on the angular separation between planets) is by astrological standards almost embarrassingly simple. Yet for 30 years Nelson was convinced he saw a correlation that in fact did not exist. So we should not be surprised that astrologers, working with generally vaguer events and far more complicated techniques, can see correlations even if none actually exists.

The astrologers' response to these five surveys, which are the only ones I know of, has not been to generate surveys of their own. Instead there has been either silence or brusque dismissal, such as that by a reviewer in the Canadian astrological journal *Fraternity News* (1986), who dismissed Culver and Ianna's entire book as "not even good objective criticism." Of course it could be argued that this is a legitimate response for the two in three Western astrologers who eschew prediction in favor of psychology and counseling. So the rest of this article is addressed to their point of view.

The Real Thing and Counseling

Here the term *counseling* is used in accordance with the following classification of astrological consultations due to Rosenblum (1983, pp. 33-44):

A. Chart reading. Usually one session only; astrologer talks, client listens.

B. Counseling. One or several sessions; client participates in a dialogue. Involves inquiry into client's life; addresses short-term problems.

C. Therapy. Regular ongoing sessions; client has major long-term problems and requires help to regain control of life. Astrologer has (or should have) orthodox qualifications in psychotherapy.

Each type blends into the next to form a continuum, so the classification is basically one of emphasis. Many astrologers consider A to be unhelpful and potentially harmful because the client is passive and dependent on the all-knowing astrologer. But if clients are merely curious about astrology, A may be all they want. In occurrence, A and B are probably roughly equal, while C is rare, probably roughly 0.5 percent of B. For these reasons I have focused on B as the real thing.

Lester (1982), a professor of psychology in the United States, visited an astrologer, talked to clients of astrologers, and surveyed astrological writings. He concluded: (1) Astrologers play a role similar to that of psychotherapists. (2) People consult astrologers for the same reason they consult psychotherapists, but without the stigma the latter may entail. (3) Clients get empathy, advice, compliments (which increase self-esteem), and positive comments about possible future traumas, all of which amounts to supportive psychotherapy.

Skafte (1969), a psychologist and counselor, tested the effect of introducing popular astrology (and palmistry and numerology) into personal and vocational counseling, for example, by saying "a person born under your sign is supposed to enjoy travel—does this sound like you?" The words were chosen to avoid implying validity and to promote dialogue. She found that: (1) this provides a focal point for discussion that often stimulates clients to talk openly about themselves, (2) mutual interest in an unconventional activity quickly creates closeness and rapport that would otherwise take many sessions to establish, (3) the focus on individual qualities (as opposed to, say, impersonal questionnaires), meets the clients need to feel special.

Clearly, when used in this way, astrology can be valuable without needing to be true. Skafte's first point about astrology's providing a focus is amply confirmed in astrology books and lectures, which often contain surprisingly little astrology. Thus an exposition by a good astrologer on the special problems faced by Neptune in each house, or how to live with a T-square or a void-of-course moon or a Splash pattern or a heavy fifth harmonic, will contain beneath the jargon a sensible and insightful commentary on human behavior that any caring person of rich experience could deliver. In such cases astrology, without needing to be true, acts as an organizing device for the otherwise unmanageable smorgasbord of human experience.

Mayer (1978), a humanistic psychotherapist and astrologer, extends Skafte's sun-sign approach to include all of astrology. His concern is to help clients confused about their identity and seeking a meaning in life. He argues that this is difficult via the orthodox personality theories used to guide therapy, but easy via the imagery and complexity of astrology without requiring it to be true. For this purpose he proposes a new kind of astrology for which no claims of validity are made, and which is contraindicated for clients opposed to nonrational approaches or overinclined to fantasy. However, therapists and clients seem unlikely to accept a tool of this complexity unless some underlying truth is assumed, in the same way that we would resist using English if it required us to speak in riddles.

Laster (1975), an educational psychologist and astrologer, makes the pragmatic point that the many people who believe in astrology can be reached on common grounds of faith by counselors familiar with astrology, just as Jews can be better reached by Jewish counselors than by non-Jewish ones. On this basis the validity of the belief—whether Jewish or astrological—should not be an issue if the belief helps to establish rapport between client and therapist. Here Laster is in effect redefining astrology as a religion, so his point becomes invalid if the client seeks earthly guidance divorced from spiritual understanding. I say more about utility vs. validity later.

Wedow (1976), a sociologist, made tape recordings of counseling sessions with eight astrologers to find out what happens when they make a wrong statement about the client. She found that they gave one or more of the following explanations:

1. Client does not know himself. } This shifts the blame from astrol-
2. Astrologer is not infallible. } ogy to the participants.

3. Another factor is responsible. } This puts the blame on the am-
4. Manifestation is not typical. } biguity of the birth chart.

Wedow notes that such explanations make the whole process nonfalsifiable, and that the participants seem to be unaware of this nonfalsifiability. Hence once the session has begun the end result can hardly fail to maintain astrology's credibility.

Note that this nonfalsifiability arises not from the chart factors themselves, which are in principle testable and therefore falsifiable, but from what astrologers do with them. The process is described so vividly by Hamblin (1982), an astrologer critical of current practice and later chairman of the U.K. Astrological Association, that he is worth quoting in full:

> If I find a very meek and unaggressive person with five planets in Aries, this does not cause me to doubt that Aries means aggression. I may be able to point to his Pisces Ascendant, or to his Sun conjunct Saturn, or to his ruler in the twelfth house; and, if none of these alibis are available, I can simply say that he has not yet fulfilled his Aries potential. Or I can argue (as I have heard argued) that, if a person has an *excess* of planets in a particular sign, he will tend to suppress the characteristics of that sign because he is scared that, if he reveals them, he will carry them to excess. But if on the next day I meet a very aggressive person who also has five planets in Aries, I will change my tune: I will say that he *had* to be like that because of his planets in Aries.

Hamblin notes that this gives astrologers an inexhaustible reserve of explanations for even the gravest difficulties. It also reduces to inutility claims like that of Metzner (1970), a psychologist and astrologer, that chart factors in combination are "probably better adapted to the complex variety of human natures than existing systems of types, traits, motives, needs, factors, or scales." More subtly, it kills off the very understanding that the real thing is supposed to promote and replaces it with tokens of understanding that have value only in an economy of free-floating, all-purpose astrobabble. We may ask how the previously cited Media Watch astrologers could believe that this kind of thing is "psychologically and intellectually more mature." As we shall see in

Part 2, due to the nature of astrology and of the human mind, the answer is: "Very easily."

The Dark Side of Astrology

Steiner (1945), a medical and psychiatric social worker, made a remarkable survey of U.S. astrologers, palmists, numerologists, Tarot readers, and similar "consultants." The survey took 12 years, during which time she posed as a consultant to find out what people's troubles were, and visited consultants (including 40 astrologers) posing as a client to find out what their advice was like. She concluded: (1) There is no agony like emotional turmoil. People will seek relief anywhere, usually quite uncritically. (2) In general, consultants were utterly untrained for professional practice. Many were unscrupulous and dishonest. (3) No technique was better or worse than the others. Yet all consultants claimed success for their particular system.

That was the situation in the United States in the 1930s and early 1940s, and it could only improve. Thus 25 years later Sechrest and Bryan (1968) consulted 18 U.S. astrologers who advertised mail-order marital advice. They found that the advice bore no discernible relationship to astrological principles but was always realistic, and was usually direct, clear, vigorous, personal, and friendly. They concluded that the advice was not likely to be damaging and, because it was friendly and cheap, was even a great bargain. In 1978 a survey of 75 astrologers found that they and their clients were mostly solidly middle class and well educated (Koval 1979). The same year, for a consultation of one to two hours, plus up to three hours of preparatory work, the average fee for 276 U.S. astrologers was $40 to $50 (American Federation of Astrologers 1978a), which per hour was about a third of the average rate for psychoanalysis. Larner (1974), an astrologer and New York businessman, divides U.S. astrologers into the following five types but without indicating their relative numbers. The costs are those of a consultation in 1974: (1) The sun-sign astrologer $2 to $10. Typically the gypsy lady with a storefront in the low-rent district or with mail-order services advertised in newsstand astrology magazines. (2) The large-volume astrologer, $5 to $10. Found mainly at parties, resorts, and fund-raising events. (3) The kitchen astrologer, $10 to $25. Typically the hobbyist, usually a homemaker, invariably conscientious and best value for money. (4) The professional astrologer, $25 to $100. Usually has training, experience, expenses, and overhead. Best judged by reputation. (5) The flamboyant astrologer, $250 to $1,000. As (4), but gives personal service and magnificently presented charts to wealthy clients like film stars.

Today most astrological organizations hold examinations, award diplomas, and have codes of ethics. What they do not have is effective regulation, which means that anyone can become an astrologer just by saying so. With or without codes of ethics, some astrologers do play God, or make irresponsible predictions, or intrude their hangups, all of which can have traumatic effect. For example, Rudhyar (1979) says, "I have received many letters from people telling me how fearful or psychologically confused they had become after consulting even a well-known astrologer and being given biased character analyses and/or predictions of illness, catastrophe, or even death." For a personal account of such an experience see Wallace (1978). For a discussion of the various sins to which astrologers are prone during a consultation, see Rosenblum (1983, pp. 120-128).

Of course people can suffer just as much from parents, teachers, and clergy, so it would be unfair to single out astrologers, especially as they are much easier to avoid. My own experience, and my canvassing of informed opinion, suggests that the proportion of astrologers who are irresponsible is something like 1 in 20. Since astrologers are about 50 times less numerous than lawyers (Dean and Mather 1977, p. 7), the problem, while distressing, is hardly of epidemic proportions. This of course may not be the case if we include fairground astrologers and newspaper columnists, whom most serious astrologers regard as irresponsible by definition.

The reasons people believe in astrology have been surveyed by Fullam (1984) using all available opinion polls from Western countries. She concluded that people believe in astrology because it is satisfying on many levels from the trivial to the profound. Some use it as entertainment. Some use it to solve problems ("Is he right for me?"). Some use it to discover the sacred meaning of life. And of course some use it to make money. In other words, different people believe in astrology for different reasons. The interesting question of how such beliefs arise in the first place will be discussed in Part 2. For the moment let us look at why *astrologers* believe in astrology.

Why Do Astrologers Believe in Astrology?

The arguments commonly put forward by astrologers to support their belief in astrology have been critically examined by Kelly et al. (1989), who concluded that none of them stood up to inspection. The arguments and (in parentheses) responses by Kelly et al. are briefly as follows.

1. Astrology has great antiquity and durability. (So has murder.)
2. Astrology is found in many cultures. (So is belief in a flat earth.)
3. Many great scholars have believed in it. (Many others have not.)
4. Astrology is based on observation. (Its complexity defies observation.)
5. Extraterrestrial influences exist. (None are relevant to astrology.)
6. Astrology has been proved by research. (Not true.)
7. Non-astrologers are not qualified to judge. (So who judges murder?)
8. Astrology is not science but art/philosophy. (Not a reason for belief.)
9. Astrology works. (The evidence suggests otherwise.)

Of these nine arguments none is more common, more simple, and more disarming of criticism than "astrology works." So let us examine this point in more detail with a look at the views of astrologers.

The late Charles Carter (1925), the leading U.K. astrologer of the 1930s and noted for exceptional clarity of expression, says: "Practical experiment will soon convince the most sceptical that the bodies of the solar system indicate, if they do not actually produce, changes in: (1) Our minds. (2) Our feelings and emotions. (3) Our physical bodies. (4) Our external affairs and relationships with the world at large."

Edith Custer (1979), editor of a U.S. quarterly magazine devoted exclusively to letters from serious astrologers, says: "Whether the scientific world accepts or rejects astrology makes it no less a valid tool for me to work with. . . . I know it works and I am satisfied with that."

Dane Rudhyar (1970), guru of person-centered astrology, says: "If, after having studied . . . his . . . birth chart, a person . . . is able to feel a direction and purpose . . . in his life . . . then astrology is 'existentially' proven to be effective in this particular case. It 'works'—for him."

Rudolf Smit (1976), founding editor of the Dutch astrological journal *Wetenschap & Astrologie* (*Wetenschap* = Science), says: "On the inevitable question 'why does astrology work?' even the most intelligent and experienced astrologers cannot give a clear-cut answer, only vague assumptions. They simply don't know. which is why virtually all serious astrologers are obliged to be pragmatic: Don't ask them how it works, because they know only that it does work, which is why they use it."

The most popular vague assumption, and the subsequent circular argument, is that everything in the universe is interrelated, so that in effect we can tell what our fingers are doing by looking at our toes. For example, Zipporah Dobyns (1986, p. 33), one of the few astrologers with a Ph.D. in psychology, says, "But, increasingly, modern astrologers are realising that the correspondences are symbolic. The sky is part of the universe, and it is visible, so it is a convenient way to see the shared order."

The idea of "shared order," more usually called *synchronicity*, is not without a certain conceit. As Mackay (1852) noted, "How we should pity the arrogance of the worm that crawls at our feet, if . . . it . . . imagined that meteors shot athwart the sky to warn it that a tom-tit was hovering near to gobble it up." Astrologers are not unaware of this but argue that, because astrology works (note the circular argument), theirs is not to reason why.

What are the views of orthodox professionals who use astrology? Dr. Edward Askren (1980), a psychiatrist who was once skeptical of astrology but who now uses it in his practice, says, "Like ethics or theology, astrology presents at its best a coherent explanation of what is, and broadly indicates how an individual does in life and how he . . . may relate to the rest of creation." He describes the benefits of using astrology in his practice of psychotherapy as follows: "[Astrology provides] me with a different view of personality—one that seems to be more congruent with the world. . . . By giving me a new set [of analogies] with which to perceive, it helps me to see things I would not see otherwise. My patients have responded—some negatively, some positively, some gradually positively."

Dr. Bernard Rosenblum (1983, pp. 3-4), a psychiatrist who uses astrology in his practice, describes his first visit to an astrologer at age 41 and incognito. He was told about his conflicts, talents, intellectual style, emotions, parental images, and much more, including the opinion that he was, or should be, a psychoanalyst or psychiatrist: "It was all pointedly meaningful to me—and surprisingly specific. The usual criticism of astrology, that it produces a variety of generalities that can refer to almost anyone, was suddenly, in my mind, relevant only to newspaper and magazine types of astrology and no longer to the experience of going to a competent astrologer. . . . Now that I have studied astrology myself, I am well aware of the excellent contributions astrology can make to human understanding."

Not all professionals come away from astrology with such glowing opinions. Dr. Anthony Stevens, a psychiatrist who assessed chart readings as part of Parker's (1970, pp. 217-219) investigation of astrology, concluded that astrology is a delusional system comparable to organized religion and is used to impose order on private chaos. For troubled people who refuse to accept personal responsiblity for their lives, a good psychiatrist or astrologer may be a necessity. But unlike psychiatrists, who free their clients from their paranoia (i.e., the attitude that events are beyond their control), astrologers reinforce it by dragging their clients into "a shared paranoia, a *folie à deux,* in which both astrologer and client subscribe to the same delusional system." He concludes:

> Astrology, in my view, is not so much anti-therapeutic as a-therapeutic, producing a psychologically sterile liaison between client and astrologer which stultifies creativity instead of making it possible: not "know thyself" but "know thy stars." At this crucial point, the similarity between astrology and psycho-analysis ends: if my own fate should bring me to the crossroads, I know to which discipline I should turn for help.

To illustrate the understanding that astrology can bring, suppose you are experiencing emotional ups and downs (or restlessness or problems at work or whatever). Your astrologer points out that your chart has Mars aspecting Venus, or the moon in a Fire sign, or a lack of Earth, or transiting Uranus in the fifth house, or any of a hundred other things, all of them indicating ups and downs and thus confirming your situation. The astrologer then explains the strengths and weaknesses of these factors and how they interact with other chart factors, and shows how any liabilities can be minimized or even turned into assets, for example, by avoiding situations abrasive to your sensitive Neptunian nature, or by concentrating on the fine communicative skills shown by your strong third and ninth houses. In effect your situation is repacked and put into coherent order by the structure of the chart, in the same way that a transactional analyst would repack it in terms of Adult, Parent, and Child. So you see your situation from a new vantage point. Since you have never heard yourself explained in such a simple and appealing way, it is a revelation. You end up reassured, self-aware, and very satisfied with the service, which the astrologer sees as yet more evidence that astrology works.

These examples (and I could have cited many more) illustrate the kind of evidence that astrologers respect most. They see that astrology gives benefit, self-understanding, and spiritual insight. They see that it helps people. They see that it works. And because seeing is believing, they don't care what the critics say—*they know*. What could be more reasonable? But phrenologists said exactly the same.

A Salutary Lesson from Phrenology

Phrenology is a system of intellectual and moral philosophy that is based on reading character from brain development as shown by head shape. Phrenology is now virtually dead, but in the 1830s it was more popular than astrology is today. Like astrology, it encourages you to assess yourself via its principles and act on the findings to achieve harmony with the world. Like astrology, it attracted people of intelligence and a vast literature wherein every criticism was furiously attacked. Like astrology, it flourished because practitioners and clients saw that it worked. For many other parallels see Dean and Mather (1985).

But the claims of phrenology are now known to be wrong. Character is not indicated by brain development because the brain does not work like that, at least not in the way and to the extent required by phrenology (Davies 1955; Flugel 1964). So a certain head-shape cannot mean what it is supposed to mean. Yet millions of people could agree that phrenology works, just as millions of people today agree that astrology works. But could millions of people be wrong? As discussed next, the answer is yes and no.

It All Depends on What You Mean by "Works"

We have seen that astrologers believe in astrology because it works. But as Eysenck and Nias (1982, p. 211) point out, it all depends on what you mean by *works*. If by *works* you mean *is helpful,* the popularity of serious astrology leaves no doubt that it does indeed work. But this is hardly surprising—after all, to most people astrological ideas have undeniable beauty and appeal, the birth chart is nonjudgmental, the interpretation is nonfalsifiable, and astrologers tend to be nice people. In a society that denies ego support to most people, astrology provides it at a very low price. Where else can you get this sort of thing these days?

But if by *works* you mean *is true,* this changes the situation entirely and brings us back to the question of utility vs. validity. It is one thing to say we can learn about ourselves by following the interaction of Mars and Venus like toy soldiers in a psychological war game and quite another to say that these interactions are related to what Mars and Venus were doing at our moment of birth. As one astrologer who recognizes the problem put it, "Any good I've done as a consultant, and I have done some good, had less to do with my being a good astrologer than with my being a good person" (Ashmun 1984).

This explains the conflict between critics and astrologers: Critics see a lack of factual evidence and conclude it doesn't work, whereas astrologers see that it helps people and conclude it does work. Both are right—and both are guilty of not wanting to know what the other is talking about. The situation is not helped by the typical astrologer's attitude toward factual evidence so well described by Levy (1982), who runs Australia's largest computerized chart calculation service: "I often get the feeling, after talking to astrologers, that they live in a mental fantasy world, a kind of astrological universe where no explanations outside of astrological ones are permitted, and that if the events of the real world do not accord with astrological notions or predictions, then yet another astrological technique will have to be invented to explain it."

In such a situation the crucial question is not whether astrology is true but whether it *needs* to be true. We have already seen that, at the trivial level, the answer is no (Skafte 1969). But what about the real thing? To find out we must first understand some more about birth charts.

The Importance of an Accurate Birth Chart

Astrology postulates a correspondence between birth chart and person. Or as above, so below. Some astrologers, like Charles Carter, hold the traditional view that the birth chart indicates character and destiny. Others, like Dane Rudhyar, see it pertaining only to the individual's potential. (Here I will ignore the problem that, because potential can never be determined, it is impossible to know whether such astrology works.) Either way, an accurate birth chart is essential—a point confirmed by astrological organizations in their codes of ethics, as shown by the following typical example from the American Federation of Astrologers (1978b): ". . . A precise astrological opinion cannot honestly be rendered with reference to the life of an individual unless it is based upon a horoscope cast for the year, month, day and time of day plus correct geographical location of the place of birth."

If an accurate birth chart is essential, then the wrong chart should ruin everything. But if the chart makes no difference, the rationale for astrology disappears—and astrology does not need to be true. So let us now put this point to the test.

Right Charts vs. Wrong Charts

Right and wrong charts have been compared in seven independent studies, nearly all of them made in the past five years, in which subjects had to decide

TABLE 1

Can Subjects Tell Right Charts from Wrong Charts?

| | | | | No. Picking Own Chart | |
Study	Note	No. of Subjects	Charts per Subject	Observed	Expected By Chance
Cummings et al. 1978	1	12	3	4	4
Neher 1980	1	18	6	3	3
Lackey 1981	2	38	2	19	19
Dwyer & Grange 1983	3	34	3	10	11
Tyson 1984	1	15	5	2	3
Carlson 1985	1	83	3	28	28
Dwyer 1986	3	30	2	15	15
Total		230		81	83

Answer: Unanimously no. The overall trend is not even in the right direction. The interpretations were prepared by (1) one or more professional astrologers; (2) the experimenter, from books; or (3) a computer. They were usually based either on the whole chart or on the whole chart minus long-term factors, such as the sign position of planets beyond Jupiter.

which of two or more chart interpretations fitted them best. One interpretation was of their own chart, the rest were those of other subjects picked at random. Care was taken to ensure the direct clues, such as birth data, were excluded and that indirect clues, such as sun-sign descriptions, were either excluded or were the same in all interpretations. According to astrology the subjects should certainly tend to pick their own charts. But the results (Table 1) show that in every study the subjects performed no better than chance. In other words, they were just as happy with wrong charts as with right ones. This suggests that the perceived validity of astrology is an illusion.

The results cannot be explained by poor interpretation. Thus, in the study by Dwyer (1986), who at the time was a tutor in the internationally known Mayo School of Astrology, the method had previously been progressively refined via a panel of 30 control subjects to maximize accuracy. And in the Carlson (1985) study the interpretations were individually prepared by experienced professional astrologers judged by their peers to be highly competent. Yet in both studies the results were at exactly chance level.

Nor can the results be explained by the subjects' not knowing themselves. Thus, in the study by Tyson (1984), the test was also given to someone who knew the subject well (usually a parent), but the results were just as negative—

3 hits vs. 3 expected by chance. Ianna and Tolbert (1984–85) tested the ability of U.S. astrologer John McCall to pick the correct chart out of four from the subject's face and build, which of course avoids the problem entirely. McCall was confident of success (he had previously put an ad in the *Washington Post* challenging scientists to test him) and was completely satisfied with the test conditions. Yet he scored only 7 hits for 28 subjects, no different from the 7 expected by chance, and scarcely better than his score of 1 hit for 5 subjects obtained in an earlier test (Randi 1983).

We may note that, if subjects do not know themselves, then valid personality questionnaires could not exist; see Cronbach (1970). Nor could astrologers ever know that astrology works. Or as one indignant correspondent to *American Astrology* put it, "I believe that I know myself better than that conceited Virgo astrologer did" (Shivers 1983).

And, indeed, when the approach used in Table 1 is applied to personality inventories, the correct profile tends to be chosen; for example, see Greene et al. (1979) and the results of Grange (1982) cited in Part 2. This shows that, while self-knowledge may not be 100 percent, it is sufficient for the present purpose. At any rate, the results of Table 1 are consistent with examples in the astrological literature where the interpretation fits the subject perfectly but the chart is subsequently found to be wrong (Dean and Mather 1977, pp. 28-31). Such examples are often very telling, as this one from the late Piet Hein Hoebens (1984) demonstrates: "In my newspaper column in *De Telegraaf* I have occasionally discussed astrological topics. Mr. Gieles, a well-known astrologer in The Hague, responded to my critical writings by publishing my horoscope, which, not surprisingly, revealed that the stars and planets had conspired to make me a critical journalist hostile to Mr. Gieles's claims. Everything fitted beautifully except one detail—poor Mr. Gieles had used the wrong birth date!"

The Problem of Words

One problem with testing astrological interpretations is that they tend to be wordy and rambling. For example, the interpretations tested by Carlson (1985) averaged just over 1,000 words each, of which the following excerpts are typical: (1) Emotions tend to be erratic especially when communications break down. (2) You want to belong and fit in, at the same time you want to be noticed. (3) You can hold jobs of singular authority when in command. (4) You have a deep mind but tend to daydream when bored and need the discipline of education to stimulate your versatility. [Total: 56 words.]

Trying to choose between three 1,000-word interpretations in such a style is conducive to mental paralysis; the mind cannot cope. This does not invalidate Carlson's study—the interpretations were prepared by highly competent astrologers, so it is a fair test of actual professional practice. But it does leave us wondering what would happen if (1) wrong charts were used in an actual consultation, and (2) the interpretations were made especially concise to facilitate detection of their wrongness. Would the client notice? I decided to find out.

References

AFAN. 1985. AFAN's Media "Watchdog Project." *AFAN Newsletter*, 3(4):3 (July 1985). And Media Watch Update. *AFAN Newsletter*, 4(1):3 (October 1985). The AFAN (= AFA Network) is an unaffiliated auxiliary organization of members of the American Federation of Astrologers and others.

American Federation of Astrologers. 1978a. AFA questionnaire summation report. *AFA Bulletin*, 40(8):6-16.

———. 1978b. Code of ethics (on reverse of membership application form to be signed by all AFA members) effective from 1 December 1978.

Ashmun, J. M. 1984. Editor's response. *Seattle Astrologer*, 16:4-5.

Askren, E. L. 1980. A psychiatrist-psychologist looks at astrology. *Journal of Geocosmic Research Monograph*, no. 1:10-15.

Carlson, S. 1985. A double-blind test of astrology. *Nature*, 318:419-423 (5 December 1985), and personal communication 1986.

Carter, C. E. O. 1925. *The Principles of Astrology*. London: Theosophical Publishing House, p. 14.

Châtillon, G. 1985. Astrology in Quebec. Letter to *Skeptical Inquirer*, 9:398-399 (Summer).

Cronbach, L. J. 1970. *Essentials of Psychological Testing*, 3rd ed. New York: Harper and Row.

Culver, R. B., and Ianna, P. A. 1984. *The Gemini Syndrome: A Scientific Evaluation of Astrology*. Buffalo, N.Y.: Prometheus Books; 250 references.

Cummings, M., Smith, M., Lovick, K., and Crosbie, P. 1978. Astrological chart interpretation: Exploring an alternative strategy for counseling. *Kosmos*, 8(2):5-26 (Winter 1978-79).

Custer, E. 1979. Editorial comment. *Mercury Hour*, no. 21:15 (April 1979).

Davies, J. D. 1955. *Phrenology Fad and Science: A 19th-Century American Crusade*. Yale University Press, pp. 140-143.

Dean, G. 1983. Forecasting radio quality by the planets. *Skeptical Inquirer*, 8:48-56 (Fall).

Dean, G., and Mather, A. 1977. *Recent Advances in Natal Astrology: A Critical Review 1900-1976*. Subiaco, Western Australia: Analogic; 1020 references. *Recent Advances* and back issues of the *Astrological Journal* are available from AA Publications, 396 Caledonian Road, London N1 1DN, England. Tel. 081-469-2828.

———. 1985. Superprize results: Six winners—and news of a new prize to challenge the critics. Part 1. *Astrological Journal*, 28(1):23-30 (Winter 1985). Contains major editorial errors (none relevent to phrenology). An error-free version appears in *FAA Journal* [Australia], 15(3-4):19-32 (September-December 1985).

Dobyns, Z. 1986. The return of Halley. *Kosmos*, 15(2):32-45 (Spring 1986).

Dobyns, Z. P., and Roof, N. 1973. *The Astrologer's Casebook*. Los Angeles: TIA Publications, p. 4.

Dwyer, T. 1986. Unpublished work performed as a potential entry for astrology superprize. For details see *Astrological Journal*, 28(3):92-93 (April 1986).

Dwyer, T., and Grange, C. 1983. Unpublished work entered for the *Recent Advances* astrological prize number 2. For details see *Astrological Journal*, 25(3):206 (Summer 1983).

Eysenck, H. J., and Nias, D. K. B. 1982. *Astrology: Science or Superstition?* New York: St. Martin's Press; 230 references.

Flugel, J. C. 1964. *A Hundred Years of Psychology*, 3rd ed. London: Duckworth, pp. 31-38. A wonderfully sympathetic but devastating critique of phrenology that applies equally to astrology.

Fraternity News. 1986. Review by Lois Strantz of *The Gemini Syndrome*. 8(1):65.

Fullam, F. A. 1984. *Contemporary Belief in Astrology.* Master's thesis, Department of Sociology, University of Chicago, July 1984.

Grange, C. 1982. *The Barnum Effect: Its Relevance for Non-Scientific Methods of Personality Assessment.* Unpublished study, Loughborough University of Technology, Loughborough, UK.

Greene, R. L., Harris, M. E., and Macon, R. S. 1979. Another look at personal validation. *Journal of Personality Assessment,* 43:419-423.

Hamblin, D. 1982. The need for doubt and the need for wonder. *Astrological Journal,* 24(3):152-157 (Summer 1982).

Hoebens, P. H. 1984. Personal communication, August 1984.

Hunter, R. N., and Derr, J. S. 1978. Prediction monitoring and evaluation program: A progress report. *Earthquake Information Bulletin* [USGS], 10(3):93-96 (May-June 1978); and personal communication, 1986. Because of lack of funds no final report was published. The program outline appeared in 8(5):24-25 (September-October 1976).

Ianna, P. A., and Tolbert, C. R. 1984-85. A retest of astrologer John McCall. *Skeptical Inquirer,* 9:167-170 (Winter).

Kelly, I. W. 1982. Astrology, Cosmobiology, and Humanistic Astrology. In P. Grim (ed.), *Philosophy of Science and the Occult.* Albany: State University of New York Press, pp. 47-69; 60 references.

Kelly, I. W., Culver, R., and Loptson, P. J. 1989. Arguments of the astrologers: A critical examination. In S. K. Biswas, D. C. V. Malik, and C. V. Vishveshwara, eds., *Cosmic Perspectives.* New York: Cambridge University Press. 45 references.

Kelly, I. W., and Krutzen, R. W. 1983. Humanistic astrology: A critique. *Skeptical Inquirer,* 8(1):62-73 (Fall).

Koval, B. 1979. Response to CAO Ethics Questionnaire. *Mercury Hour,* no. 20:40-42 (January 1979).

Lackey, D. P. 1981. A controlled test of perceived horoscope accuracy. *Skeptical Inquirer,* 6(1):26-31 (Fall).

Larner, S. P. 1974. *Astrological Assistance.* Englewood Cliffs, N.J.: Prentice-Hall, pp. 15-18.

Laster, A. 1975. *On the Psychology of Astrology: The Use of Genethliacal Astrology in Psychological Counseling.* Ph.D. thesis, School of Education, University of Pittsburgh. University Microfilms catalog number 7620183.

Lester, D. 1982. Astrologers and psychics as therapists. *American Journal of Psychotherapy,* 36:56-66.

Levy, A. 1982. Rectification and predictions. *Australian Astrologers' Journal,* 6(3):19-23 (Spring 1982).

Mackay, C. 1852. *Extraordinary Popular Delusions and the Madness of Crowds.* London: National Illustrated Library, p. 282. Reprinted by Noonday Press, New York 1977.

Mayer, M. H. 1978. *A Holistic Perspective on Meaning and Identity: Astrological Metaphor as a Language of Personality in Psychotherapy.* Ph.D. dissertation, Humanistic Psychology Institute, Calif. University Microfilms catalog number LP 00166. See especially pp. 70-75. Published in revised form as *The Mystery of Personal Identity* (San Diego: ACS, 1985).

Metzner, R. 1970. Astrology: Potential science and intuitive art. *Journal for the Study of Consciousness,* 3(1):70-91.

Miller, G. A. 1956. The magical number seven, plus or minus two: Some limits on our capacity for processing information. *Psychological Reviews,* 63:81-97.

Neher, A. 1980. *The Psychology of Transcendence.* Englewood Cliffs, N.J.: Prentice-Hall, pp. 239-242; and personal communication, 1985.

Parker, D. 1970. *The Question of Astrology.* London: Eyre and Spottiswoode.

Perinbanayagam, R. S. 1981. Self, other and astrology: Esoteric therapy in Sri Lanka. *Psychiatry,* 44:69-79.

Randi, J. 1983. A small-scale test of an astrological claim. *Skeptical Inquirer,* 7(4):6-8 (Summer).

Reverchon, J. 1971. *Value of Astrological Judgements and Forecasts.* Yerres, France: self-published. French text with English translation, 13 pp. each.

Rosenblum, B. 1983. *The Astrologer's Guide to Counseling: Astrology's Role in the Helping*

Professions. Reno, N.Y.: CRCS.

Rudhyar, D. 1970. How can astrology's claims be proven valid? *Aquarian Agent,* 1(10):6-7. In 9,000 words Rudhyar never delivers a recognizable answer to this question. The nearest we get is the excerpt cited.

———. 1979. Review of "Recent Advances in Natal Astrology." *Zetetic Scholar,* nos. 3-4, pp. 83-85.

Sechrest, L., and Bryan, J. 1968. Astrologers as useful marriage counselors. *Trans-Action,* 6:34-36.

Shivers, O. 1983. Letter to the Editor. *American Astrology,* September 1983:18.

Skafte, D. 1969. The use of palmistry in counseling. *Voices,* 5(4):38-41 (published by the American Academy of Psychotherapists). Also covers astrology and numerology.

Smit, R. 1976. *De Planeten Spreken: De Werkelijkheid achter de Astrologie* (The planets speak: The reality behind astrology). Bussum, Netherlands: Fidessa, p. 32.

Smith, R. F. 1975. *Prelude to Science: An Exploration of Magic and Divination.* New York: Scribner, p. 24.

Startup, M. J. 1984. *The Validity of Astrological Theory as Applied to Personality, with Special Reference to the Angular Separation Between Planets.* Ph.D. thesis, Goldsmith's College, London University; 350 references. University Microfilms catalog number 85-00, 985.

Stearn, J. 1972. *A Time for Astrology.* New York: New American Library, pp. 247-257.

Steiner, L. R. 1945. *Where Do People Take Their Troubles?* Boston: Houghton Mifflin.

Truzzi, M. 1979. Astrology: A review symposium (prologue). *Zetetic Scholar,* nos. 3-4, pp. 71-73.

Tyson, G. A. 1984. An empirical test of the astrological theory of personality. *Personality and Individual Differences,* 5:247-250.

Wallace, L. 1978. Dialogue: Noel Tyl interviews Lore Wallace. *Astrology Now,* no. 20, p. 59. Wallace describes going to a famous astrologer at age 17, whose prediction of things like a difficult birth and the death of a child "have damaged me probably for the rest of my life." Her comments imply that the predicted events did not occur.

Wedow, S. M. 1976. The strangeness of astrology: An ethnography of credibility processes. In W. Arens and S. P. Montague (eds.), *The American Dimension: Cultural Myths and Social Realities.* Port Washington, N.Y.: Alfred. And next reference.

———. 1974. *Perennial Wisdom on Display: The Use of a System of Knowledge* [= astrology] *in Interaction.* Ph.D. dissertation, Department of Sociology, University of California, Santa Barbara, 1974.

GEOFFREY DEAN

Does Astrology Need to Be True?
Part 2: The Answer Is No

My experiment consisted of having each of 22 subjects rate extremely concise interpretations of their astrological charts. The subjects (5 male, 17 female, mean age 31) were recruited through a local occult bookstore and ads in an occult magazine and were previously unknown to me. Each interpretation was based solely on interplanetary aspects (specified angular separations), because these generally have clear meanings, can be weighted in a strength according to exactness, and are considered by many astrologers to be the most important factors in the chart. For example, the eminent U.S. astrologer Rob Hand (1981) says, "They usually speak the loudest and yield the most reliable results." The effect of the other chart factors is discussed later.

Each interpretation consisted of a list of the closest aspects (typically 10 to 12 per chart), their exactness (range 0 to 5 degrees), their individual meanings expressed as adjectives or short statements (average 22 items per chart), and their opposite meanings (average 18 items per chart). For example, the meanings for Mars conjunct Uranus were as follows:

> Meaning: impatient, mind of own, disruptive   (3 items)
> Opposite: patient, easily led, not disruptive   (3 items)

The strength of these indications would then be strong, average, or weak, depending upon the exactness. The strength was not quantified; instead, the subject made his or her own estimate of the exactness, with verbal guidance from me. Obviously a subject who could be both x and not x would agree with anything the chart said about x, thus inflating the apparent accuracy, so opposite meanings were included to avoid this problem. The meanings and opposites were labeled so that the subject knew which was which.

The subjects were led to believe that the chart interpretations were authentic. In fact, only half the subjects received interpretations based on their actual charts. The rest received interpretations based on what I call "reversed charts." A reversed chart is one made to be as opposite to the actual chart as possible but with the same sun-sign to avoid suspicion. Thus, if the actual chart contained sun square Mars (= impetuous), the planets in the reversed chart would be juggled to give sun trine Saturn (= cautious). In this way extroverted indications were substituted for introverted, stable for unstable, tough for tender, ability for inability, and so on. The use of reversed charts is preferable to using actual charts with reversed interpretations because it allows the reader (in this case, me) to proceed normally without the need for pretense—an important consideration in a face-to-face situation. Some charts are too ambiguous to be adequately reversed; such cases were not included among the 22 subjects.

The subjects came separately for their consultations. I gave each subject a birth chart and written interpretation, explained what the chart symbols meant, and stressed the need to test the chart carefully before its indications could be accepted (this justified the next bit). The subject then rated on a 3-point scale (correct, uncertain, incorrect) each item in the interpretation, rating it as correct only if both meaning and strength were correct. This made the test as severe as possible. When finished, the subject carefully reviewed the ratings as a whole to resolve (1) any uncertainties, and (2) any conflict between one part of the interpretation and another (this accommodated the dictum that astrological factors must not be judged in isolation). Most subjects changed nothing. Each session was unhurried and occupied one to two hours. All subjects found the rating procedure to be simple and straightforward.

In this test the meanings were made as clear and concise as possible, and half the charts were made as wrong as possible. According to astrology the wrong charts should have stood out a mile. But the results (Table 1) show they were rated just as highly as right charts. In fact the results were so consistent and clear-cut that plans for further tests (which required a full day's work for each subject) were abandoned. However, I did perform a couple of similar tests in which two female subjects (known to each other but not to me) attended together and rated supposedly authentic interpretations that had in fact been switched. The results were the same: The subjects agreed with what seemed to be theirs (but were actually not theirs), and disagreed with what seemed to be not theirs (but were actually theirs). Moreover each subject agreed with the other's assessment.

However, before we can believe these results, we have to be sure that they cannot be explained by other chart factors, such as signs and houses, or by subjects preferring desirable descriptions (generous, serious) to undesirable ones (extravagant, grim). This is shown to be the case in Table 2. We also have to be sure that my personal presence did not bias the subjects' ratings. Fortunately this can be checked against an earlier published study of mine involving similar ratings done by mail and therefore free from such bias (Dean and Mather 1977, p. 39). In this study the average hit rate indicated by 44 subjects was 95 percent, or almost exactly the same as in the present study. This shows that any bias is not appreciable. So what do these results tell us?

TABLE 1

Can Subjects Tell Right Charts from Reversed Charts?

Charts	Number of items rated by 11 subjects	Number of hits*	Hits %	Range %
Authentic	261 meanings	250	96	90-100
	213 opposites	25	12	0-37
Reversed	214 meanings	207	97	90-100
	186 opposites	29	16	6-22

*Item was a hit if correct and a miss if uncertain or incorrect.

Answer: No. Subjects rated reversed charts (whose astrological indications could not have been more wrong) just as highly as authentic charts.

Reversed Charts and Cognitive Dissonance

The subjects clearly believed that their charts provided true descriptions of themselves even when, according to astrology, the descriptions could not have been more wrong. This finding is consistent with the results of previous studies (see Table 1 in Part 1) and of the German psychologist and astrologer Peter Niehenke (1984). Niehenke gave 3,150 German subjects a 500-item questionnaire designed to test astrological claims, including aspect interpretations. The results were completely negative; for example, subjects with as many as four Saturn aspects (which are supposed to indicate heavy responsibility and depression) felt no more depressed than those with no Saturn aspects.

Similar results were obtained by Neher (1980) in small-scale studies of numerology, palmistry, Tarot, and the I-Ching, by Dlhopolsky (1983) in a small-scale study of numerology, and by Blackmore (1983) in tests of Tarot interpretations involving 29 subjects. In every study the subjects were unable to pick the right interpretation at better than chance level. Similarly, Hyman (1977, p. 27) found that palmistry was just as successful when the interpretation was the opposite of what the hand indicated. I myself have given astrologers a chart that was supposedly mine, but was actually that of somebody else, and their interpretations always fitted me perfectly.

But why should subjects see the birth chart (whether right or wrong) as being valid? Possible explanations are surprisingly numerous (Table 3). Some, like the Barnum effect (Dickson and Kelly 1985), where people accept vague statements as being specific for them when in fact they apply to everybody, are well known. Others, like selective memory (Russell and Jones 1980), the

TABLE 2

Some Individual Ratings of Aspect Interpretations

AUTHENTIC CHART

Subject agreed with this (= meaning of aspect actually present)	And disagreed with this (= meaning opposite to aspect actually present)	Aspect actually present*
Self-willed, pig-headed, tense	Calm, diplomatic, not tense	Su-Ur
Imprudent, extreme, restless	Restrained, not restless	Mo-Ju
Erratic, lacks confidence	Calm, confident	Mo-Ur
Confused, overly imaginative	Methodical, not imaginative	Me-Ne
Active, overscattered	Patient, persistent	Ma-Ju
Irritable, disruptive	Even, not disruptive	Ma-Ur
Forceful, overdoes things	Moderate, not forceful	Ma-Pl

*All are hard aspects and all are exact within 3 degrees. So according to astrology their effect should be strong.

REVERSED CHART

Subject agreed with this (= meaning of aspect supposedly present)	And disagreed with this (= meaning of aspect actually present)	Aspect actually present*
Well-directed, organized	Impetuous, overactive, scattered	Su-Ma
Considerate, self-effacing	Forceful, self-centered	Su-Pl
Mind separate from feelings	Mind linked to feelings	Mo-Me
Emotionally reserved, calm	Outgoing, moody	Mo-Ju
Restless mind, innovative	Methodical mind, cautious	Me-Sa
Steady mind, overcautious	Restless mind, scattered	Me-Ur
Cautious, well-directed	Impulsive, scattered	Ma-Ju

*All have no contrary indications by sign, house, or aspect elsewhere in the chart. All except Mo-Me are hard aspects and most are exact within 2 degrees. So according to astrology their effect should be extremely strong.

The interpretations are verbatim but have been condensed where necessary to fit the space. The upper examples show that *unpleasant* interpretations can be accepted, suggesting that belief in astrology can be stronger than social desirability. The lower examples show that unambiguously *wrong* interpretations can be accepted, suggesting that the belief itself can be stronger than any astrological fact.

"Dr. Fox effect" (Naftulin et al. 1973), and hindsight bias (Marks and Kammann 1980), are less well known but can be remarkably potent. So when Rosenblum (1983, pp. 3-4), in the quotation cited in Part 1, saw pointedly specific meaning in what the astrologer said, we cannot conclude that there is necessarily something in astrology.

Most studies of actual interpretations have concentrated on generality (Barnum effect) and social desirability. For example Tyson (1984) found that the acceptance of chart interpretations prepared by a professional astrologer increased with their desirability; and Blackmore (1983) found that the acceptance of her Tarot interpretations increased with their generality and desirability, the correlation being about 0.3 in each case.

The mix of factors will of course vary with the astrologer. Thus Grange (1982) tested a professional astrologer whose interpretations happened to be clear and specific ("You have a good imagination"). They were judged by 54 subjects to be less accurate than Barnum and graphological statements (which were equally general), which in turn were judged to be less accurate than statements based on responses to the Eysenck Personality Questionnaire (which were necessarily specific and accurate). This shows that specific chart interpretations can be so wrong that the astrologer would be hard pressed to survive without support from the factors in Table 3.

In the present study, factors like generality and desirability do not apply. Hence the most likely explanation seems to be cognitive dissonance, or the need to justify our decisions and thus reduce any conflict (dissonance) between our thoughts and actions. The subjects were interested in astrology and probably believed in it, so they were motivated to avoid the pain of having their beliefs shattered. The interpretation test therefore became a search (albeit unconscious) for personal attributes to confirm their belief. Given the variability of human nature (we have all been everything at some time or another) the search could hardly fail.

This conclusion is supported by the results of Kallai (1985), who asked 101 male and female subjects aged 15 to 16 to judge the agreement between four supposedly astrological predictions and entries in a diary. Each prediction consisted of four statements, such as "You'll become more popular this week." The diary contained typically seven entries per prediction and was written so that one statement was confirmed ("I appeared on TV"), one was half-confirmed, one was disproved, and one was not referred to. The predictions were rated successful more often by believers in astrology than by nonbelievers, showing that what you believe affects what you see. Further striking but nonastrological examples are given by Marks and Kammann (1980).

If what you believe affects what you see, what happened before you believed what you believe? In other words, how do people come to believe in astrology?

TABLE 3

Twenty Ways to Convince Clients that Astrology Works

Principle	Factor	How it works.
Cues	Cold reading.	Let body language be your guide.
Disregard for reality	Illusory validity. Procrustean effect. Regression effects. Selective memory.	Sound argument yes, sound data no. Force your client to fit the chart. Winter doesn't last forever. Remember only the hits.
Faith	Predisposition. Placebo effect.	Preach to the converted. It does us good if we think it does.
Generality	Barnum effect. Situation dependence.	Statement has something for everybody. Everybody has something for statement.
Gratification	Client misfortune. Rapport.	The power of positive thinking. Closeness is its own reward.
Invention	Non-falsifiability.	Safety in numbers.
Packaging	Dr. Fox effect. Psychosocial effects. Social desirability.	Blind them with science and humor. The importance of first impressions. I'm firm, you're obstinate, he's —.
Self-fulfilling prophecies	Hindsight bias. Projection effects. Self-attribution.	Once seen, the fit seems inevitable. Find meaning where none exists. Role-play your birth chart.
Self-justification	Charing a fee. Cognitive dissonance.	The best things in life are not free. Reduce conflict—see what you believe.

This table shows that there are many nonastrological reasons that clients should be satisfied by an astrological consultation, none of which require that astrology be true. But if clients are going to be satisfied with the product offered, then astrologers can hardly fail to believe in astrology. In this way a vicious circle of reinforcement is established whereby astrologers and their clients become more and more persuaded that astrology works. An astrologer typically spends years learning to read charts and thus has ample chance to respond to such reinforcement.

How Does Belief in Astrology Arise?

For most people belief in astrology probably arises the same way: We hear or read what our sun-sign is supposed to mean, compare it with what we see in ourselves, and proceed from there. Let us look at what happens by summarizing in a single adjective the meaning of each sun-sign from Aries through Pisces, as follows—*assertive, possessive, changeable, sensitive, creative, critical, harmonious, secretive, adventurous, cautious, detached, intuitive*. Because we are interested only in our own sign, we fail to notice what astrologers aren't telling us, namely, that these traits are universal. Everybody behaves in each of these ways at various times; so, no matter what your sign is, it will agree with a trait you already possess. Lo! Astrology works—and you have started on the road to belief.

But there is more down this road than the universal validity of sun signs. Suppose astrology says that a person is extroverted, and we test this by asking the person questions. Since introverts occasionally do extroverted things, and vice versa, asking questions about instances of extroverted behavior ("Do you go to parties?") will necessarily produce extroverted answers that confirm astrology. Conversely, introverted questions ("Do you read books?") will necessarily produce introverted answers that disconfirm astrology. In other words the slant of the question can determine the outcome regardless of reality. So when testing astrology, what kind of questions do we tend to choose?

Glick and Snyder (1986) made an ingenious study to find out. They asked 12 believers and 14 skeptics to each test the validity of a brief chart interpretation (which indicated that the subject was highly extroverted) by asking the subject 12 questions chosen from a list of 11 confirmatory (extroverted) questions, 10 disconfirmatory (introverted) questions, and 5 neutral questions. The subject was in fact a confederate who gave predetermined answers matching the slant of the questions. Both believers and skeptics chose on the average just over 7 confirmatory questions, 3 disconfirmatory questions, and less than 2 neutral questions. In other words, *regardless of their stake in the outcome,* they tended to test the interpretation with questions that were bound to confirm it. This is consistent with the results of research into hypothesis-testing strategies in general (Nisbett and Ross 1980). But the surprises didn't end there.

For skeptics, the greater the number of confirmatory questions they asked (which of course increased the amount of confirmatory information they received), the more accurate they rated the chart interpretation, the correlation being an impressive 0.75—in fact two skeptics gave higher ratings than any believer. But for believers there was no correlation. All of them rated the interpretation as accurate or mostly accurate, regardless of the number of confirmatory questions asked, showing that their rating bore little or no relation to the information received.

Russell and Jones (1980) observed similar results for belief in ESP among 50 college students divided equally into believers and skeptics. For skeptics, about 90 percent accurately remembered an article on ESP regardless of whether it was favorable or unfavorable. For believers, 100 percent accurately remembered the favorable article but less than 40 percent accurately remembered the unfavorable article; 16 percent actually remembered it as favorable. The believers who read the unfavorable article were far more upset than the skeptics who read the favorable article.

So it seems that belief in astrology arises because (1) astrological interpretations tend to be universally valid, and (2) we tend to test an interpretation with strategies that are bound to confirm it. If one is basically a skeptic, one's belief will be modified by subsequent evidence. If one is basically a believer, one's belief will persist because apparently positive evidence (as in Table 3) will be remembered, whereas negative evidence (like this article) will be ignored. On this basis, regardless of the evidence, astrology is not going to go away.

Astrology and Human Inference

It would be wrong to conclude that the apparent validity of an astrological consultation (Table 3) is due to nothing more than simple-minded gullibility. As Hyman (1981) and Connor (1984) point out, words and sentences do not exist like chunks of rock but have to be interpreted before they mean anything. Thus the message received by the client is determined by his previous programming, that is, by the experiences and expectations he draws on to give it meaning. Even with a transcript you can never experience the interpretation the way the client did—what seems facile to you ("You have problems with money") may be deeply meaningful to the client. So no description given by me can possibly recreate what a chart interpretation feels like; for this you need to visit a good astrologer.

The point I am making is that, far from exemplifying gullibility, the factors in Table 3 mostly reflect the very human ways in which we cope with the world. In other words, fundamental to our understanding of astrology (and anything else for that matter) is the problem of human inference, namely, the ability of human beings to correctly judge what is going on. Because our everyday judgment is so successful most of the time, it never occurs to us that it might be grossly inadequate in certain situations, just as it never occurs to astrologers to test wrong birth charts. Thus we see that the interpretation fits and conclude that astrology works. What could be simpler and more convincing? But, as we have seen, it is not nearly that simple, and our convincing conclusion can be dead wrong.

Astrologers take every advantage of their inferential deficiencies. To them everything is a correspondence and nothing is a coincidence—an idea that casinos disprove daily. Thus a Sagittarian cavalry officer will be seen as confirming astrology (the centaur, symbol of Sagittarius, is half-man, half-horse) even though the occurrence is at chance level and everything else in his chart says he should be a banker. This confident use of glaring inconsistencies has been surveyed by Culver and Ianna (1984) and aptly named the "Gemini syndrome," after the two-faced propensities said to be typical of Gemini.

However, although our judgments may let us down, this is no reason to go to the opposite extreme—not even the most rational person makes a statistical study of dentists before deciding where to get his teeth fixed. On this basis there is no reason to suppose that when people go to an astrologer

The State of Astrology

Under this heading "The State of Astrology: Where Are We Headed?" 19 well-known U.S. astrologers recently gave views that tended to differ markedly from the usual bright optimism. According to them, astrology in the United States today is: at a dead halt, in a stormy situation of uncertain outcome, generally of decreased quality, in a very sorry state, plagued by bickering, too commercial, not accepted by society, maturing, often a waste of time, insufficiently person-centered, too person-centered, making progress, too ingrown, in trouble, in chaos, ignorant of relevant disciplines, and best in the world for its sensitive understanding of the human condition.

The main need is for: a theoretical basis, more facts and better theories, qualified people to do research, wider horizons such as application to ecological issues, reintegration of the sacred and the scientific, rigorous scientific testing, more person-centeredness, investigation of underlying mechanisms, proper accreditation, new ideas, more professionalism, better accreditation, more sophistication, thorough testing, and scientific research. (The views in each category total less than 19 because some astrologers evaded the question.)

Here the majority view is that astrology is in trouble and in need of proper testing. Perhaps the most heretical view came from John Townley, a respected, widely published astrologer with two decades of experience: "I would say that most of the accusers of astrology are probably correct. They think that astrologers are 100-percent charlatans, but I would bring it down to 90 percent. Not necessarily even intentional charlatans. But . . . they are suffering from the same failing. Maybe 50 percent of the people out there are deliberately selling hokum straight ahead." (Source: *Astro*Talk,* May/June 1986)

they are any less rational than when they go to a dentist. The point is that, if we want to know what is *really* going on, then we must be aware of our inferential deficiencies and act accordingly—which of course is what the scientific approach is all about. An excellent survey of human inference is provided by Nisbett and Ross (1980) and is essential reading for anyone wishing to understand why astrology is seen to work despite the lack of factual evidence.

Perhaps the central problem in astrology is that astrologers, like most people, including the orthodox professionals cited earlier, are not aware of their inferential deficiencies and do not act accordingly. If they were, and did, then the present shouting match between astrologers and critics might never have arisen, always assuming that the critics could make the corresponding attempt to become informed about astrology. The same is of course true of most other paranormal areas.

TABLE 4

Are the Judgments of Astrologers Using Charts Better Than
Those of Astrologers Not Using Charts?

Subjects	Judgments by 45 astrologers using subjects' birth charts		Judgments by 45 astrologers without charts, i.e., guessing	
	Judged extrovert	Judged introvert	Guessed extrovert	Guessed introvert
60 extroverts	1,472	1,228	1,401	1,299
60 introverts	1,461	1,239	1,363	1,337
Percent hits	50.2		50.7	

	Judged Unstable	Judged stable	Guessed unstable	Guessed stable
60 unstable	1,488	1,212	1,239	1,461
60 stable	1,462	1,238	1,170	1,530
Percent hits	50.7		51.3	

Answer: No. If anything the astrologers' judgments were made worse by looking at birth charts. *Source:* Dean (1985)

Charts vs. No Charts

If correct birth charts really are as essential to astrological practice as astrologers claim, then astrologers using charts should consistently outperform astrologers not using charts, i.e., simply guessing. I recently put this to the test
in a blind trial reported in detail elsewhere (Dean 1985).

From a sample of 1,198 subjects who had taken the Eysenck Personality
Inventory, I selected the most extroverted, most introverted, most stable, and
most unstable, 60 of each. Mean age was 30; 72 percent were female; and
all knew their birth times. These extreme subjects were equivalent to the top
and bottom fifteenths in the general population. Extroversion and emotional
stability were chosen because they are perhaps the most major and enduring
of known personality factors (Eysenck and Eysenck 1985) and are considered
by astrologers to be readily discernible in a birth chart (Dean 1986).

The birth charts of these subjects were given to 45 astrologers (mostly
in the U.S. and U.K.) ranging from beginners to internationally recognized
experts. Each astrologer indicated which extreme he thought each subject was
and how confident (high, medium, or low) they were in each judgment. Most
of the astrologers agreed that the test was a fair one. On the average the

judgments took each astrologer nearly 20 hours, or 4 to 5 minutes per judgment. Another 45 astrologers did the same task (circling their responses on the response sheets) but without charts; these judgments took 20 minutes.

If charts are essential to astrological practice, then astrologers using charts should have a distinct edge over those not using charts. But the results (Table 4) showed no difference; if anything, the judgments were made worse by looking at charts.

Further analysis revealed more bad news for astrology. Judgments made with high confidence were no better than those made with low confidence. Judgments on which the astrologers agreed were no better than those on which they disagreed. Supposedly crucial factors, such as experience, technique, use of intuition, and birth data accuracy, made no difference. Everything remained stubbornly at chance level.

The most damning result was the poor agreement between astrologers, the mean correlation being 0.10 for judgments and 0.03 for confidence. This indicates that 60 astrologers would on average be split 33:27 on judgments and 31:29 on confidence. (A value of 0.7 or more is generally considered satisfactory, 0.4 is poor, and 0.25 or less is useless; these correspond to a split in judgments of roughly 5:1, 5:2, and 5:3, respectively.) The agreement was little better for astrologers using much the same technique and did not improve with experience—if anything, experts showed worse agreement than beginners. Other studies have found mean correlations that are just as poor. Vernon Clark (1961), in a famous blind trial involving some of the world's best astrologers (for example, Charles Carter and Marc Edmund Jones), obtained results that on inspection reveal 0.13 for 20 astrologers matching 10 pairs of charts to case histories, and 0.12 for 30 astrologers judging 10 pairs of charts for intelligence. Macharg (1975) found 0.17 for 10 astrologers judging 30 charts for alcoholism. Ross (1975) found only 0.23 for 2 astrologers rating 102 charts on five 5-point scales of the Psychological Screening Inventory, even though both had received a similar training, both taught astrology at the same college in Miami, and both followed Rudhyar's person-centered approach. Vidmar (1979) obtained results that on inspection reveal 0.10 for 28 astrologers matching 5 pairs of charts to case histories. Fourie et al. (1980) found 0.16 for two astrologers rating 48 charts on eighteen 9-point scales of the 16PF Inventory. Steffert (1983) obtained results that on inspection reveal 0.03 for 27 astrologers judging the charts of 20 married couples for marital happiness. In other words, in none of these studies was the agreement between astrologers better than useless. If astrologers cannot even agree on what a chart indicates, then what price astrology?

Does It Matter?

If astrology does not need to be true, and if astrologers cannot agree on what a chart indicates, does it matter? The answer depends on where you are coming from. If astrology is used as entertainment or a religion, then it cannot possibly matter. Nor would it seem to matter if astrology is used like Rorschach inkblots to provide insight: Just as there is nothing really there in inkblots, so we need have no concern if there is nothing really there in celestial inkblots—at any rate we can hardly outlaw the latter while the former goes free.

But if astrology is presented as being not merely helpful but also true (and most astrology books do so present it) then on present evidence the client is being exposed to semi-institutionalized dishonesty and all the dangers that this implies. Clients seeking ways to regain control of their lives are not helped by hints that this responsibility can be passed, however slightly, to the stars. Notwithstanding the dictum that the stars incline but do not compel (and which, judging from the conversation at any astrology conference, no astrologer actually subscribes to), the remedy is simple: Astrologers wishing to be taken seriously must become more responsible. They must become aware of relevant research findings, they must desist from making claims at variance with the known facts, and they must label their product honestly so that the public is not misled. Something like CSICOP's astrology disclaimer would be a step in the right direction. Until this happens, the professional astrologer will remain a contradiction in terms. Best (1983), editor of *Correlation*, the scholarly journal of research in astrology, has put the matter bluntly: "We really have no alternative. Either we put our house in order or someone from the establishment will sooner or later take great delight in doing it for us, or, alternatively, taking it apart brick by brick." I might add that the codes of ethics adopted by astrological organizations are useless in this respect, because in effect they are concerned with skating elegantly and not with thin ice. Which brings me to the final question: If astrology does not need to be true, what is the legal position?

Astrology and the Law

Western law has traditionally (and unfairly) regarded astrologers as mere fortune tellers. And fortune telling is usually illegal. But the situation is changing. Today in many countries, including the U.K., the United States, and Australia, an astrologer is as near as the yellow pages. In the United States the legal position of astrologers was recently summarized by the AFAN (1983) as follows: "As matters stand now, there are precedents for conviction of astrologers employing any and all of the usual ways of circumventing the fortune-telling laws: religion, grandfather clause, the Evangeline Adams case, the 'truth' of astrology, and just about any other defense you can think of has been tossed out of one court or another, largely due to the lack of committed, adequate defense."

However, in 1984 there were two landmark decisions affecting astrology. The California Supreme Court of Appeals and a federal court both held that astrology and fortune telling are permitted free speech under the First Amendment of the U.S. Constitution (AFAN 1984, 1985). Among other things the First Amendment prohibits any law "abridging the freedom of speech." This was the first instance of a federal court ruling on astrology and fortune telling, the judge holding that "one need not have a scientific basis for a belief in order to have a constitutional right to utter speech based on that belief" (AFAN 1985). So even though astrology does not need to be true, the current U.S. legal view is that you have every right to practice it, just as you have every right to set up a Flat Earth society.

The Evangeline Adams case mentioned above was until recently the only instance in Western law where the details of astrological practice had been thoroughly examined in court. Because the case set a precedent and the verdict supported astrology, it is worth looking at.

The Strange Case of Evangeline Adams

The trial in 1914 of U.S. astrologer Evangeline Adams is famous among astrologers. The following account from the *Larousse Encyclopedia of Astrology* (Brau et al. 1980) is typical of those published in astrology books:

> Arrested in New York City in 1914 on a charge of fortune-telling, Adams insisted on standing trial. She came to court armed with reference books, expounded the principles of astrology, and illustrated its practice by reading a blind chart that turned out to be that of the judge's son. The judge was so impressed by her character and intelligence that he ruled in her favor, concluding that "the defendant raises astrology to the dignity of an exact science."

This gives the impression that the case was won because astrology was shown to be accurate and scientific. However, inspection of the court record tells a different story (New York Criminal Reports 1914). In the judge's opinion Adams did not pretend to tell fortunes. She merely indicated what the birth chart was supposed to mean and gave "no assurance that this or that eventually would take place." Therefore "she violated no law," and Adams was acquitted. There is only an indirect mention of the blind reading, and no mention of its subject or accuracy. Clearly the acquittal had nothing to do with astrology being accurate or scientific, and indeed the judge specifically states that the practice of astrology was "but incidental to the whole case."

Further inspection of the court record reveals a curious situation. I mentioned earlier the astrological dictum that individual factors must not be judged in isolation. Yet according to the following examples taken verbatim from the court record, this is exactly what Adams did:

Interpretation	Chart factor
Tendency to have great periods of depression.	Mercury in Capricorn
Strange fatality connected with mother's life.	Moon conjunct Neptune
Not likely to marry the first man to whom she was engaged . . . it indicated temptations.	Sun conjunct Uranus
Ambitious but lacking in confidence.	Saturn rising

Today no serious astrologer would dream of making such interpretations. For example, before finding a tendency to depression it would be necessary to examine at least the moon, Venus, Saturn, Neptune, cadent houses, triplicities, quadruplicities, and afflictions generally (Carter 1954). In other words, no astrologer could possibly claim that Adams's interpretations were accurate, let alone scientific.

So how could the judge conclude that Adams "raises astrology to the dignity of an exact science"? The answer is that he didn't, at least not in the sense implied by the quotation. The quotation is in fact incomplete and has been taken out of context, not from the final judgment, as we are led to believe, but from the introduction to the summing up. This introduction mentions that Adams had been a professional astrologer since 1897, that she had produced books in court, that her reading of a chart was "an absolutely mechanical, mathematical process," that "she claims that astrology never makes a mistake," that chart forms are used, and that "the defendant raises astrology to the dignity of an exact science—one of vibration, and she claims that all planets represent different forces of the universe." In other words, the judge is not saying that astrology is an exact science, only that Adams claims it to be so. So, in this legal case at least, contrary to what astrologers would have us believe, astrology did not need to be true.

Conclusion to Parts 1 and 2

In the past ten years, various studies have addressed astrology (the real thing, not popular nonsense) on the astrologer's terms. The results of these studies are in agreement, and their implications are clear: Astrology does not need to be true in order to work, and contrary to the claims of astrologers authentic birth charts are not essential. What matters is that astrology is believed to be true and that authentic birth charts are believed to be essential.

After surveying modern beliefs in astrology, Fullam (1984) comes to much the same conclusion:

> However, a system does not have to be real to be accepted as true as long as it is satisfying. Astrology has flourished because it is a framework within which people can discuss and look for meaning in their lives at the most superficial to the deepest levels of involvement in astrology.

Similarly, Kelly and Krutzen (1983) reached much the same conclusion after a detailed analysis of Rudhyar's humanistic astrology, which according to astrologers is the *real* real thing:

> The humanistic astrology of Dane Rudhyar is praiseworthy in its aims and shows an undoubted breadth of vision and concern for humanity. But it is dressed in obscurity and obfuscation. Worse, it . . . [requires] that no belief about anything could be false, thereby obliterating the distinction between knowledge and belief.

Thus the real thing emerges as a kind of psychological chewing gum, satisfying but ultimately without real substance. This does not deny the possibility that some as-yet-untested features of chart interpretation may work (e.g., indication of trends), or that some entirely new and valid astrological technique may be discovered, or that Michel Gauquelin's Mars effect may be eventually proved (the effect is still too weak to be considered as support for astrological practice), or that certain astrologers may achieve positive results in tests where others have failed, in which case the onus is on astrologers to demonstrate it. Nor does it deny the therapeutic utility of astrological beliefs—if invalid beliefs worked like a charm in phrenology they can do the same in astrology. What *is* denied is the essential truth of the real thing as practiced by most astrologers. As Dean and Mather (1985) note:

> Astrologers are like phrenologists: their systems cover the same ground, they apply them to the same kinds of people, they turn the same blind eye to the same lack of experimental evidence, and they are convinced for precisely the same reasons that everything works. But the phrenologists were wrong. So why shouldn't critics conclude for precisely the same reasons that astrologers are wrong?

Should astrologers wish to deny a state of affairs so contrary to their claims, all they have to do is perform appropriate tests. After all, why have a shouting match when you can have tests? However, it could be argued that the existence of mutually incompatible systems throughout astrology (for example, tropical and sidereal zodiacs), all of which are nevertheless seen as completely valid by their users, has already put this point to the test and given us convincing answers.

Acknowledgments

This paper is an updated version of one given at the Second Annual Astrological Research Conference, Institute of Psychiatry, London, on November 28, 1981. Thanks are due to Joanna Ashmun, Graham Douglas, Maureen Perkins, Austin Levy, Arthur Mather, Rudolf Smit, Drs. Rowan Bayne, Susan Blackmore, David Nias, Michael Startup, Profs. Ray Hyman, Ivan Kelly, and Andrew Neher for helpful comments, and to Profs. Ray Hyman, Ivan Kelly, and Marcello Truzzi for useful reference material. Readers who have trouble locating specific reference material may write to me at Box 466, Subiaco 6008, Western Australia. (In the three years since this article appeared, only one reader has done this.)

See pages 314-319 for the author's Postscript.

References

AFAN. 1983. AFAN to defend Shirley in Superior Court. *AFAN Newsletter,* 2(1):1 and 9 (October).
————. 1984. Major legal victory for astrology. *AFAN Newsletter,* 2(4):1-2 (July).
————. 1985. Astrology wins in Federal Court! *AFAN Newsletter,* 3(2):1-2 (January). A legal information and legislative action kit for astrologers is available for $25 from AFAN Legal Information Committee, 1754 Fell Street, San Francisco, CA 94117.
Best, S. 1983. Astrological counselling and psychotherapy: Critique and recommendations. *Astrological Journal,* 25(3):182-189 (Summer).
Blackmore, S. J. 1983. Divination with Tarot cards: An empirical study. *Journal of the Society for Psychical Research,* 52:97-101.
Brau, J. L., Weaver, H., and Edmands, A. 1980. *Larousse Encyclopedia of Astrology.* New York: McGraw-Hill, p. 21.
Carter, C. E. O. 1954. *An Encyclopedia of Psychological Astrology,* 4th ed. London: Theosophical Publishing House, pp. 124, 125, 167.
Clark, V. 1961. Experimental astrology. *In Search,* Spring: 102-112.
Connor, J. W. 1984. Misperception, folk belief, and the occult: A cognitive guide to understanding. *Skeptical Inquirer,* 8:344-354 (Summer).
Culver, R. B., and Ianna, P. A. 1984. *The Gemini Syndrome: A Scientific Examination of Astrology.* Buffalo, N.Y.: Prometheus, pp. 136-137, 149-150.
Dean. G. 1985. Can astrology predict E and N? Part 2. The whole chart. *Correlation,* 5(2):2-24 (November). *Correlation* is available from Correlation Distribution, 396 Caledonian Road, London N1 1DN, England. Tel. 081-469-2828.
————. 1986. Can astrologer predict E and N? Part 3. Discussion and further research. *Correlation,* 6(2):7-52 (December). When ranked in terms of validity and reliability against palmistry, graphology, and orthodox psychological tests, astrology comes out at the bottom. A detailed survey of the evidence with 110 references.
Dean, G., and Mather, A. 1977. *Recent Advances in Natal Astrology: A Critical Review 1900-1976.* Subiaco, Western Australia: Analogic.
————. 1985. Superprize results: Six winners—and news of a new prize to challenge the critics. Part 1. *Astrological Journal,* 28(1):23-30 (Winter). Contains major editorial errors (none relevant to phrenology). An error-free version appears in *FAA Journal* [Australia], 15(3-4):19-32 (Sept.-Dec.).
Dickson, D. H., and Kelly, I. W. 1985. The Barnum effect in personality assessment: A review of the literature. *Psychological Reports,* 57:367-382.
Dlhopolsky, J. G. 1983. A test of numerology. *Skeptical Inquirer,* 7(3):53-56 (Spring).
Eysenck, H. J., and Eysenck, M. W. 1985. *Personality and Individual Differences: A Natural Science Approach.* New York: Plenum. 1,200 refs., extremely readable, essential for anyone interested in using personality questionnaires to test astrology.
Fourie, D. P., Coetzee, C., and Costello, D. 1980. Astrology and personality: Sun-sign or chart? *South African Journal of Psychology,* 10:104-106, and personal communication 1986.
Fullam, F. A. 1984. *Contemporary Belief in Astrology.* Master's thesis, Department of Sociology, University of Chicago, July, p. 67.

Glick, P., and Snyder, M. 1986. Self-fulfilling prophecy: The psychology of belief in astrology. *Humanist*, May/June, pp. 20-25, 50.

Grange, C. 1982. The Barnum effect: Its relevance for non-scientific methods of personality assessment. Unpublished study, Loughborough University of Technology, Loughborough, U.K.

Hand, R. 1981. *Horoscope Symbols*. Rockport, Mass.: Para Research, pp. 25-26.

Hyman, R. 1977. Cold reading: How to convince strangers that you know all about them. *Zetetic* (now *Skeptical Inquirer*), 1(2):18-37 (Spring-Summer).

———. 1981. The psychic reading. In *The Clever Hans Phenomenon: Communication with Horses, Whales, Apes, and People,* edited by T. A. Sebeok and R. Rosenthal, pp. 169-181. New York: New York Academy of Sciences. In the Clever Hans effect, information unwittingly emitted by the client is fed back to the client, who is thus persuaded to see marvels where none exists.

Kallai, E. 1985. *Psychological Factors that Influence the Belief in Horoscopes*. Master's thesis, Department of Psychology, The Hebrew University, Jerusalem. In Hebrew with English abstract. And personal communication 1986.

Kelly, I. W., and Krutzen, R. W. 1983. Humanistic astrology: A critique. *Skeptical Inquirer,* 8:62-73 (Fall).

Macharg, S. J. 1975. *The Use of the Natal Chart in the Identification of Alcoholism and a Comparison of Its Diagnostic Efficacy with the MMPI*. Ph.D. dissertation (education), University of Southern California, p. 77.

Marks, D., and Kammann, R. 1980. *The Psychology of the Psychic*. Buffalo, N.Y.: Prometheus Books, pp. 176-186. The examples of subjective validation given by Marks and Kammann are examples of hindsight bias.

Naftulin, D. H., Ware, J. E., and Donnelly, F. A. 1973. The Doctor Fox lecture: A paradigm of educational seduction. *Journal of Medical Education,* 48:630-635. Dr. Fox was an actor who was coached to give a highly entertaining but otherwise meaningless one-hour lecture on games theory to various professionals, such as psychiatrists and social workers. They found his talk to be clear and stimulating, and nobody realized it was nonsense.

Neher, A. 1980. *The Psychology of Transcendence*. Englewood Cliffs, N.J.: Prentice-Hall, pp. 239-242, and personal communication 1985.

New York Criminal Reports. 1914. Vol. 32. The people ex rel Adele D. Priess vs. Evangeline S. Adams. Priess was a 49-year-old woman detective acting on behalf of the New York police. The report consists of four pages of prosecuting testimony, including one page of comment, plus eight pages of defending testimony and six pages of summing up.

Niehenke, P. 1984. The validity of astrological aspects: An empirical inquiry. *Astro-Psychological Problems,* 2(3):10-15. APP is now combined with the *NCGR Journal*, Box 34487, San Diego, CA 92103-0802.

Nisbett, R., and Ross, L. 1980. *Human Inference: Strategies and Shortcomings of Social Judgement*. Englewood Cliffs, N.J.: Prentice-Hall. A concise and more readable survey that cites Nisbett and Ross is R. M. Hogarth, *Judgement and Choice: The Psychology of Decision,* Wiley, New York, 1980.

Rosenblum, B. 1983. *The Astrologer's Guide to Counseling: Astrology's Role in the Helping Professions*. Reno, Nev.: CRCS.

Ross, L. H. 1975. *The Relationship Between the Ratings of an Established Personality Inventory and Those of Two Practitioners of Rudhyar's Person-Centered Approach*. Ph.D. thesis (education), Nova University, Fort Lauderdale, Fla., p. 44.

Russell, D., and Jones, W. H. 1980. When superstition fails: Reactions to disconfirmation of paranormal beliefs. *Personality and Social Psychology Bulletin,* 6:83-88.

Steffert, B. 1983. Marital bliss or misery: Can synastry distinguish? *Astrological Journal,* 25(3):211-213.

Tyson, G. A. 1984. An empirical test of the astrological theory of personality. *Personality and Individual Differences,* 5:247-250.

Vidmar, J. E. 1979. Astrological discrimination between authentic and spurious birthdates. *Cosmecology Bulletin,* nos. 8 and 9:56-90. Reprinted from Ph.D. dissertation (education) of the same name, University of Northern Colorado, 1978.

Postscript

RESPONSE FROM ASTROLOGERS:

Astrologers have been curiously silent. The only response was from Dr. Lee Lehman (1988), research director of the U.S. National Council for Geocosmic Research and author of *The Ultimate Asteroid Book*, or how the right asteroids (nearly 3,000 to choose from) can improve any birth chart. She did not like the article. In her view I am permanently biased against astrology, being "completely committed to negative results . . . deliberate misrepresentation . . . and wretched experimental design." So I must be stripped of refereeing duties and be denied opportunities to speak and publish. According to her, the article you have just read is so biased and wretched that it should be banned.

Such views implicated my own referees (see Acknowledgments), who were not amused—the article had gone through two preliminary versions and five rewrites before they were happy. One of them, the meticulous critic and former professional astrologer Rudolf Smit, now secretary for Professor H. J. Eysenck's Committee for Objective Research in Astrology, had carefully checked the entire text before submission. He replied: "On the contrary, I found it one of the best-balanced scientific articles on astrology I ever read" (Smit 1988).

But why have arguments when you can have tests? To resolve the issue I proposed submitting my wretched work to arbitration (Dean 1988). If Lehman's charges were substantiated then I would gladly refrain from further refereeing, speaking, and publication—provided that, if my work was acquitted, then in fairness she would do the same. That was nearly two years ago. The result has been a deafening silence.

A STATISTICAL CURIOSITY?

Table 1 in Part 1 shows that the number of subjects picking their own birth charts is *exactly* the same as chance in 5 out of 7 studies. At least six readers (none of them supporters of astrology) wrote to the *Skeptical Inquirer* pointing out that this was too good to be true (various 1987). My reply said the quoted figures were exactly as given in the original studies, so it was probably a coincidence. I also gave this breakdown of hits for the three studies that gave second and third choices:

Study	No. of subjects	Rank of authentic chart 1	2	3	No. of hits expected
Cummings et al. 1978	12	4	7	2	4
Dwyer & Grange 1983	34	10	19	5	11
Carlson 1985	83	28	33	22	28

In each case only rank 1 shows a good fit with expectancy, suggesting that the exclusion of other ranks has made the fit seem better than it is. One of the readers, Professor I. J. Good, then pointed out that the first row of results (4, 7, 2) add up to 13 when it should be 12. Just one other copying error among the other studies would make the overall results less perfect, and therefore easier to accept. On this point the original article says 4, 7, 2 despite repeatedly saying the total is 12, so it is not clear which figure is wrong (but for the record 3, 7, 2 does the least violence to their associated figures). Since then three more studies can be added to Table 1 as follows:

Study	Note	No. of Subjects	Charts per Subject	No. Picking Own Chart Observed	Expected
Krippner 1980	1	16	8	3	2
Marbell et al. 1986	2	24	3	14	8
Klein 1988	3	122	5	41	24.4

The interpretations were: (1) Prepared by a professional astrologer using the whole chart. (2) Prepared by one or more astrologers based mostly on sign and house positions. Results indicated cueing by sun sign description. (3) Computerized paragraphs based on aspects between charts describing how one partner in a relationship experiences the other. Individual partners had to choose which was theirs. There was no clear control of prior knowledge, which is important since the ad asking for volunteers was headed "Is Astrology True?" The results indicated significant collusion between partners.

The last two studies show how poor design can make the results impossible to interpret. But at least the run of exactly chance results has been broken—before we had 5 exact in 7, now we have 4 exact in 10, no longer an eyebrow-raising result. A brief review of matching tests of personality inventory profiles is included in Furnham and Schofield (1986, p. 175). Contrary to the outcome when picking their own birth charts, subjects can reliably pick their own profiles.

RECENT RESEARCH

Recent research has strengthened the argument that astrology does not need to be true. The ease of matching any birth chart to any person is illustrated by Gergen, Hepburn, and Fisher (1986). They asked university students to explain how certain traits could explain certain behaviors, which unknown to the students had been picked at random. The results showed that any trait could plausibly explain any behavior, including opposite behaviors. For example, the *hostile* person (read Mercury square Mars) *avoids social groups* because he hates people, and *seeks social groups* because he needs people to attack. The students could readily find 3 to 6 plausible explanations for any given link. No wonder that astrologers see that charts fit the person, even when the chart is later found to be wrong (see end of Part 1).

The poor agreement between astrologers noted in Part 2 is confirmed by Timm and Köberl (1986). They re-analyzed an unpublished validation study of 178 German astrologers made in 1952-55 by Hans Bender's Institute for Borderline Areas in Psychology. This is the largest sample of astrologers ever tested. Despite the traditional German concern for precision, the mean agreement was only .10, for an average of 80 astrologers matching a total of over 50 charts. For 14 studies the mean agreement was .11, or no better than useless (Dean 1986 in references to Part 2).

The judgment processes and inferential deficiencies underlying techniques like astrology, palmistry, and graphology have been recently surveyed by Dean, Kelly, Saklofske, and Furnham (1991). Included are numerous topics and artifacts not mentioned in the present article. Conclusion: Human cognitive skills are simply not equal to the task (making sense out of multiple inputs) that astrologers, palmists, and graphologists have set themselves. Their natural human biases have created false beliefs.

Here is an example. On the syndicated live TV spectacular "Exploring Psychic Powers Live" broadcast in 1989 and seen around the world, the U.S. astrologer Joseph Meriwether had previously guessed the sun signs of 12 subjects after a brief interview. The subjects were born under different signs within three years of each other. A score of 10 or more hits would win a glittering $100,000. The host, James Randi, told me that, after the interviews, the astrologer was totally confident of success. His actual score was 0 hits, worse than the 1 expected by chance.

A sidebar, "The State of Astrology," in Part 2 gave the gloomy views of 19 well-known U.S. astrologers. Quite a different view emerges from *The Future of Astrology,* a recent anthology of 14 essays from "some of the foremost international astrologers in the world today" (Mann 1987). Here there is nothing but unbounded enthusiasm. Curiously, half the essays say nothing about the future of astrology, as shown in the box on the following page.

Even more curiously, not one of the essays gives the slightest hint that hundreds of research studies exist, some of them not inconceivably pertinent to the future of astrology. The result reads like a weather forecast in which forecasters resolutely refuse to look out the window.

So here is the latest of the research studies that the world's foremost international astrologers don't want you to know about, a remarkable matching test of astrologers in Indiana conducted by McGrew and McFall (1990). The astrologers collaborated to make the test as fair as possible. And they got everything they wanted—subjects with diverse backgrounds, certified birth times accurate to five minutes or better, and case files containing results from two personality tests, responses to their own 7-page 61-item questionnaire (covering everything from height and hobbies to family deaths and favorite colors),

The Future of Astrology
Source: Mann (1987)

Author	View
Rudhyar	Astrology will bring order to the chaos of human life.
Rael	Rudhyar was a true seed man.
Hand	We must see the universe as a living conscious entity.
Addey	Astrology will become central to scientific thought.
Gauquelin	During 40 years I got positive and negative results.
Harvey	Astrology will be woven into the fabric of our lives.
Mann	Astrology will be part of medicine, education, religion.
Elwell	Every individual is a cosmic deed directed to some end.
Lewis	Growth to late 1988, then consolidation, then flowering.
Hamaker	Astrology will regain its accepted place in science.
Oken	Astrology is the purest form of occult truth.
Huber B. & L.	We must create new methods for spiritual delineation.
Lynes	Unless mundane astrology is recognized, tragedy looms.
Elliot	No future unless astrology can be shown to work.

and two photographs. The test was to match case file to birth chart for 23 subjects, all aged 30 to 32 years to avoid age cues. Six astrologers selected for competence did the test, which took them 12 to 24 hours each. On a scale of 0 to 100 percent, each astrologer's confidence in each judgment was typically 75, so the astrologers were generally confident of success. However the range of hits was 0 to 3, mean 1.33, little better than the 1 expected by chance. One nonastrologer did the matchings and scored 3 hits, as good as the best astrologer. The mean agreement between astrologers was .03, almost nil. In a follow-up study, five groups of 5 to 6 astrologers matched a subset of 5 charts and scored 2 hits, worse than the 5 expected by chance. The negative results had no noticeable effect on the astrologers' belief in astrology.

On the last point, Glick, Gottesman, and Jolton (1989) gave 216 students aged 15 to 18 a personality description. Half were told it might not apply to them, but check it anyway, and half were told it had been prepared for them by an astrologer. In fact, all the descriptions were the same two, either *favorable* (e.g., sympathetic, dependable) or *unfavorable* (e.g., unrealistic, undependable). Believers in astrology rated it more accurate if they thought it was based on astrology, even when it was unfavorable. This supports my findings with reversed charts reported in Part 2. Skeptics were less enthusiastic about unfavorable descriptions, but favorable descriptions supposedly based on astrology increased their belief even more than it did for believers. Overall the results showed that the *skeptics* were willing to change their views on astrology according to the evidence, whereas the *believers* were not—the same as observed by others (see Part 2), including McGrew and McFall (1990) above.

But we should not underestimate the cunning of psychologists. Suppose we believe that men make better bosses than women. If our belief is entrenched, contrary questions (Why do women make better bosses than men?) will have no effect. So Swann, Pelham, and Chidester (1988) use questions that are simply more extreme (Why do men *always* make better bosses than women?). Because we resist change, we think up reasons against this extreme view—and unwittingly change our belief in the opposite direction! Interestingly, the authors show that our change is a true shift in position, not just a recognition that more extreme views exist. So the next time you meet an entrenched believer in astrology, ask questions like: Why are newspaper horoscopes always so accurate? Why does astrology deny free will? When astrology has revolutionized the world, will anything be the same?

Finally, for a sympathetic but critical look at areas of astrology not covered in this article, see Kelly, Dean, and Saklofske (1990). Their survey includes validity, effect size (typically the correlation between astrology and an independent criterion is less than .10), the work of Michel Gauquelin and Suitbert Ertel, spiritual astrology (praiseworthy in its aims but contains nothing that could be true or false), and the future of astrology. Conclusions: (1) Astrology covers many areas from the trivial to the profound. Which astrology are you discussing? (2) The problems facing astrology are: effect sizes too small to justify astrological claims, poor agreement between astrologers, and obscurantism—astrologers never specify the evidence they would accept as proving their ideas are wrong. (3) Astrologers look at astrology from a value viewpoint and conclude it works. Scientists look at astrology from a factual viewpoint and conclude the opposite. Beware the difference.

References

Dean, G. 1988. Letter to the Editor. *Astro-Psychological Problems,* 6(3):31-34. Replies point by point to Lehman's allegations.

Dean, G. A., Kelly, I. W., Saklofske, D. H., and Furnham, A. 1991. Graphology and Human Judgment. In B. and D. Beyerstein (Eds.), *The Write Stuff.* Buffalo: Prometheus Books. Very relevant to astrology, somewhat longer than the present Parts 1 and 2 combined, 160 references.

Furnham, A., and Schofield, S. 1986. Accepting personality test feedback: A review of the Barnum Effect. *Current Psychological Review of Research,* 7:162-178. 61 references.

Gergen, K. J., Hepburn, A., and Fisher, D. C. 1986. Hermeneutics of personality description. *Journal of Personality and Social Psychology,* 50:1261-1270.

Glick, P., Gottesman, D., and Jolton, J. 1989. The fault is not in the stars: Susceptibility of skeptics and believers in astrology to the Barnum effect. *Personality and Social Psychology Bulletin,* 15:572-583.

Kelly, I. W., Dean, G. A. and Saklofske, D. H. 1990. Astrology: A critical review. In P. Grim (Ed.), *Philosophy of Science and the Occult,* 2nd ed., pp. 51-81. Albany, N.Y.: State University of New York Press. 80 references.

Klein, M. 1988. The accuracy of relationship description as a test of astrology. *Correlation,*

8(2):5-17. My thanks to Mavis Klein for providing additional details.

Krippner, S. 1979. Gardner Murphy and the astrology probe. In W. G. Roll (Ed.), *Research in Parapsychology 1979*. Metuchen, N.J.: Scarecrow Press, pp. 31-32. My thanks to Dr. Krippner for bringing his and Murphy's study to my attention, and for providing additional details.

Lehman, J. L. 1988. Bias in the conduct of research. *Astro-Psychological Problems*, 6(1):16-18.

Mann, A. T., ed. 1987. *The Future of Astrology*. London: Unwin Hyman. No index and (apart from Michel Gauquelin's essay) no references.

Marbell, N. Z., Novak, A. R., Heal, L. W., Fleming, L. D., and Burton, J. M. 1986. Self-selection of astrologically derived personality descriptions: An empirical test of the relationship between astrology and psychology. *NCGR Journal*, Winter 1986-87, pp. 29-43.

McGrew, J. H., and McFall, R. M. 1990. A scientific inquiry into the validity of astrology. *Journal of Scientific Exploration*, 4(1):75-83. My thanks to John McGrew for providing additional details.

Smit, R. 1988. To the defense of Geoffrey Dean. *Astro-Psychological Problems*, 6(2):32. For other independent comments see *NCGR Research Journal*, Spring 1989, p. 43, and Winter 1989, p. 44.

Swann, W. B., Pelham, B. W., and Chidester, T. R. 1988. Change through paradox: Using self-verification to alter beliefs. *Journal of Personality and Social Psychology*, 54:268-273.

Timm, U., and Köberl, T. 1986. Re-Analyse einer Validitätsuntersuchung an 178 Astrologen. *Zeitschrift für Parapsychologie und Grenzgebiete der Psychologie*, 28:33-55. (A re-analysis of a validation study of 178 astrologers.)

Various. 1987. Statistics in Dean's astrology article. *Skeptical Inquirer*, 11:418-422.

Double-Blind Test of Astrology Avoids Bias, Still Refutes the Astrological Hypothesis

Astrologers who claim they can analyze a person's character and predict a person's life course just by reading the "stars" are fooling the public and themselves, University of California physicist Shawn Carlson concluded in a unique double-blind test of astrology published in *Nature* (December 5, 1985). The controlled study was designed specifically to test whether astrologers can do what they say they can do. Carlson, a researcher at UC's Lawrence Berkeley Laboratory, found astrologers had no special ability to interpret personality from astrological readings. Astrologers also performed much worse in the test than they predicted they would, according to Carlson.

The study refutes astrologers' assertions that they can solve clients' personal problems by reading "natal charts," individual horoscopes cast according to the person's date, time, and place of birth. "It is more likely that when sitting face to face with a client, astrologers read clients' needs, hopes, and doubts from their body language," said Carlson, who has since received his doctorate in physics at UCLA and is also a professional magician who has performed "psychic ability" demonstrations.

Carlson's research involved 30 American and European astrologers considered by their peers to be among the best practitioners of their art.

The study was designed specifically to test astrology as astrologers define it. Astrologers frequently claim that previous tests by scientists have been based on scientists' misconceptions about astrology.

To check astrologers' claims that they can tell from natal charts what people are really like and how they will fare in life, Carlson asked astrologers to interpret natal charts for 116 unseen "clients." In the test, astrologers were allowed no face-to-face contact with their clients.

For each client's chart, astrologers were provided three anonymous personality profiles—one from the client and two others chosen at random—

and asked to choose the one that best matched the natal chart. All personality profiles came from real people and were compiled using questionnaires known as the California Personality Inventory (CPI). The CPI, a widely used and scientifically accepted personality test, measures traits like aggressiveness, dominance, and femininity from a long series of multiple-choice questions.

According to Carlson, the study strenuously attempted to avoid anti-astrology bias by making sure astrologers were familiar with the CPI and by incorporating many of the astrologers' suggestions. At the same time, to prevent testers from inadvertently helping astrologers during the test, the project was designed as a double-blind study where neither astrologers nor testers knew any of the answers to experimental questions.

Despite astrologers' claims, Carlson found those in the study could correctly match only one of every three natal charts with the proper personality profile—the very proportion predicted by chance.

In addition, astrologers in the study fell far short of their own prediction that they would correctly match one of every two natal charts provided. Even when astrologers expressed strong confidence in a particular match, they were no more likely to be correct, Carlson found.

Concluded Carlson:

> We are now in a position to argue a surprisingly strong case against natal astrology as practiced by reputable astrologers. Great pains were taken to insure that the experiment was unbiased and to make sure that astrology was given every reasonable chance to succeed. It failed. Despite the fact that we worked with some of the best astrologers in the country, recommended by the advising astrologers for their expertise in astrology and in their ability to use the CPI, despite the fact that every reasonable suggestion made by the advising astrologers was worked into the experiment, despite the fact that the astrologers approved the design and predicted 50% as the "minimum" effect they would expect to see, astrology failed to perform at a level better than chance. Tested using double-blind methods, the astrologers' predictions proved to be wrong. Their predicted connection between the positions of the planets and other astronomical objects at the time of birth and the personalities of test subjects did not exist. The experiment clearly refutes the astrological hypothesis.

"A lot of people believe in astrology because they think they have seen it work," Carlson observed. He believes many astrologers are successful at their art because they draw important clues about clients' personalities and lifestyles from facial expressions, body language, and conscious or unconscious verbal responses. "When magicians use the same technique, they call it 'cold reading.' " said Carlson.

Based on his scientific findings, Carlson suggests many people would do better to spend their money on trained psychology counselors. However, he disagrees with those who would like to see astrology outlawed. "People believed in astrology for thousands of years and no doubt will continue to do so no matter what scientists discover. They are entitled to their beliefs, but they should know that there is no factual evidence on which to base them."

"The astrologers' reactions so far have been pretty much what I expected," Carlson told the *Skeptical Inquirer*. "The astrologers whom I didn't test are saying that the test was not fair because I did not test them. Of course, if I had tested them instead, and they had failed, then the astrologers I actually tested would now be saying that the test was not fair because I did not test them."

Carlson's study was supported by Richard Muller, professor of physics at UC Berkeley, and paid for by a general congressional research award.

Editor's Postscript

Shawn Carlson received his doctorate and is now an astrophysicist at Berkeley working with Richard Muller on the Nemesis star project. Muller considers Carlson's *Nature* paper one of the best designed and conducted scientific tests ever done on the claims of astrology.

Did it have an effect? "I didn't expect any big-name astrologers to turn in their robes," says Carlson, "but I think some astrologers are still answering for it. Why else would they still feel the need to attack it in their writings six years later?" Nevertheless as a multimillion-dollar industry, astrology, he notes, remains unscathed. Since publishing his research Carlson has hardened his opposition to astrology. "After my study appeared, people started telling me their experiences with astrologers and I now know the terrible toll some astrologers take on their most intimate concerns. Astrological advice heeded often means a life destroyed." He says if astrologers want to be therapists "they should be held to the same standards as all other therapists; namely, needing to receive state certification in conventional therapy techniques before they counsel patients."

Although Carlson has been challenged to conduct further astrological studies, he says he now prefers more conventional research. "Astrologers have been repeatedly tested and have repeatedly failed," he says. "Scientifically, astrology is a dead issue."

Part 8: Crashed-Saucer Claims

PHILIP J. KLASS

Crash of the Crashed-Saucer Claim

A revealing indication of the credulity of many of the present leaders of the UFO movement is their widespread acceptance of the claim that the U.S. government recovered one or more flying saucers in 1947, along with bodies of the alleged occupants—a tale rejected three decades ago by leading UFOlogists. A paper on the alleged crashed saucers was featured at the 1985 conference of the Mutual UFO Network (MUFON), the nation's largest UFO organization, and at earlier MUFON conferences.

The crashed-saucer tale was first advanced in 1950, barely three years after UFOs had been "discovered" in a best-selling book by Frank Scully, then a columnist for *Variety*—the "Bible of Show-Biz." But Scully's wild claim was promptly rejected even by *True* magazine, which itself had helped launch the UFO era a few months earlier when it published an article by Donald Keyhoe claiming that the earth was being visited by extraterrestrial craft.

Scully had obtained his information on the "crashed saucers" from two men who were exposed as con-men two years later by a young reporter, J. P. Cahn, in an article published in *True*. Soon afterward, the two men were arrested and charged with selling a device called a "Doodlebug," which they claimed could find oil deposits. One of their victims had invested more than $230,000. The two men subsequently were convicted of fraud.

For almost three decades the claim of crashed saucers in New Mexico was ignored by responsible UFOlogists. Then, in 1980, it was resurrected by Charles Berlitz and William L. Moore in their book *The Roswell Incident*. Berlitz earlier achieved fame and fortune with his book on the Bermuda Triangle, which he claimed mysteriously swallowed up airplanes and ships—some of which had never existed. Moore earlier had authored the book *The Philadelphia Experiment*, which claimed that during World War II the U.S. Navy had discovered techniques that could make its ships invisible. But, according to Moore, the Navy decided not to deploy this remarkable technique because its use gave sailors a headache or made them ill.

With this heritage, one might expect the leaders of the UFO movement to treat the Berlitz–Moore claims with considerable skepticism—unless one is familiar with the incredible credulity of many UFOlogists. Even Bruce S. Maccabee, one of the most technically competent of pro-UFOlogists and head of the Fund for UFO Research, gave the Berlitz–Moore book an endorsement in a book review published in *Frontiers of Science* magazine.

It is not surprising that Berlitz and Moore intentionally omitted from their book the considerable hard evidence that denied the claim of crashed saucers. But considering the amount of time UFOlogists spend in pouring over old, once-classified documents in a desperate search for evidence of a massive government coverup, it is curious that they too have failed to note, or publicize, how this utterly demolishes the crashed-saucer hypothesis.

According to Berlitz and Moore, a flying saucer crashed on the ranch of W. W. Brazel *during the first week of July 1947,* and possibly a second crashed near Socorro shortly afterward. The Army Air Force (soon to become the U.S. Air Force) position was that the debris found by Brazel was nothing more than a balloon-borne radar reflector, a device resembling a box-kite lined with aluminum-foil, used to calibrate ground-tracking radars.

Naturally, Berlitz and Moore reject that explanation, drawing on the 30-year-old recollections of local citizens and a number of newspaper clippings dating back to 1947. One important newspaper account Berlitz and Moore omit entirely is an Associated Press dispatch dated July 9, 1947, based on an interview with Brazel himself. The article quotes Brazel as saying he discovered the debris while riding his ranch on *June 14—more than two weeks before Berlitz and Moore claim the flying saucer crashed.*

Brazel's description of what he found, quoted in the Associated Press article, confirms the government position that the object was only a balloon-borne radar reflector: "large numbers of pieces of paper covered with a foil-like substance and pieced together with small sticks much like a kite. Scattered with the materials over an area of about 200 yards were pieces of gray rubber. All the pieces were small." The article quoted Brazel as saying, "At first I thought it was a kite, but we couldn't put it together like any kite I ever saw."

According to Berlitz and Moore, the crashed saucer was promptly flown to Wright-Patterson Air Force Base, near Dayton, Ohio, for analysis. This base was the technical nerve-center for the Air Force and included its foreign intelligence operations. At the time, the base commander was Lt. Gen. Nathan Twining, who later became the USAF's chief of staff.

In September 1947, following a rash of UFO reports in the wake of the famous first sighting, reported by pilot Kenneth Arnold in June, the chief of staff of the Army Air Force had requested General Twining to provide him with a situation assessment, which Twining did in his letter of September 23, 1947. Berlitz and Moore quote extensively from this letter, including Twining's statement that "the phenomenon reported is something real and not visionary or fictitious." But the authors omit a critically important statement in the same letter, where Twining noted that there was a *"lack of physical evidence in the shape of crash-recovered exhibits which would undeniably prove the existence of these objects.*" And Twining was the commanding officer of the base where, according to Berlitz and Moore, top scientists had been analyzing the crashed saucer for more than two months.

After omitting this sentence from the Twining letter, the authors wrote: "It is understandable that the Twining memo makes no reference to the Roswell disc. . . ." It is understandable if the debris sent to Wright-Patterson AFB had turned out to be only a balloon-borne radar reflector and not a crashed saucer. The alternative explanations are that nobody thought to inform General Twining of the dramatic work under way at the base he commanded, or that Twining was intentionally lying to his own commanding officer.

Although dozens of ordinary citizens in New Mexico, without any official "need-to-know," quickly learned about the alleged crashed saucer(s), according to Berlitz and Moore word of the incident was withheld from the Army chief of staff because, as they explain, "he did not possess the necessary clearances." His name: Dwight D. Eisenhower. Even after General Eisenhower became president, according to Berlitz and Moore, he was not informed of the recovered crashed saucer(s) until more than a year later because "some of the higher-ups in the intelligence community didn't trust Ike. . . ." (Recall that Allen Dulles, director of Central Intelligence under Eisenhower, was a brother of Secretary of State John Foster Dulles and a close personal friend of Eisenhower.)

In early 1953, top officials at Air Defense Command headquarters in Colorado Springs received a briefing on the USAF's UFO-investigations program by Capt. Edward J. Ruppelt, then head of Project Blue Book. The briefing was classified "Secret," as Ruppelt explained, in case sensitive matters, such as the coverage of the nation's air defense radar network, came up during the question-and-answer period. Subsequently, Ruppelt's prepared briefing was declassified and was published a decade ago in *Project Blue Book,* a book edited by Brad Steiger.

The head of Project Blue Book told top Air Defense Command officials: "It can be stated now that, as far as the current situation is concerned, there are no indications that the reported objects are a direct threat to the U.S., nor is there any proof that the reported objects are any foreign body over the U.S. or, as far as we know, the rest of the world. This always brings up the question of space travel . . . and it is the opinion of most scientists or people that should know that it is not impossible for some other planet to be inhabited and for this planet to send beings down to the earth.

"However there is no—and I want to emphasize and repeat the word *no*—evidence of this in any report the Air Force has received. . . . *We have never picked up any 'hardware.' By that we mean any pieces, parts, whole articles, or anything that would indicate an unknown material or object. . . .*"

Other hard evidence that denies the crashed-saucer claims can be found in material once classified "Secret" obtained from Central Intelligence Agency files in late 1978 via the Freedom of Information Act. These CIA papers reveal that in mid-1952, probably sparked by highly publicized reports of UFOs on radar screens at Washington's National Airport, the White House asked the CIA to make an independent assessment of the situation. As a result, high-ranking CIA scientists went to Dayton for a USAF briefing on the findings of its Project Blue Book effort. Then, in mid-August, these top CIA scientists briefed the director of Central Intelligence.

In one of these briefing papers, dated August 14 and originally classified "Secret," the briefer discussed the possible explanations for UFO reports, including the possibility that some might be generated by extraterrestrial craft. But the briefer added that "there is no shred of evidence to support this theory at present. . . ." Another once-"Secret" briefing paper, dated August 15, states: "Finally, no debris or material evidence has ever been recovered following an unexplained sighting."

Recently, using the Freedom of Information Act, UFOlogists obtained an Air Intelligence Report, dated December 10, 1948, originally classified "Top Secret." It was considered such an important "find" that the *MUFON UFO Journal* devoted almost its entire July 1985 issue to reproducing this report, prepared jointly by the USAF's Directorate of Intelligence and the Office of Naval Intelligence. The objective of the report was to provide a best-estimate of the UFO situation as of 1948.

Although this once "Top Secret" report was prepared more than a year after Berlitz and Moore claim that at least one crashed saucer was recovered in New Mexico by defense officials, *there is not a single mention of any such evidence*. Instead, the report focuses its speculation on the possibility that UFO reports might be generated by Soviet reconnaissance overflights, possibly using advanced vehicles built with the help of captured German scientists.

This 1948 report concludes: "IT MUST [*sic*] be accepted that some type of flying objects have been observed, although their identification and origin are not discernible. In the interest of national defense it would be unwise to overlook the possibility that some of these objects may be of foreign origin."

Presumably this report will be studied by MUFON's international director, Walter Andrus, and by many other leading UFOlogists. Will they recognize its obvious implications (and the other hard evidence cited above) in terms of the Berlitz-Moore claim of crashed saucers? Or will Moore continue to spin his tales at future MUFON conferences, prompting his audience to believe that somewhere in some secret government vault lies the debris, and perhaps even the bodies, that could at long last confirm UFOlogists' fondest hopes?

The MJ-12 Crashed-Saucer Documents.
Part 1

On May 29, 1987, William L. Moore and two associates, Stanton Friedman and Jaime Shandera, released what purport to be "Top Secret" government documents that are either the biggest news story of the past two millennia or one of the biggest cons ever attempted against the public and the news media.

If authentic, the documents show that the U.S. government recovered a crashed flying saucer in mid-1947, and four extraterrestrial-creature bodies, much as Moore claimed in his 1980 book, *The Roswell Incident* (coauthored with Charles Berlitz), and that the government also recovered the remains of another saucer, which crashed on December 6, 1950, near the Texas-Mexico border.

Further, these documents indicate that on September 24, 1947, President Harry S Truman authorized Defense Secretary James Forrestal and Dr. Vannevar Bush, president of the Carnegie Institution, to create a top-secret panel of 12 scientists, military leaders, and intelligence officials—called Operation Majestic-12 (MJ-12). Its function, presumably, was to analyze the crashed saucer to determine its technological secrets and to make recommendations for a suitable U.S. response to extraterrestrial visitors whose intentions might prove to be hostile.

The papers released by Moore, Friedman, and Shandera consisted of three elements, purporting to be the following: (1) a "Top Secret" memorandum from President Truman to Defense Secretary Forrestal, dated September 24, 1947, authorizing him and Dr. Bush to proceed with Operation Majestic-12; (2) a seven-page "Top Secret/Eyes Only" Majestic-12 document used to brief President-Elect Eisenhower, dated November 18, 1952; (3) a "Top Secret" memorandum from Robert Cutler, special assistant to President Eisenhower, to General Nathan Twining, USAF chief of staff, dated July 14, 1954.

According to Moore, the Truman/Forrestal memo and the Eisenhower briefing document were received in mid-December 1984 by Moore's friend Jamie Shandera, a Los Angeles television writer-producer, on an undeveloped roll of 35-mm film.

As Moore described the circumstances in his banquet speech at the 1987 MUFON UFO conference in Washington in late June, the package containing the film was wrapped in plain brown paper "taped with official-looking brown tape on all seams. The address label was carefully typed, with no return address. Inside the [brown] wrapper was a second one, similarly sealed, inside of which was yet another white envelope, inside of which was a cannister, inside of which was a roll of *unprocessed* film." (Moore has not replied to my repeated requests that he send me a photocopy of the postmark, showing city and date of mailing.)

If the MJ-12 documents film is authentic, it is odd that it was not sent to Moore, whose book and numerous MUFON conference papers have made him world famous as *the* leading crashed-saucer proponent and researcher—or to Stanton Friedman, who has been Moore's closest collaborator on crashed-saucer research for almost a decade. As Moore explained at the MUFON conference, in recent years he has focused his efforts on trying to establish contacts within the intelligence community "to find out what happened to the wreckage after it came into custody of military authorities."

Why would the film be sent to Shandera, who had never published any papers on UFOs or crashed saucers and does not even consider himself a UFOlogist? How would the sender of the 35-mm film even know that Shandera and Moore were friends and that the contents would find their way to Moore?

Even before the film was developed and the MJ-12 papers became visible, Shandera demonstrated "psychic powers" in "knowing" that the undeveloped roll of 35-mm film in the plain brown wrapper from an unknown sender would be of interest to Moore. This explains why he promptly called Moore even before the film was processed and why Moore was present when it was being developed, according to Moore's report to MUFON.

According to Moore, the person who made the 35-mm film had photographed the MJ-12 documents in two duplicate sequences, seemingly to ensure that there would be at least one good set of imagery. But the sender had not thought to process the film himself for final assurance before sending it to Shandera.

The film's seven-page Eisenhower briefing document indicated that the briefing officer was Admiral Roscoe H. Hillenkoetter, who had been head of the Central Intelligence Agency in 1947 when MJ-12 allegedly was created and thus would logically be a member. But in 1950 Hillenkoetter left the CIA to return to sea duty as commander of the Seventh Task Force in Formosan Waters and did not return for duty in the United States until late 1951—the year of the alleged briefing—to become commander of the Third Naval District in New York.

It would have been more logical for Eisenhower to have been briefed

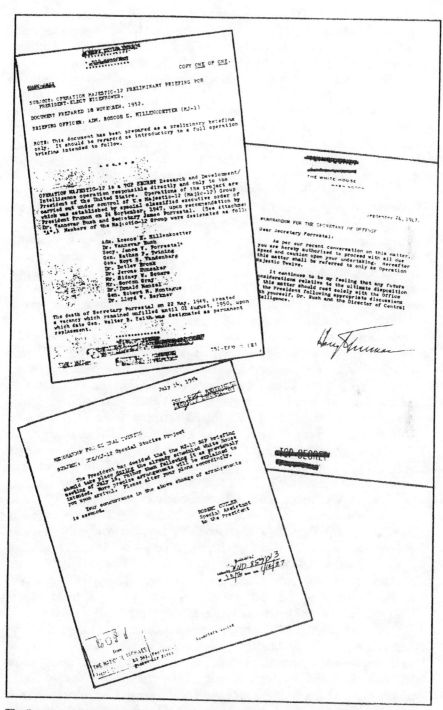

The "top secret/eyes only" Hillenkoetter briefing document, the "top secret" Cutler memorandum, and the "top secret" Truman memorandum to Defense Secretary Forrestal.

by the chairman of MJ-12, who had remained in the United States, close to the committee's activities, since 1947. Presumably this would have been Dr. Bush, who allegedly organized MJ-12 and is shown as one of its original members. (Although the Hillenkoetter briefing document lists the 12 members of the group, it does not indicate who was chairman but identifies Hillenkoetter as "MJ-1.")

While there are many such substantive anomalies in the contents of the alleged Hillenkoetter/Eisenhower briefing documents, which will be discussed in a subsequent article, the most revealing is the format used to write dates. (I am indebted to Christopher Allan, Stoke-on-Trent, England, who first brought these very significant anomalies to my attention.)

Whoever typed the Hillenkoetter briefing document used a peculiar style for writing dates—an erroneous mixture of civil and military formats. In the traditional civil style, one would write: November 18, 1952. Using the standard military format, one would write: 18 November 1952. But whoever typed the Hillenkoetter briefing document used a military format with an unnecessary comma: "18 November, 1952." *Every date* that appears in this document uses this erroneous military format, with the "unnecessary comma." By a curious coincidence, this is precisely the same style used by William L. Moore in *all* of his many letters to me since 1982, when our correspondence began.

Another curious anomaly in the Hillenkoetter document is the use of a "zero" preceding a single-digit date, a practice that was not used in 1952, when the briefing document allegedly was written, and which has come into limited use only in very recent years. Examination of numerous military and CIA documents written during the 1950s, 1960s, and 1970s shows the standard format was to write: "1 August 1950." Yet the Hillenkoetter document contains the following: "01 August, 1950" and "07 July, 1947," and "06 December, 1950."

My files of correspondence from Moore show that he used a single digit *without* a zero until the fall of 1983—roughly a year before the Hillenkoetter document film reportedly was sent to Shandera—when he then switched to the same style used in the Hillenkoetter briefing document.

The other document contained on the 35-mm film is what purports to be a "Top Secret/Eyes Only" memorandum, dated September 24, 1947, on White House stationery signed by President Truman. There is no question of the authenticity of the signature, but thanks to invention of the Xerox machine, it is easy to substitute bogus text on a photocopy of an authentic original, obtained, for example, from the Truman Library, in Independence, Missouri, which both Moore and Friedman visited prior to late 1984.

The format of the September 24 memorandum to Defense Secretary Forrestal differs significantly from that used by the president's secretary in other memoranda written to Forrestal, and others, during the same period. The typewriter used for the September 24 document was a relatively inexpensive one with a worn ribbon and keys that had not been recently cleaned, in contrast to the more elegant typeface, fresh-ribbon appearance of authentic Truman memoranda written at about the same time.

Furthermore, Truman was a blunt-spoken man whose letters reflect that style. Yet the second paragraph of the two-paragraph September 24 memo is filled with "un-Truman-like" gobbledygook: "It continues to be my feeling that any future considerations relative to the ultimate disposition of this matter should rest solely with the Office of the President following appropriate discussions with yourself, Dr. Bush and the Director of Central Intelligence." There was no need for Truman to be vague for security reasons, because the September 24 letter is stamped "Top Secret/Eyes Only."

If the letter were authentic, I'm confident it would have read more like the following: "Let's find out where in the hell these craft are coming from, whether they pose a military threat, and what in the hell we can do to defend the country against them if they should attack. I trust you will place all our forces on alert status and inform me if you need additional funds or other resources to protect this nation."

Moore told his MUFON audience that for two and a half years "we sat on the [MJ-12] material and did everything we could with it" to check its authenticity. He noted that all of the persons listed as being members of MJ-12 are now dead. Moore added: "If I was going to pick a panel at that time, capable of dealing with a crashed UFO, I would certainly want to consider [those on] that list." In other words, the members of MJ-12 were persons whom Moore himself would probably have selected for such a committee.

In mid-1982, more than two years before learning of Bush's key role from the MJ-12 papers, Moore demonstrated remarkable psychic abilities in a paper presented at a MUFON conference in Toronto. Moore said that Bush would be "the logical choice for an assignment to set up a Top Secret project dealing with a crashed UFO." Two years later, the MJ-12 papers confirmed Moore's judgment.

In the spring of 1985, Friedman learned that more than a hundred boxes of once Top Secret USAF intelligence documents from 1946 through 1955 were being reviewed by USAF representatives for declassification at the National Archives, in Washington, and he informed Moore of this. In July, Moore and Shandera flew to Washington and were the first persons—according to Moore—to gain access to those more than one hundred cartons of once Top Secret documents.

Lady Luck smiled, enabling Moore and Shandera to discover a sorely needed sheet of paper that could authenticate the MJ-12 documents on the 35-mm film. This key document" purports to be a brief, two-paragraph memorandum, dated July 14, 1954, to USAF Chief-of-Staff Twining written by Robert Cutler, then special assistant to President Eisenhower. The subject of Cutler's memo was "NSC/MJ-12 Special Studies Project," and it informed General Twining that "the President has decided that the MJ-12 SSP briefing should take place during the already scheduled White House meeting of July 16, rather than following it as previously intended."

Moore explained the importance of the July 1985 discovery of the Cutler memo to his MUFON audience in these words: "For the first time we had an official document available through a public source [National Archives] that talked about MJ-12." One might logically have expected that Moore would promptly "go public" with his remarkable MJ-12 papers, which now seemingly were authenticated by the Cutler memo. Yet, curiously, Moore did not do so, *for nearly two years!*

In the April 30, 1987, issue of a newsletter Moore publishes, he first released *three* of the *seven* pages of the Hillenkoetter briefing document, but in *heavily censored* form—*censored by Moore himself.* There was no mention of the Truman memo of September 24, 1947, nor of the Cutler memo of July 14, 1954, nor of the 35-mm film. Instead, Moore implied that the three heavily censored pages of the Hillenkoetter document had been provided by his "well-placed contacts within the American intelligence community" and said that "assurances have been given that additional information can be made available to us over the next several months."

This suggests to me that Moore planned to "dribble out" the MJ-12 material, in his possession since late 1984, in subsequent issues of his newsletter. This could generate more paid subscribers. If this was Moore's plan, it was thrown into disarray in mid-May when British UFOlogist Timothy Good met with the press to promote his new book, which claims a global UFO coverup. Good told British news media about the MJ-12 documents, which he said he had obtained "two months ago from a reliable American source who has close connections with the intelligence community. . . ."

Shortly afterward, Moore went public with the MJ-12 documents, including the Truman and Cutler memoranda, crediting them to the Moore-Shandera-Friedman Research Project. His release said: "Although we are not in a position to endorse its authenticity at this time, it is our considered opinion, based upon research and interviews conducted thus far, that the document and its contents *appear* to be genuine. . . . One document was uncovered at the National Archives which unquestionably verifies the existence of an 'MJ-12' group in 1954 and definitely links both the National Security Council and the president of the United States [Eisenhower] to it. A copy of this document, with its authenticating stamp from the National Archives, is also attached for your examination."

Stanton Friedman, nuclear physicist turned full-time UFO lecturer, who recently has returned to his original field, has been Moore's principal researcher-collaborator on crashed-saucer matters. Moore and Friedman continued to collaborate even after Friedman moved from California to New Brunswick, Canada, in 1980, as evidenced by their jointly authored paper on crashed-saucers presented at the 1981 MUFON conference in Cambridge, Massachusetts.

Thus one would think that, immediately after discovering the MJ-12 papers on the 35-mm film in late 1984, Moore would have sent a copy to Friedman. Yet it was not until late May 1987 that Friedman obtained a set of the documents, according to Friedman. In view of Moore's claim that he and Shandera spent more than two years trying to verify the authenticity of the MJ-12 papers,

Exhibit "A." At top are Hillenkoetter MJ-12 documents. Below these are examples of authentic Military/CIA-document format of the 1950s.

On 06 December, 1950, a
origin, impacted the e
Guerrero area of the e

On 07 July, 1947, a
recovery of the wrec
During the course of

until 01 August, 1950, upon
g designated as permanent

UNITED STATES GOVERNMEN

SECRET
SECURITY INFORMATION

DATE: 4 April 1958

elligence

UNITED STATES GOVERNMENT

DATE: 9 December 1952

DEPARTMENT OF THE AIR FORCE

Exhibit "B." At top are Hillenkoetter MJ-12 documents. Below are authentic Military/CIA documents showing correct date format, without comma following the month.

Hillenkoetter MJ-12 documents:

red by Dr. Bron
roup (30 November, 1947) th
human-like in a ance, t

DOCUMENT PREPARED 18 NOVEMBER, 1952.

ANALYSIS OF FLYING OBJECT INCIDENTS IN THE U. S.

Air Intelligence Division No. 203
10 December 1948

Directorate of Intelligence and Office of Naval Intelligence

DEPARTMENT OF THE AIR FORCE
WASHINGTON 18 AUGUST 1954

AFR 200-2
1-5

INTELLIGENCE
Objects Reporting (Short Title: UFOR

Exhibit "C." William L. Moore letters showing same incorrect "mixed military-civil format as that used in the Hillenkoetter MJ-12 briefing paper.

il Klass
04 "E" St. SW
C 20024

14 May, 1984

WILLIAM L. MOORE PUBLICATIONS & RESEARCH
4219 WEST OLIVE ST., SUITE № 247, BURBANK, CA. 91505

WILLIAM L. MOORE
WAYNE W. "DAVID" DUKE
PHONE: (818) 506-8365

Date 17 June, 1985

WILLIAM L. MOORE
WAYNE W. "DAVID" DUKE
PHONE: (818) 506-8365

Date 06 June, 1985

Klass
SW
24

WILLIAM L. MOORE PUBLICATIONS & RESEARCH
4219 WEST OLIVE ST., SUITE # 247,
BURBANK, CALIFORNIA, 91505

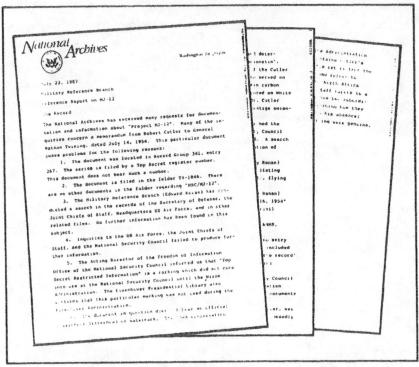

National Archives memorandum casting doubt on the authenticity of the "Cutler-Twining" memo.

one would have expected that Moore would promptly have sent the MJ-12 papers to Friedman to enlist his help in trying to authenticate them.

Friedman told me that Moore first informed him by telephone of the MJ-12 papers in late 1984 or early 1985. But, as Friedman explained in a recent letter, at that time one of his sons was fatally ill and Friedman was preoccupied with buying a new house and preparing to leave for a long UFO lecture tour. So it did not occur to Friedman to ask that Moore send him a copy of the MJ-12 papers, nor did he request a copy during the subsequent two years. That Moore did not send Friedman a copy on his own seems a most curious oversight since the documents, if authentic, were world-shaking in their importance. It is especially odd in view of Timothy Good's claim that his unidentified American source had supplied him with a copy of the MJ-12 papers earlier than Moore supplied a copy to Friedman, his closest collaborator on the case. (Moore has not responded to my repeated queries as to whether he was the American source who supplied Good with the MJ-12 papers.)

On July 22, 1987, Jo Ann Williamson, chief of the military reference branch of the National Archives, wrote a three-page memorandum summarizing the results of its own investigation into the Cutler/Twining memo, which played a key role in "authenticating" the MJ-12 papers. The National Archives memo pointed out that every other Top Secret document in the boxes of material in which the Cutler/Twining memo allegedly was found was stamped with an individual "register number"—a protocol used by the USAF reviewers to assure that each is properly accounted for and none is mislaid. The National Archives memo notes that the Cutler/Twining memo "does not bear such a number."

The Cutler/Twining memo purported to be a carbon copy on onionskin paper—which understandably would not carry the White House logo and would not necessarily be signed or initialed by Cutler. But the National Archives memo noted that "the Eisenhower Library has examined its collection of the Cutler papers. All documents created by Mr. Cutler while he served on the NSC staff have an eagle watermark in the onionskin paper." The Cutler/Twining memo found by Moore and Shandera did *not* have such a watermark. Furthermore, typewriter-key impressions protruded from the backside, suggesting it was an "original" and not a carbon copy as it appeared to be.

The National Archives memo quoted Eisenhower Library officials as stating that even when President Eisenhower had "off-the-record" meetings, his appointment books "contain entries indicating the time of the meeting and the participants. . . ." But "President Eisenhower's Appointment Books contain no entry for a special meeting on July 16, 1954, which might have included a briefing on MJ-12."

More significant, Robert Cutler could not possibly have written the memo on July 14, 1954, telling of last-minute changes in the president's schedule, *because Cutler had left Washington 11 days earlier (July 3)* to visit major military facilities in North Africa and Europe and did not return to Washington until July 15. This is shown by his subsequent trip report to the president, dated July 20, housed in the Eisenhower Library.

On August 20, 1987, the Committee for the Scientific Investigation of Claims of the Paranormal issued a four-page press release that characterized the MJ-12 papers as "clumsy counterfeits." It cited some of the discrepancies discussed above and attached a copy of the National Archives memo of July 22, 1987.

Several weeks later, the J. Allen Hynek Center for UFO Studies responded with a press release that said the question of the authenticity of the MJ-12 documents "is still open." CUFOS quoted Moore as saying that CSICOP "failed to raise a single issue which cannot be explained by further examination of the evidence." Moore charged that CSICOP's appraisal was "not only premature, but unscientific and emotional."

Shortly afterward, Citizens Against UFO Secrecy, a group that often accuses the government of a UFO-coverup, distributed the September issue of its newsletter *Just Cause*. The entire issue was devoted to the MJ-12 papers. Editor Barry Greenwood said that he remains open-minded to the possibility that a flying saucer crashed in New Mexico in 1947. But, based on his own investigation into the MJ-12 papers, Greenwood characterized them as "a grand deception and, consequently, a giant black eye on the face of UFOlogy. . . . The deeper we looked, the worse it became."

The MJ-12 Crashed-Saucer Documents: Part 2

Numerous flaws and inconsistencies in the "Top Secret/Eyes Only" MJ-12 document, allegedly written by Rear Admiral R. H. Hillenkoetter on November 18, 1952, to brief President-elect Dwight D. Eisenhower about crashed flying saucers, provide further evidence, beyond that discussed in Part 1 of this report (preceding chapter), that the document is a counterfeit.

The document, made public in late May 1987 by UFOlogist William L. Moore, supports his earlier claims of a 1947 crashed saucer made in the book *The Roswell Incident,* which he coauthored with Charles Berlitz in 1980, based on research conducted by Moore and UFOlogist Stanton T. Friedman.

One obvious anomaly is that the introductory portion of the Hillenkoetter/Eisenhower briefing document is written as if Eisenhower had never been informed that a crashed saucer (allegedly) had been recovered in New Mexico on July 7, 1947, by Army officers while Eisenhower was Army chief of staff. Moore made a similar claim in his book.

Yet historical records turned up by Moore himself show that only two days after the alleged crashed-saucer recovery, Eisenhower spent more than two hours with Lt. Gen. Lauris Norstad, a top Army Air Force (AAF) official. Surely General Norstad would not withhold word of so momentous an event from his commanding officer, who would be responsible for defending the nation if the crashed saucer was the precursor of an extraterrestrial attack.

Later that same day, July 9, Lt. Gen. Hoyt Vandenberg, AAF vice chief of staff, who Moore claims knew of the crashed saucer, attended a two-hour meeting of the Joint Chiefs of Staff in which Eisenhower participated. Yet the contents of the Hillenkoetter document imply that General Vandenberg also withheld news of recovery of a crashed saucer from the nation's top military officials.

Stanton Friedman has attempted to explain this anomaly by pointing out that a month before the (alleged) recovery of the crashed saucer Eisenhower had announced that he would retire later that year to become president of Columbia University. Although Eisenhower remained on duty as one of the nation's top military officers for another six months, Friedman claims that information on the crashed saucer was withheld from him because of his pending retirement.

According to the Hillenkoetter briefing document, "Numerous examples of what appear to be a form of writing were found in the wreckage. Efforts to decipher these have remained *largely* unsuccessful. Equally unsuccessful have been efforts to determine the method of propulsion or the nature or method of transmission of the power source involved. Research along these lines has been complicated by the *complete absence* of identifiable wings, propellers, jets, or other conventional methods of propulsion and guidance, as well as a *total lack* of metallic wiring, vacuum tubes, or similar recognizable electronic components." (Emphasis added.)

As a result, the document reports, "it is virtually certain that these craft do not originate in any country on earth. . . ." (Based on the described characteristics, it should have been *absolutely* obvious and certain that the craft could not have been built on Earth.)

The briefing document notes that the "implications for the National Security are of continuing importance in that the motives and ultimate intentions of these visitors remain completely unknown. In addition, a significant upsurge in the surveillance activity of these craft beginning in May [1952] and continuing through the autumn of this year has caused considerable concern that new developments may be imminent."

The document completely fails to mention that UFOs seemingly had been spotted near the nation's capital quite recently—July 19-20 and again on July 26-27—on radar scopes at Washington's National Airport, only a few miles from the White House and the Pentagon. One should expect that Hillenkoetter would have begun his briefing by citing these specific and potentially alarming incidents that could indicate possible hostile intent.

Instead, Hillenkoetter began his briefing by discussing historical trivia, such as the first UFO sighting report by pilot Kenneth Arnold. Another example: the document describes the (alleged) history of Majestic-12:

OPERATION MAJESTIC-12 is a TOP SECRET Research and Development/Intelligence operation responsible directly and only to the President of the United States. Operations of the project are carried out under control of the Majestic-12 (Majic-12) Group which was established by special classified executive order of President Truman on 24 September, 1947, upon recommendation by Dr. Vannevar Bush and Secretary [of Defense] James Forrestal. (See Attachment "A.") Members of the Majestic-12 Group were designated as follows:

Adm. Roscoe H. Hillenkoetter	Dr. Jerome Hunsaker
Dr. Vannevar Bush	Mr. Sidney W. Souers
Secy. James V. Forrestal	Mr. Gordon Gray
Gen. Nathan F. Twining	Dr. Donald Menzel
Gen. Hoyt S. Vandenberg	Gen. Robert M. Montague
Dr. Detlev Bronk	Dr. Lloyd V. Berkner

A footnote states: "The death of Secretary Forrestal on 22 May, 1949, created a vacancy which remained unfilled until 01 August, 1950, upon which date Gen. Walter B. Smith was designated as permanent replacement."

The Hillenkoetter document never identifies who was chairman of MJ-12, although his being selected to give the briefing and a designation "MJ-1" after his name (where he is listed as the briefer) implies he was its chairman. (No such designation is shown for others.) Yet, if Dr. Bush played a key role in organizing such a group, logically he would have been named chairman.

The names are not listed in alphabetical or any other logical sequence. Hillenkoetter seemingly lists himself as "Adm. *Roscoe* H. Hillenkoetter," even though examination of authentic letters that he wrote while director of the CIA, obtained from the Truman Library, show that he used his initials "R. H." and not "Roscoe." Further, in these authentic letters he correctly showed his rank as "Rear Admiral," not "Admiral," whereas the briefing document implies he was a "full admiral" with four-star rank.

Five of those listed as original members of the alleged MJ-12 were highly respected scientists—Bush, Bronk, Hunsaker, Menzel, and Berkner—presumably selected because of their technical expertise. Thus, if MJ-12 existed, they would be likely to remain on the committee (except possibly for Bush, for reasons to be discussed shortly).

General Vandenberg, as vice chief of staff in 1947 of the newly created U.S. Air Force, who would soon become chief of staff and hold that position until he retired in mid-1953, would also be a logical original member who would likely remain on MJ-12 because of the USAF's vital role in protecting the nation's air space and operating its network of early-warning radars.

But others listed as original members of MJ-12 would have been named to the group because of the positions they held as of September 1947, when MJ-12 was created. When they left these posts for other assignments or retired from government, their successors in those posts would logically have replaced them on MJ-12. Yet that did not occur, according to the briefing document.

For example, James Forrestal as secretary of defense would logically be named to MJ-12 in 1947. But when Forrestal resigned in the spring of 1949 (shortly before committing suicide on May 22) the new defense secretary, Louis A. Johnson, would have been his logical replacement on MJ-12.

Instead, the Hillenkoetter document claims that Forrestal's vacancy was not filled for more than a year and that his replacement was Gen. Walter B. Smith. But at that time General Smith was replacing Hillenkoetter as director of the CIA, and thus logically should have taken Hillenkoetter's place on

MJ-12, not Forrestal's.

When Hillenkoetter left the CIA in the fall of 1950, he returned to active duty in the Pacific as commander of the Seventh Fleet in Formosan Waters, a post he held for more than a year. Thus he would be far removed from Washington and could not readily direct the important work of an MJ-12 committee—if he originally had been its chairman. If he was not its chairman, why would Hillenkoetter be selected in late 1952 to brief Eisenhower? In the intervening year since he returned to the United States in the fall of 1951, he had served as commander of the Brooklyn Navy Yard and later of the Third Naval District, in New York. (The reason for Hillenkoetter's being chosen by the document's counterfeiter will be disclosed shortly.)

In July 1947, when the alleged crashed saucer was sent for analysis to the AAF's large technical center at Wright Field (now Wright-Patterson AFB, Ohio), its commander was Lt. Gen. Nathan Twining, who is shown as an original member of MJ-12. But within several weeks after MJ-12 (allegedly) was created, Twining was transferred to Alaska to head the Alaskan Command, and he remained there for more than two years before being transferred back to Washington. Surely if Twining was selected for MJ-12 because of his technical expertise, he would have stayed on at Wright-Patterson AFB to monitor analysis of the crashed saucer. Or his successor there should logically have taken his place on MJ-12.

Gordon Gray, who had become *assistant* secretary of the army about the time that MJ-12 allegedly was created, would have been a most unlikely member based on background and position. Gray was trained as a lawyer and had spent the previous ten years as publisher of two newspapers, and he did not hold a top-ranking Pentagon position. In mid-1949, Gray was named Secretary of the Army, but he resigned April 12, 1950, to return to civilian life and was succeeded in that post by Frank Pace, Jr. Yet more than two years later, when the briefing document allegedly was written, Gray was still listed as an MJ-12 member.

Sidney W. Souers might have been a logical choice as an original member of MJ-12 because on September 26, 1947, he was named executive secretary of the president's newly created National Security Council. Souers, a naval reservist, had risen to the rank of rear admiral during World War II to become deputy chief of Naval Intelligence and played a role in organizing the then-new CIA. Souers retired from his NSC position in early 1950 to return to civilian life, but he, rather than his successor at NSC, is listed in late 1952 as an MJ-12 member.

Thus, according to the briefing document, as of late 1952, MJ-12's membership included two civilians who were not scientists and had since retired to civilian life. But MJ-12 failed to include the current secretary of defense, Robert A. Lovett, who should certainly have been a member because of "implications for the National Security. . . ."

At roughly the same time as the briefing document allegedly was written, the CIA was preparing to secretly convene a panel of five top scientists, headed by Dr. H. P. Robertson, of the California Institute of Technology, for a four-day study of the USAF's best UFO cases. The CIA sought this assessment because it was then considering launching its own UFO investigation. (How strange that CIA director Gen. Walter B. Smith, who would have known of the crashed saucer through his membership in MJ-12, would allow the CIA's top scientific officials to waste their time, and his own time in briefings, when he already knew the "truth" about UFOs.)

By a curious coincidence, one of the five scientists selected for the CIA's UFO assessment was Lloyd V. Berkner—allegedly also a member of the MJ-12 group. If MJ-12 existed and Berkner was a member, it was a waste of his valuable time to participate in a "make-believe" exercise for which he already knew the answer. The final report of the "Robertson Panel," which was classified "Secret" and signed by Berkner, concluded that there was no evidence that any UFOs were extraterrestrial craft or posed any potential threat to national security.

Although two of the five members of the Robertson Panel convened in early 1953 are still alive—Dr. Luis Alvarez and Dr. Thorton Page—every person listed as being a member of MJ-12 in late 1952 was deceased when Moore, Friedman, and Shandera released their documents. Hillenkoetter died in 1982, two years before Moore says he received the MJ-12 papers.

The briefing document, allegedly found on a roll of 35-mm film received in the mail by Moore's friend Jaime Shandera on December 11, 1984, also contains what purports to be a "Top Secret/Eyes Only" memorandum from President Truman to Defense Secretary Forrestal, dated September 24, 1947, asking that Forrestal and Dr. Vannevar Bush create the MJ-12 group.

Although the White House letterhead and the Truman signature appear authentic, the memorandum's format and typewriter typeface differ significantly from that of copies I received from the Truman Library of other Truman memoranda written to Forrestal and other cabinet members during that same time period. The library has no record of the alleged September 24, 1947, memorandum. (A counterfeit can easily be made by obtaining a photocopy of an authentic Truman memo from the Truman Library, superimposing a spurious message, making a photocopy, touching it up with white typewriter correction fluid, and making another copy.)

Moore, Friedman, and Shandera stress that Forrestal and Bush met with the president on the date of the purported Truman memorandum calling for the creation of MJ-12, which is true. But in reality the meeting was held to prevail upon Bush to serve as chairman of the new Defense Research and Development Board, authorized under the then-recent National Security Act of 1947. The next day, September 25, Truman announced that Bush had accepted the appointment, and Bush replied in a letter dated September 26 acknowledging his acceptance.

In his letter, Bush emphasized "my wish ultimately to be free of governmental duties in order to return more completely to scientific matters, and I am happy to have your assurance that you will look to the Secretary and me for a suggestion of a successor to take over my responsibilities after the new organization is well launched."

Roughly a year later, on October 1, 1948, Bush wrote to the president to submit his resignation so he could "give more complete attention to the affairs of Carnegie Institution of Washington." Yet the MJ-12 briefing document indicates that as of four years later Bush still was an active member.

If President Truman knew that a crashed saucer had been recovered in July 1947, and the remains of another in late 1950, as claimed in the briefing document, certainly he would have been gravely concerned when he read the headlines in Washington newspapers on July 20, 1952, reporting that mysterious blips had been spotted the previous night near Washington, D.C., on the scopes of an old military radar recently installed at National Airport.

Although later government investigation would show the blips were the result of anomalous radar propagation conditions, surely Truman would immediately have called for a White House briefing by General Vandenberg, USAF chief of staff and member of the alleged MJ-12 group, to determine what was being done to protect the nation's capital against possible UFO attack. And certainly General Vandenberg would promptly have beefed up Washington's air defenses by rapidly shifting large numbers of interceptors to nearby military airfields.

Yet one week later, when mysterious blips again appeared on the radar displays at Washington's National Airport shortly after 9:00 P.M., it was not until two hours later that the USAF's New Castle County air base in Delaware was alerted and dispatched a mere two interceptor aircraft. Two hours later, another pair of interceptors took off to replace the first two, which were running low on fuel. Thus a total of only four interceptors were launched that night to defend the nation's capital!

If the president knew of crashed saucers, surely the two incidents would prompt him to quickly seek a briefing on July 27 by his MJ-12 group, and an explanation from General Vandenberg of why he had failed to promptly provide much more air cover for the nation's capital in view of the similar incident the previous week. Yet instead, President Truman simply asked his Air Force aide, General Landry, to make a telephone inquiry to USAF intelligence at the Pentagon.

If General Vandenberg had been so grossly derelict in his duty to try to defend the nation's capital, surely he would have been promptly replaced both as USAF chief of staff and as a member of MJ-12. Yet the briefing document indicates that General Vandenberg remained on MJ-12, and he held his top USAF post until normal retirement in mid-1953.

The briefing document claims that MJ-12, with President Truman's concurrence, opted to keep the crashed saucer secret "to avoid public panic at all costs." It is interesting to check this claim against Truman's reaction when he learned on September 19, 1949, that the USSR had exploded its first nuclear weapon several weeks earlier. The potential implications for the U.S. public were far more terrifying. Only four days later, on September 23, after allowing time to brief the Joint Chiefs of Staff and congressional leaders, President Truman publicly announced the first Soviet nuclear test. He quite correctly sized up public reaction and there was no panic.

The launch of the first small Soviet Sputnik in October 1957, and the launch of a much larger satellite a few weeks later, prompted U.S. concern that the Soviets might someday be able to use satellites to deliver nuclear weapons from space. To enable the United States to continuously monitor all objects in space, Navy scientists developed and deployed an "electronic fence"—stretching across the southern portion of the United States from the Atlantic to the Pacific. This Navy Space Surveillance System was so sensitive it could detect a six-inch-long metal strap released from one of our satellites.

To speed deployment, the system was built using commercially available television-station transmitters and large numbers of slightly modified television-set antennas. Such a system could have been deployed in the late 1940s or early 1950s, if the United States had captured one or more crashed extra-terrestrial craft and wanted to be able to detect and monitor any future intrusions. Yet such a space surveillance fence was not built until the late 1950s, following the launch of Sputnik.

Hard evidence that the MJ-12 papers are a fraud can be found in a once "Top Secret" document that was declassified in 1985 at the National Archives and featured in the July 1985 issue of the *MUFON UFO Journal.* The document, dated 10 December 1948, was prepared by the USAF's Directorate of Intelligence and the Office of Naval Intelligence and is titled: "Analysis of Flying Object Incidents in the U.S."

The opening page of the top-secret report begins: "PROBLEM: TO EXAMINE pattern of tactics of 'Flying Saucers' (hereinafter referred to as flying objects) and to develop conclusions as to *the possibility of existence.*" (Emphasis added.) In other words, more than one year after the U.S. government allegedly captured a crashed saucer in New Mexico, top USAF and Navy intelligence officers were uncertain whether there really were extraordinary craft in our skies.

The report states, "THE ORIGIN of the devices is not ascertainable," and considers two possible explanations. One is that "the objects are domestic devices." The other alternative is that the "objects are foreign, and if so, it would seem most logical to consider that they are from a Soviet source." The report then proceeds to focus on the possibility that UFOs are Soviet craft, possibly built with the aid of captured German scientists.

This 1948 top-secret report, after citing a number of seemingly credible UFO-sighting reports, concludes: "It is not known at this time whether these observations are misidentifications of domestically launched devices, natural phenomena, or foreign unconventional aircraft."

The reason Hillenkoetter was selected by the counterfeiter as the alleged briefer of President-elect Eisenhower was to exploit Hillenkoetter's known involvement in the UFO movement shortly after he retired from the Navy in 1957. At that time he became a member of the board of governors of

NICAP (National Investigations Committee for Aerial Phenomena), a newly formed group that would soon become the largest and most influential pro-UFO group.

NICAP's official position was that some UFOs were extraterrestrial craft and that the U.S. Air Force was resorting to coverup tactics to keep the truth from the public. In 1960, Hillenkoetter "went public" in support of NICAP's coverup charges. For example, the February 28, 1960, edition of the *New York Times* carried a feature article that quoted Hillenkoetter, identified as a former CIA director, as saying "it is time for the truth [about UFOs] to be brought out in Congressional hearings."

In an article on Hillenkoetter in the November/December 1986 issue of *International UFO Reporter,* published by the J. Allen Hynek Center for UFO Studies (CUFOS), Bruce S. Maccabee wrote: "His [pro] UFO activities seem to make no sense in the context of official policy of the government for which he had worked for many years. Yet his actions do make sense if he sought to persuade the public that UFOs are real—without revealing government secrets about UFOs." (The article was written some months before release of the MJ-12 papers, and I am confident that Maccabee had no involvement in their preparation.)

In naming Hillenkoetter as the MJ-12 briefer, the counterfeiter sought to provide an added irony: that Hillenkoetter actually knew about two crashed saucers and seemingly joined NICAP and issued his public statements to try to break loose the secret. Yet if Hillenkoetter knew of crashed saucers through his work at the CIA and in MJ-12, he would never have dared make such public statements, lest the government terminate his retirement benefits, or worse. Surely a man skilled in intelligence would be wise enough to use "leaks" to a major news media reporter who would have protected his identity and retirement benefits.

The addition of the name of Dr. Donald Menzel, a world-famous astronomer and leading UFO-debunker, is an attempt at revenge by the MJ-12 counterfeiter. Menzel was hated and maligned by the "UFO-believers" during the first two decades of the UFO era. In the eyes of a UFO-believer, there could be only one thing worse than being a UFO-debunker—and that is a debunker who *knowingly* resorted to falsehood. The counterfeiter tried to heap this final indignity on a world-famous scientist, now deceased, by listing him as a member of MJ-12. (Recently I was told by one UFOlogist that he suspected that I had replaced Menzel on MJ-12 following his death.)

Prior to the release of the MJ-12 papers, some members of the UFO movement considered Moore and Friedman to be among the most rigorous of researchers and investigators. Moore claims that he and his associates did not release the MJ-12 papers for more than two years while they investigated their authenticity. Moore said, in his lengthy banquet speech at the 1987 MUFON conference in Washington, that he and his associates had done "every conceivable thing we could do." Moore added: "What we found was that nothing seemed inconsistent."

Moore said he recognized that "maybe someone was out to embarrass us" and that the briefing document and Truman memo (allegedly) received from an anonymous source might be "disinformation." And, Moore added, "we felt if we didn't do anything with it, maybe somebody might urge us to and [this] might give us a hint as to where it had come from." However, Moore said, "we never got urged to" make public the MJ-12 papers.

This provides useful insights. If the MJ-12 papers were *authentic*, the person who sent them to Moore's friend Shandera in late 1984 so they could be made public might logically become impatient after a year or two and submit them to the *New York Times* or the *Washington Post*, or one of the sensationalist tabloid newspapers that feature wild UFO stories. But that did not occur. Nor did Moore or Shandera ever receive an anonymous letter urging them to make the MJ-12 papers public.

If, on the other hand, the MJ-12 papers are counterfeits intended for public disinformation, how could the hoaxer be confident that Moore and his associates would not spot the discrepancies noted above and in Part 1 of this article? After so many months had elapsed since the papers (allegedly) were mailed to Shandera without being made public, the counterfeiter might logically decide that Moore et al. had concluded they were frauds and that the MJ-12 papers should now be sent to less discriminating UFOlogists or the news media. Yet this did not occur either.

If the MJ-12 papers are counterfeits intended to embarrass Moore and his associates by demonstrating their inability to spot a hoax, then the counterfeiter clearly succeeded. Shortly after the MJ-12 papers were made public, Jerome Clark, editor of the *International UFO Reporter*, made an informal survey of UFO "researchers whose judgments I trust" to obtain their appraisal of the authenticity of the documents, as he reported in the May/June 1987 issue of *IUR*. Clark said he found only two UFOlogists who believed the MJ-12 documents were authentic: William Moore and Stanton Friedman, plus Shandera, who does not consider himself a UFOlogist.

Clark admitted that "Moore has been much less than forthcoming about his sources and his information, leading critics to charge that whatever his intentions, he has become for all practical purposes a part of the cover-up— a point that Moore himself probably would concede." Clark said that "Moore's promise that further (and, we would hope, better-substantiated) revelations would follow the MJ-12 release has not been fulfilled at this writing," i.e., mid-1987.

But in the next issue of *IUR*, mailed in October, Clark sharply criticized CSICOP (and me) for a news release, issued August 20, 1987, that characterized the MJ-12 papers as a forgery. He noted that CUFOS had promptly issued a refutation "prepared in cooperation with [Stanton] Friedman and William Moore." Clark admitted that the MJ-12 "document may well be a forgery. Yet if it is, it is hardly a 'clumsy' counterfeit, as CSICOP would have us believe, but an extraordinarily skilled one, and there is no reason whatever to accuse Moore of responsibility for it." (CSICOP's release did not make such an accusation. Instead, it invited Moore to cooperate with CSICOP by supplying information requested earlier that could help pinpoint the hoaxer. Moore has not responded.)

Clark concluded: "What is especially troubling is that many of those who have been swept up in the wave of anti-Document [*sic*] hysteria are among ufology's finest and ordinarily most sensible people." Clark predicted that it would be possible to resolve the authenticity question "within a few months."

Useful insights are provided by the one-page top-secret memo that mentions MJ-12, allegedly written to General Twining by President Eisenhower's special assistant, Robert Cutler. Moore and Shandera claim they found this Cutler/Twining memo in mid-1985 in one of more than a hundred boxes of recently declassified USAF intelligence documents in the National Archives to which they were the first to gain access.

The memo purports to be an unsigned carbon copy written by Cutler on July 14, 1954, but Cutler had left the country 11 days earlier and did not return until July 15. When this and numerous other discrepancies raising doubts about authenticity appeared in a National Archives memo of July 22, 1987, Friedman's attempted explanation was that the July 14 memo had been written by Cutler's deputy, James Lay, using Cutler's name.

If Friedman had checked with the Dwight D. Eisenhower Library he could have discovered that its files include a "Top Secret" memo to the Secretary of Defense and Chairman of the Atomic Energy Commission, written by James Lay on the *same date*—July 14, 1954—*over his own name, not Cutler's*. Thus, if the Twining memo had been written by Lay, as Friedman suggests, his name would appear at the bottom, not Cutler's.

One of the many discrepancies in the Cutler/Twining memo noted by the National Archives is that it does not bear a registration number as do other top-secret papers declassified and made public in the massive collection in which Moore and Shandera said they found the Cutler/Twining memo. Friedman's explanation is that the Cutler/Twining memo was "planted" by one of the USAF officers involved in reviewing and declassifying the many thousands of documents. The careless officer simply forgot to stamp a registration number on the "planted" document.

But how would the officer know that Moore and Shandera had the MJ-12 papers and sorely needed what Moore has called "an official document available through a public source that talked about MJ-12" to "authenticate" the MJ-12 papers? And that Moore and Shandera would promptly visit the Archives and spot the one-page memo among the many thousands of papers in more than a hundred boxes of newly released files?

At MUFON's 1982 conference, held in Toronto in early July, Moore presented a paper on new evidence to support the Roswell crashed-saucer story and said it would be included in his new book, *The Roswell Evidence*, to be published in the spring of 1983. But the book was never published, perhaps because public interest in UFOs then was at an all-time low.

In early January 1987, a UFO-sighting report by a Japan Air Lines 747 pilot attracted considerable media attention. Soon afterward, Whitley Strieber's book *Communion,* describing his alleged abduction by UFOnauts, quickly climbed to the best-seller list. By the spring of 1987, publishers were interested in books on the subject—at least those that were pro-UFO.

The April 30, 1987, issue of a UFO newsletter that Moore publishes contained the first three pages of the MJ-12 document, *which had been heavily censored by Moore himself.* In the newsletter, Moore said the "information is being shared with the UFO community, and through them, ultimately, with the public at large, in anticipation that the resulting controversy will prove useful in further verifying the authenticity of its contents." He implied that the material had been supplied by his "well-placed contacts within the American intelligence community."

Several weeks later, the *London Observer* broke the story on the MJ-12 papers, which had been supplied to it by British UFOlogist Timothy Good, author of a new book entitled *Above Top Secret: The Worldwide UFO Coverup,* scheduled for publication that July. Good was quoted as saying he had obtained the MJ-12 papers "two months ago from a reliable American source who has close connections with the intelligence community." Shortly thereafter, Moore, Friedman, and Shandera made public the full and uncensored version of the MJ-12 papers, including the Cutler/Twining memo. (Moore has never replied to my several letters asking if he was Good's "American source.")

Moore, Friedman, and Shandera have publicly said they are "90-95 percent" confident of the authenticity of the MJ-12 papers, but they have offered no reason for their not being 100 percent confident. If they discovered any disturbing discrepancies in their more than two years of investigation, they have *never revealed them publicly.* Nor have they acknowledged the validity of any of the many anomalies turned up by others, even by members of the UFO movement.

Instead, Moore chooses to ignore critics and speak of additional evidence to be released at some future date, while Friedman sharply attacks the critics. One possible explanation is an understandable reluctance to admit they are victims of a hoax because of their superficial investigation. The alternative explanation is too obvious to require statement here. Neither alternative is likely to enhance their reputations in the UFO movement.

PHILIP J. KLASS

New Evidence of MJ-12 Hoax

A "smoking gun" recently has been discovered that confirms beyond any doubt that the alleged "Top Secret/Eyes Only" MJ-12 documents, which seemingly showed that the U.S. government had captured at least one crashed flying saucer and the bodies of several extraterrestrials in 1947, are counterfeit.

The MJ-12 documents were made public on May 29, 1987, by William L. Moore and two associates, Jaime Shandera and Stanton T. Friedman. If authentic, the documents would confirm claims made in a 1980 book, *The Roswell Incident,* authored by Moore and Charles Berlitz, of "Bermuda Triangle" fame.

The MJ-12 papers include what purports to be a one-page memorandum from President Harry Truman to Defense Secretary James Forrestal, dated September 24, 1947—several months after the alleged crashed-saucer recovery in New Mexico. The letter authorized Forrestal and Vannevar Bush to create a top-level Majestic-Twelve (MJ-12) group to analyze the crashed saucer and alien bodies. The other MJ-12 document is a lengthy status report on MJ-12's crashed-saucer research efforts, seemingly intended to brief President-elect Eisenhower, dated November 18, 1952. The briefing paper seemingly was written by Rear Admiral R. H. Hillenkoetter, who had earlier headed the Central Intelligence Agency and allegedly was a member of MJ-12.

A roll of 35-mm film, together with photocopies of these two "Top Secret/Eyes Only" documents, reportedly arrived at the home of Shandera by mail from an unknown sender on December 11, 1984. Moore, Shandera, and Friedman claim that they spent the next two and a half years investigating the authenticity of the MJ-12 papers before making them public in May 1987.

Moore and his associates said that their lengthy investigation had failed to turn up anything that would cast doubt on the authenticity of the MJ-12 papers. My own investigation revealed many reasons to suspect the MJ-12 papers were counterfeit. (See the previous three chapters in this section.)

Recently, I discovered hard physical evidence that demonstrates that these documents are counterfeit. This is based on the fact that a person's handwritten signature is like a snowflake—no two are ever *identical.*

Before the advent of the "Xerox era" and "signature machines," the very existence of two identical signatures was considered to be *"very strong evidence of forgery,"* according to the book *Questioned Documents,* by Albert S. Osborn, published in 1978. Osborn notes that "the fact that two signatures are very nearly alike is not alone necessarily an indication of forgery of one or both but the question is whether they are *suspiciously alike."* (Emphasis added.)

The "Harry Truman" signature on the MJ-12 Truman memorandum of September 24, 1947, is *suspiciously like* the signature on the letter that Truman wrote to Vannevar Bush on October 1, 1947. I found the October 1 letter in the Bush collection in the Manuscript Division of the Library of Congress and made several photocopies of it there.

In signing the authentic letter to Bush, Truman's pen accidentally skidded slightly, creating a small extraneous mark on the left upper part of the right-hand vertical stroke in the letter "H." *The same "skidmark" appears on the Truman signature of the MJ-12 memo of September 24, 1947.* It is slightly heavier on the MJ-12 memo because of the multiple photocopying operations used to make the hoax document.

(Photocopies of both signatures are shown in Figure 1. Readers who are sufficiently interested can make photocopies and superimpose them before a strong light to confirm that the two are identical.)

If the Truman signature is a counterfeit, then so is the alleged Hillen-koetter MJ-12 briefing paper, contained on the same 35-mm film, which makes specific reference to this "special classified executive order of President Truman on 24 September, 1947. . . ."

To obtain an expert corroboration of my own findings, I called David Crown, a professional "document examiner" in the Washington, D.C., area, who previously headed the Central Intelligence Agency's questioned documents laboratory. Crown informed me that the Truman memo had already been exposed as a hoax because it was written on a typewriter that "did not even exist in 1947." He told me that this discovery had been made by a highly respected document examiner, whose name and telephone number he provided. (I will refer to the latter document examiner as PT because of his reluctance to become a public figure in the MJ-12 controversy.)

When I called PT, he expressed great interest in obtaining a copy of the authentic Truman-Bush signature of October 1 because he had earlier been drawn into the MJ-12 controversy through a friend, also a professional document examiner. PT's earlier analysis of the typeface of the machine used to prepare the MJ-12 Truman memo indicated that it was a Smith-Corona machine that first appeared in 1963—more than 15 years *after* the September 24, 1947, date on the memo.

PT asked me to send the October 1 memo to him by overnight mail because he was leaving in two days for a meeting of professional document examiners in San Francisco, and I did so. In our first conversation, I mentioned

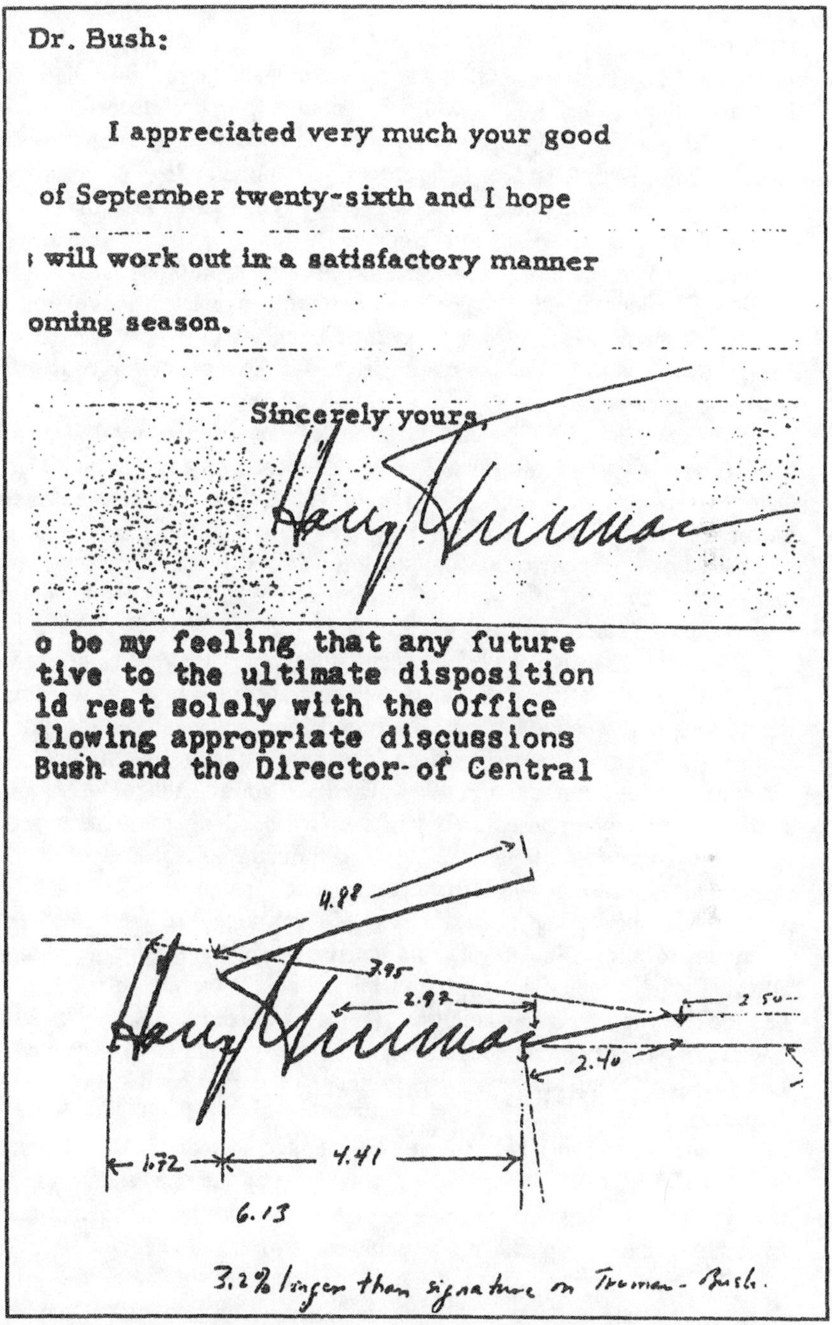

Dr. Bush:

I appreciated very much your good

of September twenty-sixth and I hope

will work out in a satisfactory manner

oming season.

Sincerely yours,

o be my feeling that any future
tive to the ultimate disposition
ld rest solely with the Office
llowing appropriate discussions
Bush and the Director of Central

FIGURE 1. Authentic Harry Truman signature from letter of Oct. 1, 1947, to Vannevar Bush (*top*). This signature and the one on MJ-12 document (*bottom*) are "suspiciously alike"—indicating MJ-12 memo is a forgery. The MJ-12 skidmark on the "H" is heavier because of multiple photocopies used to create counterfeit document.

that the MJ-12 Truman signature was approximately 3.6 percent longer than the one on the October 1 letter, which I attributed to optical distortion during the several photocopying operations needed to produce a counterfeit. PT explained that Xerox, and its competitors, intentionally do not reproduce a thin border around the outside of a document to be copied—to avoid creating unwanted lines at the edges. To compensate for this, the original copy is enlarged by roughly 1.2 percent—which is imperceptible to the casual reader.

Thus, if a counterfeiter had needed three photocopying iterations to produce the MJ-12 memo—as my own experiments suggested—this would account for the fact that the MJ-12 signature is about 3.6 percent larger than the October 1 signature.

Eight days later, PT called and informed me that the MJ-12 signature was "a classic signature transplant," i.e., a photocopy forgery. In the authentic October 1 signature, a portion of the top of the "T" in "Truman" barely intersected the "s" at the end of "Sincerely your*s*." When the counterfeiter had used typewriter correction fluid to retouch out the "Sincerely yours," he had slightly "thinned" the width of the top of the "T." This retouching, PT told me, is the "kind of coup de grâce we look for."

PT told me he had made overhead projector transparencies of the MJ-12 and October 1 signatures and taken to San Francisco to show at the meeting of professional document examiners. He first showed his audience the MJ-12 Truman memo typeface, pointing out that the Smith-Corona machine used did not exist in 1947. Then PT showed the MJ-12 Truman signature and superimposed a copy of the October 1 signature—enlarged by about 3.6 percent—and pointed out the "thinning" of the top of the "T." PT said his audience gave a verbal endorsement—"a chorus of 'Ah-haa!' "

PT told me he had already called Moore's longtime associate Stanton Friedman to inform him of PT's findings because "he had [earlier] sent me all this [MJ-12] material . . . [and] I felt I owed it to him to tell him that he should just wash his hands of this." (Friedman opted to ignore PT's advice. The next week Friedman spoke at a MUFON regional conference near St. Louis and repeated his earlier endorsement of the authenticity of the MJ-12 papers.)

Friedman, who has been the most outspoken defender of the authenticity of the MJ-12 papers, knew at least shortly after their release—more than two years ago—that the Truman signature on the MJ-12 memorandum "match[ed]" the one on a letter Truman wrote to Bush in October 1947.

Friedman reported this fact in his article published in the September/October 1987 *International UFO Reporter* claiming that this "match" confirmed the authenticity of the MJ-12 document. In fact, it really revealed just the opposite. (I am indebted to Christopher D. Allan of the United Kingdom for bringing Friedman's claim to my attention, and to Joe Nickell for supplying references from the book *Questioned Documents*.)

Earlier this year, Friedman requested and received a $16,000 grant from the Fund for UFO Research (FUFOR) for further investigation into the authenticity of the MJ-12 papers. Ironically, he already had in his possession the "smoking gun." Friedman, in an interim report on his FUFOR funded research, published in the September 1989 *MUFON UFO Journal*—prior to receiving PT's call—said his research had found nothing to question the "legitimacy" of the MJ-12 papers.

Others have earlier pointed out another suspicious flaw in the alleged Truman memo to Forrestal. This is the fact that the numerical portion of the date—"24, 1947"—was typed using *a different machine* from the one used to type "September."

The logical explanation for this flaw is that the counterfeiter used an old-vintage machine to make it appear that the memo was written in 1947. But the machine's numerical keys were inoperative, forcing the counterfeiter to type the numerical part of the date on a different machine and paste it in. If this were an authentic Truman memo, it would indicate that the President's secretary did not have access to a fully operable typewriter—which is highly unlikely.

Friedman and Moore visited the library to peruse the Bush collection in 1981-1982, prompted by a 1950 memorandum written by Wilbert B. Smith, a Canadian engineer. Smith's memo claimed that the U.S. government was conducting a highly classified investigation into "flying saucers," directed by Bush.

In Moore's paper presented at a MUFON conference in early July 1982, he reported that he and Friedman had "spent considerable time in Washington, D.C. over the past year locating and researching dusty files and records. . . ." This enabled him to report that Vannevar Bush and Defense Secretary Forrestal had met with President Truman on September 24, 1947—the date of the MJ-12 memo—after Bush had agreed to head the Pentagon's new research and development board.

A third document made public by Moore, Shandera, and Friedman in the spring of 1987 was what purported to be a "Top Secret" memo from President Eisenhower's special assistant, Robert Cutler, to USAF chief-of-staff Gen. Nathan Twining. The memo, dated July 14, 1954, informed Twining of a slight change of plans for a White House meeting of the "NSC [National Security Council]/MJ-12 Special Studies Project" to be held on July 16.

Moore and Shandera said they found the unsigned carbon copy when they visited the National Archives in mid-1985. As Shandera explained to me, because the memo was found in the National Archives it seemed to officially confirm the existence of MJ-12. However, the Cutler memo lacked a registration number, which all other Top Secret documents in the same files had. Nevertheless, Friedman claimed the memo was authentic because it concluded with "your concurrence in the above change of arrangements is assumed"—almost identical language to that used by Cutler in an earlier memo to Twining, dated July 13, *1953*. Friedman and Moore had found this authentic memo in 1981 in the collection of Twining's papers at the Library of Congress.

Curiously, the MJ-12 Cutler memo was found in recently declassified

USAF intelligence material—an unlikely place for a carbon copy seemingly intended for White House files. Also, it had been folded as if it had been carried in the breast pocket of a man's suit. Subsequent investigation by the National Archives revealed that Cutler could not possibly have written the letter because he was out of the country on July 14, 1954. This and other questionable aspects of the document were detailed by a National Archives official in a three-page memorandum.

Did Twining attend an NSC meeting at the White House, as instructed by the MJ-12 Cutler memo? When I checked Twining's official log for July 16, 1954, it showed many appointments but no NSC briefing. When I pointed out this discrepancy to Friedman, he argued that the White House MJ-12 meeting was so secret that it would not be listed in Twining's official log.

If Friedman's logic were valid, then Twining's official log ought not show him attending the "Extraordinary Meeting of the National Security Council" referred to in the authentic Cutler memo of July 13, *1953*. Cutler's memo explained that "special security precautions" should be taken "to maintain absolute secrecy regarding participation" in the NSC meeting. For example, Cutler explained that Twining was to enter the White House grounds via a special entrance and his Pentagon limousine should not remain parked near the White House. No such security precautions were prescribed in the MJ-12 Cutler memo.

When I checked Twining's official log in the Library of Congress it *did* show that Twining attended the very secretive NSC conference in 1953. His log showed: "National Security Council at White House all day"—demolishing Friedman's claim. By a curious coincidence, this secret July 16, 1953, NSC meeting was held one year to the day of the alleged MJ-12 NSC meeting.

Ironically, in the introduction to a paper on crashed-saucer claims authored by Moore and Friedman, presented at the 1981 MUFON conference, they quoted Albert Einstein as follows: "The right to search for the truth implies also a duty; one must not conceal any part of what one has recognized to be the truth." This recalls the admonition by French philosopher Charles Peguy: "He who does not bellow the truth when he knows the truth makes himself the accomplice of liars and forgers."

Editor's Note: William L. Moore was informed of the investigation and conclusions reported above. In a letter (October 16, 1989), Moore acknowledged that the document examiner referred to as PT had indeed made his (hoax) findings available "some time ago" and "we have not yet published them." But, he said, PT was only one of four document examiners he and his colleagues had consulted and claimed the opinions of the four about the issues involved with the Truman document are "mixed." He did not name the other examiners.

MILTON A. ROTHMAN

Cold Fusion: A Case History in "Wishful Science"?

Creative science requires an interplay between two opposing modes of thought: imagination and skepticism (T. Rothman 1989). New ideas, concepts, theories, and inventions come into being by the use of free imagination, for only in this manner can elements of reality be arranged into new and unexpected patterns. However, the unfettered imagination, if not linked to reality by observation and experimentation, has a tendency to fly off into the realm of pure fantasy. Good science requires a balance between the opposing impulses of creative fantasy and reality testing. A scientist too much wedded to fantasy is prone to believe in untested hypotheses. On the other hand, too much fixation on mundane reality produces a dogmatic skeptic, a naysayer, the kind who "proves" that people cannot fly, that computers cannot think.

You might think that experimental scientists would tend to be more realistic than theoreticians. After all, their instruments determine what they see, and machines can do only what they must do. Yet Albert Einstein, the theoretical physicist, was the most realistic of philosophers. Many experimenters, on the other hand, are unable to locate the fine line separating fantasy from reality.

While instruments do not lie (although they can make mistakes), perceptual and conceptual hazards beset the experimentalists whenever they must interpret data so as to extract meaning from an experiment. Sometimes they interpret the numbers so that they will agree with a preconceived theory. At this point it is possible for errors to arise if care is not taken.

I have, on more than one occasion, listened to physicists describe results that were clearly beyond the capabilities of their equipment. One nuclear physicist, measuring the energies of gamma rays emitted by materials when bombarded by neutrons, claimed to have detected a large number of different energies in a narrow energy band. He simply ignored the fact that his instrument could not separate energies so close together. The most famous case of this kind is the observation of canals on Mars by the astronomers Giovanni Schiaparelli, Nicolas Flammarion, and Percival Lowell. They wanted to see canals, so they did, even though their telescopes were inherently unable to resolve such structures. (Photographs never showed them.) These experiences illustrate how the experimenters' expectations can color their judgment. Over-whelmingly strong expectations are the chief cause of what we might call "pathological science" or "wishful science."*

A seemingly prize example of wishful science has recently been reported. I refer to the controversy concerning the alleged discovery of "cold fusion," the release of energy from the fusion of deuterium nuclei within a palladium electrode at room temperature. If these reports had been true, then it would have been a discovery of the greatest importance; for the oceans of the world contain enough deuterium to supply civilization with power for many millions of years (Bishop 1958). Therefore a method to extract this power using relatively simple and inexpensive equipment would reward the discoverer with enormous wealth and honor.

Fusion reactions have been studied intensively by nuclear physicists since the 1930s, using particle accelerators to bombard a target with energetic ions. When a deuterium target is bombarded with deuterium nuclei, two different reactions take place, with equal probability. One of these reactions forms a helium-3 nucleus plus a neutron plus 3.2 MeV of energy; the other forms a tritium nucleus plus a proton plus 4.0 MeV. Therefore, whenever fusion takes place in deuterium, neutrons must be emitted. Indeed, this reaction is often used as a source of energetic neutrons.

Research aimed at using the fusion of hydrogen isotopes (deuterium and tritium) to generate useful power has been going on since the mid-1950s. Fusion-power research is based on the idea that to make two nuclei fuse it is necessary to overcome the mutual electrostatic repulsion that prevents them from getting close together. All methods considered in the past have depended on making the nuclei move fast enough to get past the energy barrier. This is done by heating the deuterium gas to extremely high temperatures. An offspring of this effort has been the development of plasma physics, the study of the properties of gases at temperatures so high that they are completely ionized—their atoms stripped of all their electrons.

My own experience in this endeavor was as a member of the research team at the Princeton Plasma Physics Laboratory, where I did experiments with very large, complex, and expensive machines to learn how to heat the

---

*In a famous lecture in 1953, the Nobel laureate chemist Irving Langmuir coined the term *pathological science*. He used it to refer to cases in which scientists "perfectly honest, enthusiastic over their work . . . completely fool themselves." These are cases "where there is no dishonesty involved but where people are tricked into false results by a lack of understanding about what human beings can do to themselves in the way of being led astray by subjective effects, wishful thinking, or threshold interactions. These are examples of pathological science." Perhaps a better term is *wishful science*. (A transcription of Langmuir's lecture, circulated among academics for years, has at long last been published in full, in the October 1989 *Physics Today*, pp. 36-48.)

plasma to the required temperature. These machines, called *stellarators* (predecessors of the modern tokamaks), cost many millions of dollars to build and required dozens of physicists, engineers, and technicians to operate.

Therefore, when I first read of "cold fusion" in the newspapers, my first reaction was one of incredulity. For these reports claimed release of fusion energy in an apparatus built by two chemists on an ordinary lab table for a cost of less than $100,000. The claims were totally in opposition to the experience of previous fusion research. Yet the question had to be faced: Was it really possible that these people had discovered a reaction overlooked by all others during the past four decades?

The news reports told the following story: The two chemists, B. Stanley Pons and Martin Fleischmann, working at the University of Utah, claimed that they had electrolyzed heavy water (deuterium oxide) by passing electric current between a platinum anode and a palladium cathode immersed in the water. The released deuterium concentrated itself inside the solid palladium, which has long been known to be a good absorber of hydrogen. When the deuterium within the palladium was sufficiently concentrated—so the story went—the deuterium nuclei fused together, and energy was released. The evidence for this was a measurement of more thermal energy coming out of the device than was put in by the electric current.

Another independent researcher, physicist Steven Jones, of Brigham Young University, had already been working along somewhat the same lines for about two years. His observations showed the emission of a very small number of neutrons, presumably from fusion reactions, but claimed no unusual production of heat. Jones became aware of Pons and Fleischmann's results when he was asked to review a grant proposal they had submitted to the U.S. Department of Energy. Since Jones had facilities for neutron measurements and the University of Utah group specialized in calorimetric measurements, Jones suggested that the two groups collaborate. Pons and Fleischmann were not receptive to the idea, but after some discussion it was agreed that they would submit research papers simultaneously to the prestigious British journal *Nature*. The date chosen for the submission was March 24, 1989 (Pool 1989).

On March 23 the University of Utah group called a news conference, at which they announced the results of their experiments. This was the day before they sent their paper to *Nature*. University officials stated that the reason for the premature press conference was that too many rumors and publicity leaks were already circulating and that it was important to claim priority for patent purposes. Jones felt that he had been sandbagged. *Nature* did not publish the Pons-Fleischmann paper because it wanted more details, and the experimenters—by that time up to their necks in controversy—were too busy to comply. Therefore they simply withdrew the paper ("Fusion Illusion?" *Time*, May 8, 1989, p. 72). In the meantime, the governor of Utah announced that he would ask the state legislature to provide $5 million for a fusion research lab at the University of Utah.

Following the initial Pons-Fleischmann news conference there was a stampede by dozens of labs to replicate the cold-fusion experimental results. Even though many physicists were dubious about the claims, they felt it necessary to duplicate the experiment in order to be sure they were not missing something new. Initial reports were conflicting. Some laboratories measured heat and no neutrons; others measured neutrons but no heat. Some of the experiments exhibited symptoms of hasty planning and execution. A group from the Georgia Institute of Technology reported copious neutrons issuing from the apparatus. Later they said their neutron detectors were curiously sensitive to temperatures. Finally they retracted their results entirely, blaming faulty detectors.

Theoreticians also had their day. Most prominent was Peter Hagelstein, of MIT, who tried to explain how a new kind of fusion reaction could produce heat without generating neutrons. This theory claimed that two deuterons joined to form a helium-4 nucleus, depositing the extra energy directly into the lattice vibrations of the palladium crystal. MIT promptly filed patent applications. Others tried to explain how the heat could be produced by reactions other than nuclear. Later, Hagelstein retracted his theory.

After the initial period of stumbling about, the more cautious labs had their say. Caltech, MIT, Yale, Brookhaven, Oak Ridge, and others said their measurements had produced no evidence of either heat generation or emission of neutrons. As far as they were concerned, "cold fusion" was not taking place in beakers of heavy water. A meeting of the American Physical Society held in May unanimously rejected the Utah claims.

Whereupon members of the Utah contingent issued rude statements about "the mean bullies from the Eastern establishment," and things degenerated into a name-calling bout between the chemists and the physicists.

The negative statements from the various physics labs failed to stop the controversy. Those hoping for some important new discovery continued to explore various exotic aspects of the experiments. Some thought that electric fields established within cracks inside the palladium electrode might be accelerating deuterium ions so as to create a small number of fusions.

On June 15, an announcement from the Harwell Laboratory, the British government's major fusion laboratory, dealt a death blow to hopes for cold fusion. Working with the full cooperation of Martin Fleischmann, ten Harwell scientists had spent three months and a half a million dollars trying to replicate the original Utah experiment. After trying eight different types of palladium metal, they failed to find either production of fusion byproducts (helium-3, helium-4, or tritium), generation of heat, or emission of neutrons. A few weeks later, a committee formed by the Department of Energy in Washington concluded that there was no persuasive evidence of a new nuclear process called cold fusion and recommended the DOE not fund any new facilities or research efforts to find cold fusion.

What is the average citizen to make of this confusion? If even experts disagree, what can the layman do? What he can do is to sit tight and adopt an attitude of skepticism. An attitude of skepticism is not the same as an attitude of cynicism or disbelief. With proper skepticism one applies simple rules for judging newspaper stories (M. Rothman 1988). Some of these rules and their application to the cold-fusion controversy are:

1. *Don't believe everything you read or hear.*

The history of science is littered with theories that have fallen by the wayside and discoveries that have turned out to be illusory. Two examples are the claims of the detection of gravity waves and magnetic monopoles. While these discoveries were reported in legitimate scientific journals, most physicists doubt that anything was actually observed, because nobody but the original observer was able to get the same results. Nevertheless, the work was properly carried out, possible errors were thoroughly analyzed, and procedures and results were published for all to see.

Contrast this protocol with the cold-fusion proceedings, starting with the announcement to news media of a scientific discovery before it was published in a refereed scientific journal. The purpose of refereed publication is to ensure that the paper gives all the essential details of an experiment so that others can duplicate it. If an experiment cannot be duplicated, then it cannot be trusted, particularly if there are independent reasons for doubt (see Rule 3 below). Pons and Fleischmann violated this fundamental rule of research etiquette. Therefore skepticism was an appropriate response to their claims.

2. *Cast a cold eye on studies and experiments from which different workers elicit different answers.*

Contradictory results are endemic to studies in which one looks for small signals within a noisy background. Currently in the headlines are questions about the health effects of 60-cycle electromagnetic (EM) fields. The effects, if any, are so small that they are not easily detected, and so some of the studies say EM fields are bad for you, and other studies say there is nothing to worry about. A similar phenomenon is currently taking place among researchers looking for new kinds of gravitational fields. Some people say they have found a new, weak gravitational field that acts as a repulsion. Others claim a new field that does not follow the usual inverse-square law. The measured effects are so small that they are almost imperceptible. There is a possibility that the anomalous effects arise simply from errors in accounting for the distribution of mass in the earth's crust.

The research described above is legitimate. The cold-fusion experiments, on the other hand, are not dealing with tiny effects hidden by noise. If the process is to be useful as an energy source, cold fusion should generate easily detectible amounts of heat. Yet the reported results were as contradictory as possible. Some workers claimed generation of heat without emission of neutrons; others claimed detection of neutrons but little heat. Some who claimed neutrons then withdrew their claims because their detectors were not working properly. Then, when the more cautious labs came in with their results, their unanimous finding was that nothing was happening. This sequence of events is evidence of wishful science.

3. *If a claim is made for a phenomenon that violates one or more of the laws of nature, be doubly cautious.*

A law of denial is a law of nature that forbids the performance of certain actions (M. Rothman 1988). Two examples: (1) conservation of energy says that no reaction can take place that changes the amount of energy in a closed system; (2) conservation of momentum says that no reaction can take place that changes the total momentum of a closed system. We use these laws to decide between possibility and impossibility. Thus, when trying to judge an anomalous claim or a new theory, we must ask whether any law of denial is being violated.

But, in dealing with cold fusion we must first ask what kind of reaction is taking place. If known nuclear fusion reactions were taking place within the bottles of deuterium oxide, the neutron flux emitted would have been hazardous to the health of those in the lab. For each watt of power generated, the neutron flux at a distance of one meter would have amounted to about 4 rem/hour. This is far more than is allowable. Was any shielding seen around the apparatus in the Utah lab? No exceptional numbers of neutrons were detected by any of the later experiments.

For this reason some people have theorized that a new kind of nuclear reaction must have been taking place, perhaps one in which two deuterium nuclei fuse to form helium-4, giving the energy resulting from the reaction directly to vibrations of the palladium crystal lattice, without the emission of neutrons.

Two objections can be raised to this theory. First, it would be extremely odd if the environment *outside* the deuterium nucleus could *suppress* the normal nuclear reactions and substitute a previously unknown reaction. Therefore we must be given a reason for believing that the well-known and normal fusion reactions were *totally* replaced by a new and strange reaction. But even more important is that any reaction proposed must obey conservation of momentum. If the result of a reaction moves in one direction, something else has to move with equal momentum in the opposite direction.

How does this law apply to fusion? When two deuterium nuclei fuse together, the resulting "compound nucleus" has an excess energy of 23.6 MeV. Normally it gets rid of this energy by splitting into smaller particles that move off in opposite directions. There is also the possibility that a high-energy gamma ray photon might be given off, with the helium-4 nucleus recoiling in the opposite direction. There is no mention of 23 MeV photons being observed. Instead, we are asked to believe that the 23 MeV energy is imparted directly to the crystal lattice, which then twangs like a tennis racquet. If you claim this mechanism, then you must explain what kind of force does the pushing. There is no nuclear force in existence that can cause this kind of action.

The entire theory is highly implausible and was rightly withdrawn soon after being proposed.

4. *Be skeptical of the opinions of experts outside their areas of expertise.*

The difference of opinion between the physicists and the chemists about the worth of the cold-fusion experiments is going to fuel many dissertations by sociologists of science. The case illustrates vividly how scientific opinions are fashioned by subjective social causes as well as by objective evidence. It would be natural for chemists to side with their compatriots. However, it is also significant that the experiments of the physicists tended to show the absence of cold fusion, in disagreement with the more positive results of the chemists. In my opinion the reason for this difference is that the physicists were more experienced than the chemists in this kind of work and also more skeptical in their attitude. Also, the physicists are more accustomed to thinking in terms of particle and nuclear reactions.

It is in the area of neutron measurements that differences were most apparent. The initial newspaper accounts of Pons and Fleischmann's experiment did not even mention neutron detection, even though neutron emission is the prime indicator of fusion reactions. Later reports indicate that Pons and Fleischmann did use an indirect method for counting neutrons and found a flux a billion times smaller than would be expected if the heat production was a result of known fusion reactions. Jones's neutron count (obtained with a proper neutron spectrometer) was a hundred thousand times smaller (Levi 1989). Yale and Brookhaven found ten times fewer neutrons than Jones.

Careful measurements of neutrons require expertise. When Georgia Tech announced that copious neutrons had been detected coming from a cold-fusion reaction, then confessed that their neutron detector was strangely sensitive to temperature changes, and finally admitted that their results had been obtained with a malfunctioning detector, it was apparent that nonexperts were involved. When a physicist sets out to measure neutron flux, he first uses a radioactive neutron source to calibrate his detector so that he knows how many neutrons he is measuring. He then makes a background count in which he measures the neutron flux in the absence of a source. He then turns on the reaction of interest and counts the neutrons coming out of the reaction, if any. This procedure would have avoided the announcements and retractions seen in the press. The statement that the neutron counter was sensitive to temperature changes was absurd. There is no reason for a proper neutron counter to be temperature sensitive.

5. *Be wary of scientists (and economists and theologians) who fall madly in love with their own theories.*

Readers may well be familiar with this phenomenon. Except for cases of outright fraud, claims of the paranormal are invariably made by persons obsessed with their theories. This obsession interferes with scientific research, since it encourages the scientist to make errors in judgment and procedure that tend to reinforce his or her own beliefs. The potential importance of the cold-fusion research made it essential that the most stringent controls be used and that particular care be taken in analyzing possible sources of error. Jones compared neutron fluxes obtained using ordinary water with those obtained with heavy water and found a difference. There is no mention of this sort of control in the University of Utah experiments.

Those of us on the outside have no way of getting inside the minds of Pons and Fleischmann to determine how much wishful thinking was involved. We have no way of divining how much responsibility should be laid at the feet of Chase Peterson, president of the University of Utah. We do know that the decision to hold a press conference before publication of the research results was motivated entirely by the anxiety to establish priority for purposes of obtaining patents and research grants. We know that Chase Peterson was at the side of Pons and Fleischmann when they testified before the House Committee on Science, Space, and Technology and asked for $25 million to set up a fusion research center at the University of Utah. Considering the enormous implications of fusion for the future of humankind, it would be surprising if psychological pressures did not play a part in distorting the interpretation of the experimental results.

The irony of the situation is that the betrayal of Steven Jones by beating him to the press would not have done Pons and Fleischmann any good so far as obtaining a patent is concerned. Jones's dated notebooks proved that he had been already working on cold fusion for two years.

Consider the precedent of the scandal revolving around the invention of the digital computer. The ENIAC, built by John W. Mauchly and J. Presper Eckert at the University of Pennsylvania in 1946, was long accepted as the first automatic electronic digital computer. However, between 1937 and 1942 a man named John V. Atanasoff had developed and built an electronic digital computer on his own. Furthermore, Mauchly had observed Atanasoff's computer and knew how it worked. Because of Atanasoff's retiring personality, he made no claim on the patent until much later. Finally, in 1973 a court decision gave Atanasoff proper credit and patent rights. As any lawyer will tell you, a patent is mainly useful as a license for going to court.

One sign of wishful science was Pons and Fleischmann's refusal to disclose important details of their work that would have assisted others in replicating it. Their paper in the *Journal of Electroanalytical Chemistry and Interfacial Electrochemistry* was too sketchy to be of much use. Another sign was their insistence that their work was still valid even after many other labs had failed to show evidence of cold fusion.

Many man-hours of work will be expended to explain the anomalous results that were obtained by several observers. It is possible that something unusual was actually happening in those experiments that showed emissions of small numbers of neutrons. These experiments are still hard to explain and deserve explaining, but there is essentially no chance that cold fusion is going to become a source of energy in the future.

It was exaggerated belief in a theory that tilted the cold-fusion work into disaster. Without that psychological factor the case would simply have been a matter of experimental error or misinterpretation of results, unfortunate circumstances that can happen to anybody. If the results obtained by Pons and Fleischmann had been sent to *Nature* without the initial publicity, the paper would have been reviewed according to normal procedure. Perhaps more information would have been requested by the journal. Perhaps a little later it would have been published. Eventually the published results would have been challenged by others, but there would not have been the feeling that something awful had happened. There would have been simply a minor embarrassment, something that could happen to anybody.

But overenthusiasm and apparent greed and hubris changed a minor event into a major embarrassment for all of science. The manner in which scientists are perceived by the public has been diminished as a result of this affair. Fortunately, science is a self-correcting enterprise. The community of scientists responded in a responsible manner by trying the cold fusion experiment in many independent labs before passing judgment.

Maybe the next college president tempted by fame and fortune will think twice before he encourages his professors to rush toward a conclusion that should be based on hard science rather than on politics. Perhaps in the future congressmen will allow scientists to decide among themselves who is doing valid work.

In the long run the controversy may have been for the best. The publicity helped the public see that science is not simple and that scientists are human beings. When millions or billions of dollars hang in the balance, scientists can get tempted into folly as easily as stock speculators can.

Postscript

Much has happened in the world of cold fusion since the fall of 1989, none of which is encouraging to those expecting the cold-fusion process to become a cheap new energy source. In November 1989 a report of the Energy Research Advisory Board to the U.S. Department of Energy concluded that there was no significant evidence for the generation of power by cold fusion, and a great amount of evidence against it. In the same month a paper in *Science* from a Caltech group analyzed the raw data obtained by Pons and Fleischmann and concluded that their results could be caused by errors made in computing the heat output of their electrolysis cells (Miskelly et al. 1989). A study by a group from Sandia National Laboratories showed how low-level neutron measurements could be sufficiently in error to account for some of the anomalous claims (Ewing et al. 1989). Much research concentrated on the appearance of tritium in the palladium electrodes, supposedly the product of d-d reactions.

One researcher, Kevin Wolf, finally admitted contamination of his palladium source. In another lab, at Texas A&M, there were hints of tampering (Taubes 1990), and although an investigation by the university administration produced no evidence of scientific fraud, it did criticize the carelessness and the behavior of many of the scientists involved (Pool 1990b).

In the meantime, research on cold fusion continued at the National Cold Fusion Institute, financed by a $5-million grant from the state of Utah. Finally, just as Utah's Fusion Energy Advisory Council was about to review the Institute's work and decide whether to renew the grant, Stanley Pons, initiator of the cold fusion work, disappeared (Broad 1990a). His house was put up for sale, and his phone was disconnected. Later Pons was seen in Utah, but refused to answer questions, saying, "People who want to see us do not seem to have any problems." This was in contradiction to university officials who found him unreachable except through his attorney. Pons said he had written two new papers that he would submit "as soon as their release does not negate granting of patent protection" (Associated Press 1990).

The unorthodox behavior of Pons was sure to dampen the spirits of that small band of researchers who refused to give up the hope that some new nuclear phenomenon was taking place within the palladium electrodes. While tritium production has been discounted as due to contamination, reports continued to appear concerning the production of small amounts of neutrons, and, as reported by *Science,* ". . . like the grin on the Cheshire Cat, this area shows signs of lingering long after the rest has gone" (Pool 1990a).

One of the symptoms of pathological science is the tendency of some researchers to persevere after everybody else has decided there is nothing real to find. Such was the case in the polywater flap of the 1970s. Another symptom is the manner in which investigators subvert the normal procedures of peer review and publication by going straight to the media, as described in an extensive review of the cold-fusion controversy in the *New York Times* (Broad 1990b). The most outrageous assault on the academic system took place when M. H. Salamon, a Utah physicist, reported in *Nature* (in March 1990) that he could find no gamma rays emitted by Pons's cold-fusion apparatus. This was strong evidence against a nuclear reaction. Whereupon Salamon was threatened with a lawsuit by Pons and Fleischmann's attorney. For an attorney to presume to criticize a published scientific paper is the most pathological behavior of all. The lawyer later apologized.

For a firsthand account of the cold-fusion controversy by a physicist who is a harsh critic of the University of Utah work, see Frank Close, *Too Hot to Handle: The Race for Cold Fusion,* published in May 1991 by Princeton University Press (and in Britain by W. H. Allen). An accessible article by Close is "Cold Fusion I: The Discovery That Never Was," *New Scientist,* January 19, 1991, pp. 46-50; a follow-up exchange between Close and Martin Fleischmann appeared in the April 20 issue, p. 12.

References

Associated Press. 1990. *New Mexican* (Santa Fe, N. Mex.), November 9.

Bishop, A. 1958. *Project Sherwood: The U.S. Program in Controlled Fusion.* Reading, Mass.: Addison-Wesley.

Broad, W. J. 1990a. Utah to start search for cold fusion scientist. *New York Times*, October 26, A15.

———. 1990b. Cold fusion still escapes usual checks of science. *New York Times*, October 30, C1.

Ewing, R. I., et al. 1989. Negative results and positive artifacts observed in a comprehensive search for neutrons from "cold fusion" using a multidetector system located underground. *Fusion Technology*, 16:404, November.

Levi, B. G. 1989. Doubts grow as many attempts at fusion fail. *Physics Today*, June, p. 17.

Miskelly, G. M., et al. 1989. Analysis of the published calorimetric evidence for electrochemical fusion of deuterium in palladium. *Science*, November 10, p. 793.

Pool, R. 1989. Fusion followup. *Science*, April 7, p. 27.

———. 1990a. Cold fusion: Only the grin remains. *Science*, November 9, p. 754.

———. 1990b. Cold fusion at Texas A&M: Problems but no fraud. *Science*, December 14, pp. 1507-1508.

Rothman, M. 1988. *A Physicist's Guide to Skepticism.* Buffalo, N.Y.: Prometheus Books.

Rothman, T. 1989. *Science à la Mode: Physical Fashions and Fictions.* Princeton, N.J.: Princeton University Press.

Taubes, G. 1990. Cold fusion conundrum at Texas A&M. *Science*, June 15, p. 1299.

MARTIN GARDNER

Water With Memory?
The Dilution Affair

"Experimenter effect" has two meanings. Outside psychic research circles it refers to the way a strongly held mind-set can unconsciously bias an experimenter's work. Among parapsychologists it also refers to the supposed unconscious influence of an experimenter's PK (psychokinetic) powers on the research.

Putting aside the second meaning (if such an effect is real it would throw doubt on all empirical findings since Galileo), a bizarre, almost comic instance of the experimenter effect came to light in the summer of 1988. It involved a group of scientists at INSERM U200, a medical-research institute in a Paris suburb. Their findings were widely publicized (*Newsweek*, July 25; *Time*, August 8), not merely because they were so astounding, but because for the first time they seemed to provide strong empirical support for the fringe medicine of homeopathy.

The century's most notorious instance of an experimenter effect that sparked a vigorous scientific controversy also occurred in France. In 1903 René Prosper Blondlot, a respected French physicist, claimed to have discovered a new kind of radiation, which he called N-rays after the University of Nancy, where he worked. Scores of papers confirming the reality of N-rays had appeared in French journals before a skeptical American physicist, Robert Wood, visited Blondlot's laboratory and played a dirty trick on him. Wood secretly removed from Blondlot's apparatus a prism that was claimed to be essential to the observation of N-ray spectra. Blondlot went right on describing the lines he fancied he was seeing. After Wood reported this in the British science journal *Nature* (vol. 70, 1904, p. 530), N-rays vanished from physics, but poor old Blondlot never acknowledged his self-deception.

In June 1988, physicists and chemists around the world were incredulous over a paper in *Nature* (vol. 333, June 30, p. 816) titled "Human basophil degranulation triggered by very dilute antiserum against IgE." The report was signed by 13 biologists—two from Israel, one from Italy, one from Toronto, and the others part of a team at INSERM headed by biochemist Jacques Benveniste. The phrase "very dilute" in the title is a whopping understatement. As the editors of *Nature* pointed out in an unusual disclaimer accompanying the article, the dilution of the French group was so extreme that not a molecule of the antiserum was left in its solvent. The editors considered the results unbelievable, but said they were publishing the paper for two reasons: It purported to give an accurate account of work that had been widely trumpeted in France by popular articles, and it provided other scientists with an opportunity to confirm or falsify the extraordinary claims.

What were these claims? In essence the French researchers were convinced that, after all the molecules of a certain antibody were removed from distilled water, the water somehow "remembered" the antibody's chemical properties. Although such a claim violates fundamental laws of physics, it lies at the very heart of homeopathy, a medical pseudoscience that flourished in the United States in the nineteenth century and is now enjoying a modest revival. Homeopaths maintain that, if a drug produces symptoms of a disease in a healthy person, inconceivably small quantities of that same drug will cure the disease. Moreover, the smaller the amount of the drug—including its total absence—the more potent its curative power.

Thousands of homeopathic drugs are listed in the cult's *materia medicas*—handbooks that vary widely from time to time and from country to country. If a drug is soluble—bee venom, for example—it is mixed with water or alcohol in repeated dilutions. The mixture must be shaken violently for about ten seconds after each dilution, otherwise the medicine won't work. If a drug is not soluble, it is ground into a fine powder and diluted by repeated mixing with powdered lactose (milk sugar). A moderate homeopathic dose, called "30c," is arrived at by first diluting the drug to a hundredth part and then repeating the process 30 times. As someone pointed out, it is like taking a grain of a substance and dissolving it in billions of spheres of water, each with the diameter of the solar system.

Benveniste claims that the antibody he used is still potent when dilutions are even more extreme—one part to 10^{120} parts of water! As science writer Malcolm Browne remarked in his *New York Times* account of the French claims (June 30, 1989), astronomers estimate the number of stars in the universe as a mere 10^{20}. Benveniste said the potency of his dilutions is comparable to swirling your car key in the Seine, going some hundred miles downstream, taking a few drops of water out of the river, and then using them to start your car. It is easy to show mathematically that when such extreme dilutions are made of homeopathic drugs, as they are constantly, the chance of a single molecule remaining in the solvent or powder is vanishingly small.

Certain white blood cells, called "basophils," have granules that stain a reddish color when treated with a blue dye. Incubating these cells with a strong solution of an antibody causes them to lose those granules, a process known as "degranulation." When a solution of the antibody has been diluted to the point at which no molecules of the antibody remain in the distilled water, one would expect the cells to retain their red-staining granules. Not so. According to Benveniste, about half the basophils continued to degranulate when so treated.

How do homeopaths explain this supposed potency of infinitesimal doses, even when the dilution removes all molecules of a drug? They invoke mysterious vibrations, resonances, force fields, or radiations totally unknown to science. Benveniste suggests in his paper that antiserum molecules may somehow cause water molecules to rearrange their hydrogen atoms in some inexplicable fashion that mimics the action of the antibody even when it is no longer there. In other words, water can remember the properties of a missing substance.

This magic memory water is even weirder than polywater, a conjectured new type of water that caused an enormous flap among chemists in the 1960s. Boris Derjaguin, a Soviet chemist, announced that when water collects in hairlike capillary tubes it acquires all sorts of strange properties. John D. Bernal, a noted British physicist and historian of science (he was also a dedicated communist and a great admirer of Soviet science), hailed it as the "most important physical-chemical discovery of this century."

Because polywater, as it was called, could have great military uses, the Army, Navy, and other U.S. agencies began tossing out generous grants. A flood of papers about polywater popped up everywhere. Derjaguin even wrote a nontechnical article about the water for *Scientific American* (November 1970). *Nature* (224, 1969, p. 198) published a warning from an American scientist that research on polywater should proceed with extreme caution because it might polymerize the earth's oceans, destroy all life, and change the earth into a planet like Venus.

It turned out that the miraculous water was just ordinary water contaminated by dirty test tubes. Derjaguin himself threw in the towel by announcing that for ten years he had wasted his time studying nothing more than dirty water. Meanwhile millions of dollars had been squandered on polywater research. You can read all about this remarkable farce in *Polywater* (MIT Press, 1981), a fine book by Felix Franks. He faults government agencies for premature funding, technical journals for overpermissiveness, experimenters for repeated self-deception, and the mass media for irresponsible hype.

It is too early to know if Benveniste's homeopathic water will survive as long as polywater did, or if the French biochemist will eventually withdraw his paper. *Nature*, highly suspicious of so outrageous a claim, asked a team of unpaid volunteers to fly to Paris to devise and observe a replication of Benveniste's experiments in his own laboratory. (The visit and investigation were preconditions for publication of the original article.) Benveniste readily agreed, and even planned a celebration with champagne when the replication was over and his results were vindicated. The team consisted of John Maddox, editor of *Nature*, who has a background in physics; Walter Stewart, an organic chemist and a specialist in scientific fraud from the National Institutes of

Health in Bethesda, Maryland; and the indomitable magician and psi detective James Randi.

Their blistering report in *Nature* (334, July 28) opens: "The remarkable claims made . . . by Dr. Jacques Benveniste and his associates are based chiefly on an extensive series of experiments which are statistically ill-controlled, from which no substantial effort has been made to exclude systematic error, including observer bias, and whose interpretation has been clouded by the exclusion of measurements in conflict with the [claims]. . . . The phenomenon described is not reproducible in the ordinary meaning of that word. We conclude that there is no substantial basis for the claim. . . . The hypothesis that water can be imprinted with the memory of past solutes is as unnecessary as it is fanciful."

The popular French magazine *Science et Vie* (Science and Life) in its August issue was disturbed by the fact that Benveniste had announced as early as May, at a national conference on homeopathy, that his paper would be appearing in *Nature*. On July 1, journalists in France received a thick press release about the forthcoming paper, and in July the French stock exchange did a brisk business in Boiron shares. *Science et Vie* wondered if French newspapers and television stations would give as much publicity to the debunking of Benveniste's work as they did to its promotion. If not, "water memory will remain an established fact for believers in homeopathy."

The key person in all the French experiments, as well as in their "confirmation" by a laboratory in Israel, was Dr. Elizabeth Davenas, a young woman in her twenties and a good friend of Bernard Poitevin, one of the two homeopathic doctors in the French group. She is the observer who looks through the microscope to count the red-staining granules that remain. Randi listed 15 different pretexts on which she accepted "good" cases and rejected "bad" ones; Stewart's list contained 19 such items. It is not clear whether she is deceiving herself in a manner similar to Percival Lowell's famous self-deception when he peered through telescopes and drew pictures of intricate canals on Mars, or whether some cells actually lose color occasionally because of contaminants. On this point the *Nature* investigators write:

> In circumstances in which the avoidance of contamination would seem crucial, no thought seemed to have been given to the possibility of contamination by misplaced test-tube stoppers, the contamination of untended wells during the pipetting process and general laboratory contamination (the experiments we witnessed were carried out at an open bench). We have no idea what would be the effect on basophil degranulation of the organic solvents and adhesives backing the scotch tape used to seal the polystyrene wells overnight, but neither does the laboratory.

The original *Nature* report was understandably greeted with loud hosannas by homeopaths around the world. Readers interested in the wild history of this once most popular of all alternative medicines can consult Chapter 16 of my *Fads and Fallacies* (Dover, 1952), or "Homeopathy: Is It Medicine?" by Stephen Barrett (this volume); see also comments in the letters section

of the 1988 Spring and Summer issues of the *Skeptical Inquirer.* Dr. Barrett
is also the author of a hard-hitting paper in *Consumer Reports* (January
1987) about a yearlong investigation of homeopathy. His report concludes:

> Unless the laws of chemistry have gone awry, most homeopathic reme-
> dies are too diluted to have any physiological effect. . . . CU's [Consumers
> Union's] medical consultants believe that any system of medicine embracing
> the use of such remedies involves a potential danger to patients whether
> the prescribers are M.D.'s, other licensed practitioners, or outright quacks.
> Ineffective drugs are dangerous drugs when used to treat serious or life-
> threatening disease. Moreover, even though homeopathic drugs are essen-
> tially nontoxic, self-medication can still be hazardous. Using them for a
> serious illness or undiagnosed pain instead of obtaining proper medical
> attention could prove harmful or even fatal.

I find in my files a sad clipping from the *New York Post* (July 25, 1954)
about Jerold Winston, a Long Island boy, age 4, who died of leukemia. For
16 months he had been treated only with a homeopathic remedy by his mother,
the daughter of a homeopathic doctor. The parents were facing a possible
manslaughter charge for child neglect. Who knows how many tragedies like
this occur when gullible people rely solely on worthless medicines?

Homeopathy had almost died in the United States by 1960, though it
continued to be popular in France, Germany, Russia, India, England, Mexico,
Argentina, Brazil, and other countries. But in the New Age climate of the
seventies and eighties it experienced a surprising upsurge among those who
are attracted to holistic medicine, natural foods, herbal remedies, acupuncture,
reincarnation, and the paranormal. There are now several hundred homeopathic doctors in the United States, about half with orthodox medical degrees.
The others are mostly chiropractors, naturopaths, dentists, veterinarians, and
nurses. This is a small number compared to some 14,000 such physicians
in 1900, when more than 20 schools in the United States taught the art and
there were more than 100 homeopathic hospitals.

New books on homeopathy are appearing on general trade lists. Jeremy
Tarcher, a publisher of New Age literature (including books on Spiritualism),
has two homeopathic volumes in his current catalog: *Everybody's Guide to
Homeopathic Medicines,* by Stephen Cummings and Dana Ullman, and
Homeopathic Medicine at Home, by Dr. Maesimund Panos and Jane Heimlich. Heimlich is the wife of Dr. Henry Heimlich, orginator of the famous
"Heimlich maneuver," used to aid persons choking on food. In 1980 she was
quoted in the *New York Times* (November 19) as saying she took great pride
in converting her father, the dancing teacher Arthur Murray, to homeopathy.

The Complete Book of Homeopathy, by Michael Weiner and Kathleen
Goss, was issued by Bantam Books in 1981. With all the media publicity about
Benveniste, and the continuing growth of New Age nonsense, more such books
are surely on the way. Nothing stimulates a fringe medical cult more than
attacks by skeptics, or by "allopaths," the homeopathic term for orthodox doctors.

Cummings and Ullman, in their book on homeopathy, claim there are more than 6,000 homeopathic doctors in France today, and 18,000 pharmacies that sell their medicines. In India, they tell us, more than 70,000 doctors practice the art. (In an article in the January 1984 issue of *Fate* magazine, the nation's sleaziest occult periodical, Ullman upped this number to 200,000.) England's royal family, according to the Queen's physician, has been under homeopathic care for more than 150 years. Dozens of famous nineteenth-century American writers, political leaders, and businessmen patronized homeopathic physicians, including Washington Irving, who died under the care of a homeopathic family doctor. Ullman, who holds a master's degree in public health from the University of California, Berkeley, is the nation's top homeopathic journalist. He was arrested in California in 1976 for practicing medicine without a license.

How will homeopathic doctors and true believers react to *Nature*'s debunking? There is not the slightest doubt they will take their cues from Benveniste's angry reply, which ran in the same issue of *Nature* as the critique. His invective is unprecedented in a science journal. Members of the *Nature* team are branded "amateurs" who created "hysteria" in the French laboratory. Their investigation is called a "mockery of scientific inquiry." Benveniste likens it to the Salem witch-hunts and the McCarthy persecutions. He told the *Wall Street Journal* by phone (July 17, 1988) that the *Nature* report was a sinister plot to discredit him. In Paris he told *Le Monde* (July 27, 1988) that Walter Stewart was incompetent and the investigation was a "scientific comedy . . . conducted by a magician and a scientific district attorney who worked in the purest . . . Soviet ideology style" to install a "scientific gulag."

"During the whole week I was tempted to kick them out," he said to the French newspaper *Le Figaro* (July 27, 1988). "We never could imagine the extent to which these 'experts' were going to shuffle the cards." He attacked *Science et Vie* for calling him a "new Lysenko," adding "you should know that I am the most important researcher in the world and the one most in demand at colloquia."

It is obvious from Benveniste's fury that he learned absolutely nothing from *Nature*'s careful, restrained investigation. Its lessons had the same effect on his mind as a vanishing substance has on distilled water.

Consider the egotism and folly of this man. He rushes into print with a claim so staggering that if true it would revolutionize physics and medicine, and guarantee him a Nobel prize. Yet he did this without troubling to learn the most elementary techniques for conducting truly double-blind tests or for supervising self-deceiving observers. When Randi mentioned N-rays to him, he said he had never heard of them! Does he remember, one wonders, the story of polywater?

A few scientists and science journalists criticized *Nature* for publishing the original article. Daniel Koshland, Jr., editor of *Science,* agreed that a responsible journal should "encourage heresy" but added that it also should "discourage fantasy." It is one thing, he told science writer Walter Sullivan

(*New York Times,* July 17, 1989), to publish unorthodox work that may turn out to be wrong, but the French claim about water with a memory was too far on the fantasy side, like an account of the successful construction of a perpetual-motion machine.

Other scientists have faulted *Nature* both for publishing the French paper in the first place and for later investigating the claims. Arnold Relman, editor of the *New England Journal of Medicine,* said *Nature* should have required confirmation by an independent group of biochemists before running the article, and that it was not its function to serve as an investigative body. This view was shared by Henry Metzger, a colleague of Stewart at the NIH. He said he had urged this approach when he refereed the paper for *Nature.* Immunologist Avrion Mitchinson, at University College London, thought the French paper not worth publishing. However, he did not believe it would do much harm. "Anyone who thinks the great ship of science can be damaged in such a way is greatly mistaken." (On such criticisms see "More Squabbling Over Unbelievable Result," in *Science,* 241, August 5, and "The Ghostbusters Report from Paris," in *New Scientist,* August 4.)

Nature's correspondence section (August 4, 1989) ran four letters from scientists who proposed conventional explanations for the French results. In the same issue the editors defend themselves in an editorial headed "When to Publish Pseudo-science." When one-fourth of French doctors prescribe homeopathic medicines, they argue, "there is plainly too much at stake for the issue to be dropped."

The editorial recalls an earlier publication in *Nature* (238, 1972, pp. 198-210) of the claim that, when the protein scotophobin is extracted from the brains of rats trained to run a maze and then injected into untrained rats, there is a transfer of maze-running ability. The paper was followed by a "devastating critique" by Stewart, and that ended the matter. "Is not a little of the 'circus atmosphere' inescapable on these occasions?" Why did *Nature* not withhold the French report until they made their investigation, then publish the two reports side by side? The editors have replied elsewhere that they did not do this because Benveniste had leaked information about his paper to the French press and, had they withheld his paper until after their investigation, he would have refused to allow *Nature* to print it.

INSERM (the letters stand for Institut National de la Santé et de la Recherche Médicale) has refused to take sides in the controversy. "*Nature* sends a magician to check my research," Benveniste declared, "and INSERM doesn't even protest. It's the limit!" I quote from Peter Coles's article, "Benveniste Controversy Rages in the French Press" (*Nature,* August 4, 1988). He also reveals that Boiron, a 51-percent shareholder in another firm, Laboratoires Homéopathiques de France, has purchased all remaining shares.

"Look," Randi said to the French group, "if I told you I keep a goat in the backyard of my house in Florida, and you happen to have a man nearby, you might ask him to look over my garden fence. He would report, 'That man keeps a goat.' But what would you do if I said, 'I keep a unicorn in my garden?"

The point of course is that no extraordinary verification is needed to establish the existence of a goat. But a unicorn? As the *Nature* authors write sadly at the close of their indictment, "We have no way of knowing whether the point was taken."

Editor's Postscript

In July 1989 two committees reporting to the French national institute of health and medical research (INSERM) strongly criticized Jacques Benveniste's work on the "remembering water." In a statement to the press, the INSERM directorate said the two committees offered "a very favorable opinion" of the overall activities of Benveniste's laboratory, INSERM Unit 200, but were "extremely reserved regarding the studies of high dilutions." The statement criticized Benveniste for "an insufficiently critical analysis of the results he reported, the cavalier character of the interpretations he made of them, and the abusive use of his scientific authority *vis-à-vis* his informing of the public." The scientific committee's report said Benveniste's interpretations were "out of proportion with the facts" and appear as "a laboratory curiosity to which satisfactory explanations have not yet been given and whose import will remain limited." It recommended INSERM not continue funding high-dilution research.

However, Philippe Lazar, INSERM director-general, did not endorse that latter recommendation, suggesting to do so would interfere with the freedom accorded a laboratory director. In a four-page letter, Lazar asked Benveniste to look for sources of experimental error to explain his "unusual results" but also strongly criticized *Nature* for its handling of the matter, starting with its decision to publish an "insufficiently founded" paper, for the "oddness" of its visiting panel, and for the "offensive content" of its conclusions. He appealed to the media to "let Dr. Benveniste get on with his work."

Two years later Benveniste was back at it again. In March 1991 he and his INSERM colleagues published a report in the journal of the French Academy of Sciences on fresh experiments they had conducted. The experiments seemingly supported their theory that solutions of antibody diluted to the point where no antibody molecules remain continue to evoke an immune response. Benveniste said the experiments had added key controls not present in the earlier ones. Nevertheless, according to *New Scientist* (March 16, 1991), both *Science* and *Nature* had rejected the same paper upon the advice of reviewers, one of whom said the paper contained "crippling flaws." Among the flaws was that Benveniste discarded results from certain cells, including cells that were either completely unresponsive to antibody or cells that responded spontaneously, even to solutions that had never contained antibody. *Nature*'s referee called this "throwing out data because they don't fit the conclusion." Benveniste described it as standard practice. "It is not an error. It is merely checking that the cells are working properly."

MARTIN GARDNER

Science, Mysteries, and the Quest for Evidence

A man is a small thing, and the night is very large and full of wonders.

Lord Dunsany, *The Laughter of the Gods*

Parapsychologists and psi journalists are fond of an argument that goes like this: Orthodox science is making such colossal strides, putting forth such bizarre theories, that no one should hesitate to accept the reality of psi. It is a theme that pervades Arthur Koestler's influential *Roots of Coincidence*. As parapsychology becomes "more rigorous, more statistical," Koestler writes on the very first page, theoretical physics becomes

> more and more "occult," cheerfully breaking practically every previously sacrosanct "law of nature." Thus to some extent the accusation could even be reversed: parapsychology has laid itself open to the charge of scientific pedantry, quantum physics to the charge of leaning towards such "supernatural" concepts as negative mass and time flowing backwards.
>
> One might call this a negative sort of rapprochement—negative in the sense that the unthinkable phenomena of ESP appear somewhat less preposterous in the light of the unthinkable propositions of physics.

It is true that modern science is making discoveries and formulating theories that contradict experience and boggle the mind, but this has always been the case. I suspect that most people are less boggled today by the wonders of science than they were boggled in the past by the notion that the earth rotates and goes around the sun. Indeed, all the evidence of the senses suggests that the earth is immovable and the heavens rotate. The centuries that elapsed before the Copernican theory became entrenched in the common beliefs of the civilized world—including the beliefs of Catholics and Protestants, who fought the theory as long as they could—testify to the cultural shock of such a monumental paradigm shift, to use Thomas Kuhn's fashionable phrase.

Today the public is much less bewildered by the paradoxes of relativity and QM (quantum mechanics), not just because it has grown accustomed to the surprises of science but because the paradoxes are too technical to understand. If a twin takes a long space trip at fast speeds and returns to earth, he will be younger than his stay-at-home twin. If he goes far enough and fast enough, he could return to find that centuries on earth had gone by. Most nonphysicists, unless they read science fiction, have never heard of the paradox.

The same can be said of recent confirmations of the notorious EPR paradox that Einstein and two friends (E, P, and R are the initials of the three last names) devised to show that QM is incomplete. Two particles, separated by vast distances, can under certain circumstances remain "correlated" in the sense that, if one particle is measured for a property, the other is altered even though there is no known causal connection between the pair. Who is troubled by what Einstein called the "telepathy" of this paradox except physicists and philosophers of science?

The Big Bang, black holes, pulsars, and other awesome aspects of modern cosmology have been dramatic enough to reach the general public, but I see no evidence that the public is disturbed. If *Time* reports that some physicists now think all particles are made of inconceivably tiny "superstrings," vibrating in spaces of ten dimensions, it is not likely to be a topic of cocktail-party chatter except in science circles. The only establishment claim now arousing strong public emotion is evolution, and that is because of the astonishing revival of Protestant fundamentalism.

From the beginning, science has been upsetting and drastically modifying history. It does not, however (as Koestler writes), progress by breaking sacrosanct laws. No laws of science are sacrosanct, and "breaking" is a poor word for the meandering process by which laws are refined. Great paradigm-shifts build on what went before. Ancient astronomers were good at predicting the motions of planets long before astronomy accepted a central sun. Let $1/c$, where c is the speed of light, reduce to 0 in the formulas of relativity, and you have Newton's formulas. Let Planck's constant equal 0, and QM becomes classical mechanics. The great revolutions of science are better described as benign evolutions. They refine what was known before by placing that knowledge within new theoretical frames that have superior power to explain and predict.

There are other reasons that the progress of science is cumulative and increasingly rapid. Every decade the number of working scientists increases. In Galileo's day you could count the number of physicists on your fingers. Today tens of thousands of journals report the latest scientific discoveries and conjectures, many of the conjectures (as Koestler rightly perceived) more outlandish than the claims of parapsychology. Instruments of observation get better and better. Galileo's telescope was a child's toy. Microscopes using particles other than photons have greatly increased the range of observation of the small. Giant particle accelerators provide empirical underpinnings for strange new theories of matter that could not possibly have been devised even in Einstein's day. Space probes have disclosed more facts about the planets in the past 20 years than in the previous 200.

Koestler is right in one sense. The results of science should instill in all of us a strong awareness of how mysterious and complex nature is. In the words of J. B. S. Haldane, which occult journalists love to quote, the universe is queerer than we can suppose. Every scientist and every layperson should be open to any scientific claim no matter how preposterous it may seem. If it turns out that the human mind can view a remote scene by clairvoyance, or influence a falling die or a random-number generator, this surely would be no more surprising than thousands of well-confirmed natural phenomena.

Does it follow from such admirable open-mindedness, from what the American philosopher Charles Peirce called the "fallibilism of science," that we should all accept the ability of psychics to bend paperclips with their psi powers? It no more follows than it follows from modern cosmology that (as Velikovsky maintained) the moon's craters are only a few thousand years old, or (as Jerry Falwell firmly believes) that the earth was created in six literal days and dinosaurs were beasts that perished in Noah's flood.

We can now say what is wrong with Koestler's rhetoric. The extraordinary claims of modern science rest on extraordinary evidence. No physicist today would be bothered in the least by the seemingly paranormal aspect of the EPR paradox if it did not follow inescapably from firmly established laws of QM, and from carefully controlled laboratory tests. But the extraordinary claims of parapsychology are *not* backed by extraordinary evidence.

For reasons that spiritualists have never been able to explain, the great mediums of the nineteenth century could perform their greatest miracles only in darkness. The equivalent of that darkness today is the darkness of statistics, and why psi phenomena flourish best in such darkness is equally hard to comprehend. If a mind can alter the statistical outcome of many tosses of heavy dice, why is it powerless to rotate a tiny arrow, magnetically suspended in a vacuum to eliminate friction? (J. B. Rhine's laboratory, by the way, made many unsuccessful experiments of just this sort, but they were never reported.) The failure of such direct, unequivocal tests is in my opinion one of the great scandals of parapsychology.

Why is it that the most respectable evidence today for PK, the work of Helmut Schmidt and Robert Jahn, involves sophisticated statistical analyses of thousands of repeated events? The skeptic's answer is that, when a supposed PK effect is so weak that it can be detected only by statistics, many familiar sources of bias creep into the laboratory. In the case of S. G. Soal, once hailed as England's top parapsychologist, we now know that the bias was outright fraud. Even when researchers are totally honest, it is as difficult to control the effect of passionate desires on methods of getting and analyzing data as it is to keep sealed flasks free of bacterial contamination.

No skeptic known to me rules psi forces outside the bounds of the possible.

They are merely waiting for evidence strong enough to justify such extraordinary claims. Their skepticism is not mollified when they find the raw data of sensational experiments sealed off from inspection by outsiders or when failures of replication by unbelievers are blamed on unconscious negative vibes.

I am convinced that today's skeptics would have not the slightest difficulty —*I* certainly would not—accepting ESP and PK the instant evidence accumulates that can be reliably replicated. Unfortunately, for 50 years parapsychology has rolled along the same murky road of statistical tests that can be repeated with positive results only by true believers. Psi forces have a curious habit of fading away when controls are tightened or when the experimenter is a skeptic—sometimes even when a skeptic is just there to observe.

Surely every parapsychologist worthy of respect now knows (even though he won't say so) that psychics are unable to bend spoons, move compass arrows, or produce thought photographs if a magician is watching. As for the more responsible and more modest claims that rest on statistics, they are too often obtained solo or by a small band of researchers who will not let an outsider monitor what is going on. Raw data is often kept, as is most of it at SRI International, permanently under wraps.

Parapsychologists are forever accusing establishment psychologists of wearing blindfolds that make it impossible for them to see the results of the new Copernican Revolution. If the results are as claimed, it is indeed a paradigm shift more sensational than most of the great shifts of the past, and Rhine deserves to rank with Copernicus, Newton, Einstein, and Bohr. Alas, the claims remain as poorly verified as nineteenth-century claims that character traits correlate with bumps on the head.

It would be good for every parapsychologist to study the history of phrenology. The number of scholarly journals devoted to this "science" once far exceeded the number of journals that are today devoted to parapsychology; and, at one time, the number of distinguished scientists who believed that phrenology had been strongly confirmed far exceeded the number of distinguished psychologists today who believe that parapsychology has established the reality of the phenomena it studies.

MARTIN GARDNER

Relativism In Science

In recent decades there has been a growing trend among a small group of sociologists and humanities professors, even among a few scientists and philosophers, to deny that science moves closer and closer to objective "truth." This bizarre view is closely linked to an anti-realist trend that has been stimulated by the paradoxes and mysteries of quantum mechanics. The properties of particles and quantum systems are, in a sense, not "real" until they are measured. The measurements can be made by apparatus, but the apparatus itself is a quantum system, so it too seems to be in an "indefinite" state until it has been observed by a person. Alas, the observer also is a quantum system. Is he indefinite until someone observes *him?* And how can we escape from this seemingly endless regress?

A few physicists, notably Eugene Wigner, argue that the quantum world, which of course is the entire universe, has no reality until observed by conscious minds. This view runs into grave difficulties over the question of how high on the evolutionary scale a mind has to be to make an object real. As Einstein, who was repelled by this kind of social solipsism, liked to ask: Is the moon nonexistent until a mouse observes it? And how about observation by a butterfly? Evolution seems to entail, for someone like Wigner, that reality is a matter of degree; that as life evolved on (at least) the earth, the universe slowly developed from some sort of featureless fog to the complicated mechanism it is today. And what would happen to the universe if all life became extinct? Would it fade back into the gloom?

If the universe has no reality without human observers, it is an easy step to suppose it is *we* who shape the structure of the outside world. If you and I are the creators of its laws, it follows easily that science should be regarded as similar to art, poetry, music, philosophy, and other products of human culture. Because folkways change in time and vary from culture to culture, and because science clearly is part of culture, one can look upon the history of science in the way one looks upon the history of fashions.

Women's skirts are up in one decade, down in the next, then up again. The height of the skirt is a cultural preference. We cannot say a particular height is "true" and the others "false."

It is hard to believe that some intelligent people not only see the history of science as a series of cultural preferences but even write books about it. The Harvard astronomer Bruce Gregory, for example, recently produced a volume entitled *Inventing Reality: Physics as a Language* (Wiley, 1989). His wild theme is that physicists do not discover laws of nature. They invent them. Newton didn't discover the law of gravity. He invented it. J. J. Thompson didn't discover the electron. He made it up the way one makes up a tune. "The universe is made of stories," Gregory quotes the poet Muriel Rukeyser, "not of atoms."

Gregory's views are in the tradition of pragmatists who put human experience in the center of what is "real." They don't deny that there is an outside world with which we interact; but because we can know nothing about it except what we experience, they are unable to take seriously any talk about structures "out there" independent of human minds. Following in the footsteps of such pragmatists as Karl Pearson and Benjamin Lee Whorf, Gregory focuses on human language (including, of course, the language of mathematics) as the principal shaper of what scientists like to think is out there. "The stubbornly physical nature of the world we encounter every day is obvious," he writes. "The minute we begin to talk about this world, however, it somehow becomes transformed into another world, an interpreted world, a world delimited by language. . . ."

Since the world we talk about is the only one we can know, it follows that "as our vocabulary changes, so does the world." Again: "When we create a new way of talking about the world, we virtually create a new world." Books are real "not because of some mystical connection between language and the world, but because you can ask me to bring you a book and my action can fulfill your expectation."

Consider unicorns. Ordinary people would say they are unreal because there *are* no such animals. But Gregory claims that "unicorns are not 'real' because our community has no expectations about living or dead unicorns that can be fulfilled. . . ." Moreover, our language can even alter the past. When we stopped talking about unicorns, they ceased to be real. "History is not as immutable as we might think; language can apparently transform the past as readily as it shapes the present and the future." Shades of Orwell's *1984*, in which communist historians continually rewrite history!

It is a short step from such human-centered hubris to the belief of Shirley MacLaine and other New Agers that we have the power to create our own realities. There may be some sort of timeless world out there; but if so, as Kant maintained, its ultimate structure is forever beyond our grasp. "The laws of physics," Gregory bluntly puts it, "are *our* laws, not nature's." *We* are the gods who shape reality.

It is not surprising to learn that Gregory is a devotee of the early New

Age cult of *est*. "I owe my appreciation of the immense power of the myth of 'is,' " he writes, "to Werner Erhard's relentless commitment to making a difference in my life. Absent his unremitting efforts to uncover the role of speaking in shaping experience, this book never would have been written."

Let's try to clear up some confusions involving subjectivity and relativism. First, the notion that science is always fallible is an ancient one, ably defended by the Greek skeptics, that no scientist or philosopher today denies. The very term *fallibilism* was coined by the American philosopher Charles Peirce to emphasize the way scientific statements differ from theorems in mathematics and formal logic. In logic and mathematics there are ironclad proofs inside formal systems. For example, you can prove the Pythagorean theorem within the system of Euclidian geometry—a proof that remains undamaged by the non-Euclidian structure of space-time. Given the axioms of Euclidian geometry, the theorem is true in all possible worlds. Science, on the other hand, has no infallible proofs.

Although all scientific statements are corrigible, it does not follow that they can't be placed in a continuum of probabilities that range from virtual certainty to almost certain falsehood. No one doubts, for instance, that the earth is shaped like a ball, goes around the sun, rotates, has a magnetic field, and has a moon that circles it. It is almost certain that the universe is billions of years old and that life on earth evolved over millions of years from simple to more complex forms. The big bang origin of the universe is not quite so certain. The inflationary model of the universe is still less certain. And so on. Science at present lacks any technique for applying precise probability values, or what Rudolf Carnap liked to call "degrees of confirmation," to its statements. That doesn't mean, however, that a scientist is not justified in saying that evolution has been strongly confirmed or that a flat earth has been strongly disconfirmed.

The title of Nancy Cartwright's book *How the Laws of Physics Lie* (Oxford University Press, 1983) seems to suggest that she agrees with Gregory, but on careful reading it turns out otherwise. What she does maintain—and who can disagree?—is that the phenomenological laws of physics (laws based on direct observations) have a much higher degree of confirmation than theories. We can be sure that all elephants have trunks because we can verify the statement by direct observation. Cartwright says she "believes" in the phenomenological laws, and also in such theoretical entities as electrons, even though their observation is indirect. If electrons don't make tracks in bubble chambers, she asks, what does? But when you turn to theoretical laws, such as the laws of relativity and quantum mechanics, she doesn't "believe" in them in the same way because they are too far from strong confirmation, and too subject to change. It is in this sense that science "lies."

Where does this leave us? Surely it does not leave us with a relativism in which competing scientific theories are "incommensurable"—that is, without standards by which they can be ranked. Science is like an expanding region with a solid core of truths that are very close to certainty. As you move outward from the core, assertions become progressively more tentative. In no way can one deny that science progresses in a manner quite different from the "progress" of music, art, or fashions in clothes.

Like almost all scientists, philosophers, and ordinary people, Peirce was a hard-nosed realist. Science, he wrote, is a method "by which our beliefs are determined by nothing human, but by some external permanency—by something upon which our thinking has no effect."

Here is how the eminent Harvard physicist Sheldon Glashow said the same thing in a mini-essay in the *New York Times* (October 22, 1989):

> We believe that the world is knowable, that there are simple rules governing the behavior of matter and the evolution of the universe. We affirm that there are eternal, objective, extrahistorical, socially neutral, external and universal truths and that the assemblage of these truths is what we call physical science. Natural laws can be discovered that are universal, invariable, inviolate, genderless and verifiable. They may be found by men or by women or by mixed collaborations of any obscene proportions. Any intelligent alien anywhere would have come upon the same logical system as we have to explain the structure of protons and the nature of supernovae. This statement I cannot prove, this statement I cannot justify. This is my faith.

It is important to understand that, when a theory becomes strongly confirmed by repeated observations and experiments, it can move across a fuzzy boundary to become recognized by the entire scientific community as a fact. That planets go around the sun was once the Copernican theory. Today it is a fact. That material objects are made of molecules was once a conjecture. Indeed, for many decades it was ridiculed by many physicists and chemists. Today it is a fact. In Darwin's day there was a theory of evolution. Today, only ignorant creationists refuse to call it a fact. It is also important to understand that so-called revolutions in science are not revolutions in the sense of overthrowing an earlier theory. They are benign refinements of earlier theories. Einstein didn't discard Newtonian physics. He added qualifications to Newtonian physics.

"The history of physics makes it hard to sustain the idea that we are getting closer to speaking 'nature's own language,' " Gregory naively writes. On the contrary, the history of physics makes it easy. Who, except academics smitten by relativism, can deny that science steadily improves its ability to explain and predict? Absolute truth may indeed be forever unobtainable, but if theories are not getting closer to accurate descriptions of the universe, why do they work so amazingly well? How is it we can build skyscrapers, hydrogen bombs, television sets, spacecraft, and other wonders of modern technology? Why is quantum mechanics able to predict with accuracies of many decimal places the outcomes of thousands of sophisticated experiments?

Surely it is insane to suppose that the enormous predictive power of science is nothing more than the power to predict the behavior of a world fabricated inside our tiny skulls. Of course all predictions are tested by human experience, but since everything we do is human experience, to say this is to say something obvious and trivial. Wigner wrote a now-famous essay on "The Unreasonable Effectiveness of Mathematics." To those who believe in a mathematically structured universe, independent of you and me, what could be more reasonable than the way mathematics fits the universe?

Nobody denies that science is a human tool, or that its history is influenced by cultural forces in all sorts of interesting ways. Nobody denies that scientists invent theories by creative acts similar to those of poets and artists. But once a theory is formulated, it is tested by a process that, in the long run, is singularly free of cultural biases. False theories are not shot down by a change of language, but by the universe.

James Trefil, in his stimulating book *Reading the Mind of God* (Scribner, 1989), recalls a lecture by a young sociologist on the history of the now-popular conjecture that dinosaurs were killed off by climatic changes that followed the impact of an extraterrestrial object striking the earth. She was good in describing the infighting among geologists, but she had no interest whatever in the evidence pro and con. From her perspective, her only task was to describe the conflict as if it were a battle between two rival art critics, with no mechanism for ever deciding who was right. A frustrated senior paleontologist in the audience finally burst out with the question, "Is it really news to sociologists that evidence counts?"

After all, Trefil concludes, "gravity pulls on the Bushman as well as on the European." Reading Shirley MacLaine, you might decide to create your own reality by jumping off a high building and soaring like Superman. Are we not assured by transcendental meditators that with training one can suspend gravity and levitate? Did not Jesus, that great super-psychic, walk on water? Last year a Russian psychic stood on a railroad track and tried to suspend the law of momentum (mass times velocity) by stopping a train. The poor man is no longer with us. Here is how Stephen Crane, in one of his short poems, reminded us that we are not the measure of all things:

> A man said to the universe:
> "Sir, I exist!"
> "However," replied the universe,
> "The fact has not created in me
> A sense of obligation."

RECOMMENDED READING

Alcock, James E. *Science and Supernature: A Critical Appraisal of Parapsychology.* Prometheus Books, Buffalo, N.Y., 1990.

Barrett, Stephen, M.D., and the Editors of Consumer Reports. *Health Schemes, Scams, and Frauds.* Consumer Reports Books, Fairfield, Ohio, 1991.

Blackmore, Susan. *Adventures of a Parapsychologist.* Prometheus Books, Buffalo, N.Y., 1986.

Culver, Roger B., and Philip A. Ianna. *Astrology: True or False?* Prometheus Books, Buffalo, N.Y., 1988.

Feder, Kenneth L. *Frauds, Myths, and Mysteries: Science and Pseudoscience in Archaeology.* Mayfield Publishing, Mountain View, Calif., 1990.

Frazier, Kendrick, ed. *Science Confronts the Paranormal.* Prometheus Books, Buffalo, N.Y., 1986.

Gardner, Martin. *Science: Good, Bad, and Bogus.* Prometheus Books, Buffalo, N.Y., 1981, 1989.

Gardner, Martin. *The New Age: Notes of a Fringe-Watcher.* Prometheus Books, Buffalo, N.Y., 1988, 1991.

Hines, Terence. *Pseudoscience and the Paranormal: A Critical Examination of the Evidence.* Prometheus Books, Buffalo, N.Y., 1988.

Hyman Ray. *The Elusive Quarry: A Scientific Appraisal of Psychical Research.* Prometheus Books, Buffalo, N.Y., 1989.

Klass, Philip J. *UFOs: The Public Deceived.* Prometheus Books, Buffalo, N.Y., 1983.

Klass, Philip J. *UFO-Abductions: A Dangerous Game.* Prometheus Books, Buffalo, N.Y., 1988, 1989.

Kurtz, Paul. *The Transcendental Temptation.* Prometheus Books, Buffalo, N.Y., 1986, 1991.

Kurtz, Paul, ed. *A Skeptic's Handbook of Parapsychology.* Prometheus Books, Buffalo, N.Y., 1985.

Neher, Andrew. *The Psychology of Transcendence.* Prentice-Hall, Englewood Cliffs, N.J., 1980.

Nickell, Joe. *Inquest on the Shroud of Turnin.* Prometheus Books, Buffalo, N.Y., 1983, 1987.

Nickell, Joe. *Secrets of the Supernatural.* Prometheus Books, Buffalo, N.Y., 1988.

Paulos, John Allen. *Innumeracy: Mathematical Illiteracy and Its Consequences.* Hill and Wang, New York, N.Y., 1988.

Radner, Daisie, and Michael Radner. *Science and Unreason*. Wadsworth Publishing, Belmont, Calif., 1982.

Randi, James. *Flim-Flam!* Prometheus Books, Buffalo, N.Y., 1982.

Randi, James. *The Faith-Healers*. Prometheus Books, Buffalo, N.Y., 1987, 1989.

Reed, Graham. *The Psychology of Anomalous Experience,* rev. ed. Prometheus Books, Buffalo, N.Y., 1988.

Rothman, Milton. *A Physicist's Guide to Skepticism*. Prometheus Books, Buffalo, N.Y., 1988.

Ruchlis, Hy. *Clear Thinking*. Prometheus Books, Buffalo, N.Y., 1990.

Sagan, Carl. *Broca's Brain*. Random House, New York, N.Y., 1979.

Williams, Stephen. *Fantastic Archaeology: The Wild Side of North American Prehistory*. University of Pennsylvania Press, Philadelphia, 1991.

Zusne, Leonard, and Warren H. Jones. *Anomalistic Psychology*. Lawrence Erlbaum Associates, Hillsdale, N.J., 1982.

The articles in this anthology appeared in the *Skeptical Inquirer (SI)*, a magazine devoted to the critical investigation of pseudoscience from a scientific viewpoint, and are reprinted by permission.

Part 1: Understanding the Human Need
"The Burden of Skepticism," by Carl Sagan. *SI*, Fall 1987, pp. 38–46.
"The Perennial Fringe," by Isaac Asimov. *SI*, Spring 1986, pp. 212–214.
"Reflections on the 'Transcendental Temptation,' " by Paul Kurtz. *SI*, Spring 1986, pp. 29–32.
"The Uses of Credulity," by L. Sprague de Camp. *SI*, Spring 1986, pp. 215–217.
"The Appeal of the Occult," by Phillips Stevens, Jr. *SI*, Summer 1988, pp. 376–385.

Part 2: Encouraging Critical Thinking
"A Field Guide to Critical Thinking," by James W. Lett. *SI*, Winter 1990, pp. 153–160.
"Assessing Arguments and Evidence," by Ray Hyman. *SI*, Summer 1987, pp. 400–404.

Part 3: Evaluating the Anomalous Experience
"The Brain and Consciousness: Implications for Psi Phenomena," by Barry Beyerstein. *SI*, Winter 1987–88, pp. 163–173.
"The Aliens Among Us: Hypnotic Regression Revisited," by Robert A. Baker. *SI*, Winter 1987–88, pp. 147–162.
"The Varieties of Alien Experience," by Bill Ellis. *SI*, Spring 1988, pp. 263–269.
"Past-Life Hypnotic Regression: A Critical View," by Nicholas P. Spanos. *SI*, Winter 1987–88, pp. 174–180.
"Past-Tongues Remembered," by Sarah G. Thomason. *SI*, Summer 1987, pp. 367–375.
"The Myth of Alpha Consciousness," by Barry Beyerstein. *SI*, Fall 1985, pp. 42–59.
"Pathologies of Science, Precognition, and Modern Psychophysics," by Donald D. Jensen. *SI*, Winter 1989, pp. 147–160.

Part 4: Considering Parapsychology

"The Elusive Open Mind: Ten Years of Negative Research in Parapsychology," by Susan J. Blackmore. *SI*, Spring 1987, pp. 244–255.

"Parapsychology, Miracles, and Repeatability," by Antony Flew. *SI*, Summer 1986, pp. 319–325.

"Ganzfeld Studies," by Kendrick Frazier. *SI*, Fall 1985, pp. 2–7.

"Improving Human Performance: What About Parapsychology?" by Kendrick Frazier. *SI*, Fall 1988, pp. 34–45.

"PSI Researchers' Inattention to Conjuring," by Martin Gardner. *SI*, Winter 1985–86, pp. 116–120.

"The Obligation to Disclose Fraud," by Martin Gardner. *SI*, Spring 1988, pp. 240–243.

Part 5: Examining Popular Claims

"The Hundredth Monkey Phenomenon" and "Watson and the Hundredth Monkey Phenomenon," by Ron Amundson. *SI*, Summer 1985, pp. 348–356, and *SI*, Spring 1987, pp. 303–304.

"An Investigation of Firewalking," by Bernard J. Leikind and William J. McCarthy. *SI*, Fall 1985, pp. 23–34.

"Incredible Cremations: Investigating Spontaneous Combustion Deaths," by Joe Nickell and John F. Fischer. *SI*, Summer 1987, pp. 352–357.

"Write and Wrong: The Validity of Graphological Analysis," by Adrian Furnham. *SI*, Fall 1988, pp. 64–69.

"Graphology and Personality: 'Let the Buyer Beware,' " by Robert Basil. *SI*, Spring 1989, pp. 241–243.

"A Study of the Kirlian Effect," by Arleen J. Watkins and William S. Bickel. *SI*, Spring 1986, pp. 244–257.

"The Moon Was Full and Nothing Happened," by Ivan W. Kelly, James Rotton, and Roger Culver. *SI*, Winter 1985–86, pp. 129–143.

"Testing Psi Claims in China," by Paul Kurtz, James Alcock, Kendrick Frazier, Barry Karr, Philip J. Klass, and James Randi. *SI*, Summer 1988, pp. 364–375.

Part 6: Medical Controversies

"The Psychopathology of Fringe Medicine," by Karl Sabbagh. *SI*, Winter 1985–86, pp. 154–164.

"Folk Remedies and Human Belief-Systems," by Frank Reuter. *SI*, Fall 1986, pp. 44–50.

"Chiropractic: A Skeptical View," by William Jarvis. *SI*, Fall 1987, pp. 47–55.

"Homeopathy: Is It Medicine?" by Stephen Barrett. *SI*, Fall 1987, pp. 56–62.

Part 7: Astrology

"Does Astrology Need to Be True? Part 1: A Look at the Real Thing," by Geoffrey Dean. *SI*, Winter 1986–87, pp. 166–184.

"Does Astrology Need to Be True? Part 2: The Answer Is No," by Geoffrey Dean. *SI*, Spring 1987, pp. 257–273.

"Double-Blind Test of Astrology Avoids Bias, Still Refutes the Astrological Hypothesis," from News and Comment. *SI*, Spring 1986, pp. 194–196.

Part 8: Crashed-Saucer Claims

"Crash of the Crashed-Saucer Claim," by Philip J. Klass. *SI*, Spring 1986, pp. 234–241.

"The MJ-12 Crashed-Saucer Documents," by Philip J. Klass. *SI*, Winter 1987–88, pp. 137–146.

"The MJ-12 Papers: Part 2," by Philip J. Klass. *SI*, Spring 1988, pp. 279–289.

"New Evidence of MJ-12 Hoax," by Philip J. Klass. *SI*, Winter 1990, pp. 135–140.

Part 9: Controversies Within Science

"Cold Fusion: A Case Study in 'Wishful Science,' " by Milton A. Rothman. *SI*, Winter 1990, 135–140.

"Water With Memory? The Dilution Affair," by Martin Gardner. *SI*, Winter 1989, pp. 132–141.

"Science, Mysteries, and the Quest for Evidence," by Martin Gardner. *SI*, Summer 1986, pp. 303–306.

"Relativism in Science," by Martin Gardner. *SI*, Summer 1990, pp. 353–357.

CONTRIBUTORS

JAMES E. ALCOCK is professor of psychology at Glendon College, York University, Toronto, and author of *Parapsychology: Science or Magic?* (1981) and *Science and Supernature* (1990).

RON AMUNDSON is professor of philosophy at the University of Hawaii at Hilo, specializing in the history of philosophy of science, especially evolutionary biology.

ISAAC ASIMOV has published more than 460 books (so far), fiction and nonfiction, and is one of the foremost explainers and popularizers of science of our time.

ROBERT A. BAKER, professor emeritus of psychology at the University of Kentucky, Lexington, spends most of his spare time working with the Kentucky Association of Science Educators and Skeptics (KASES) in an attempt to improve science education in Kentucky and increase skepticism worldwide. His most recent books are *They Call It Hypnosis* (1990) and *Hidden Memories: Voices and Visions from Within* (forthcoming).

STEPHEN BARRETT, M.D., a practicing psychiatrist and consumer advocate, edits *Nutrition Forum Newsletter* and has produced 26 books on health topics, including *The Health Robbers*. He is a board member of the National Council Against Health Fraud and is co-chairman of CSICOP's Paranormal Health Claims Subcommittee.

ROBERT BASIL edited *Not Necessarily the New Age: Critical Essays* and co-edited *On the Barricades: Religion and Free Inquiry in Conflict.*

BARRY BEYERSTEIN is professor of psychology at Simon Fraser University, Burnaby, British Columbia, and a researcher in the university's Brain Behavior Laboratory.

WILLIAM S. BICKEL is a physicist in the Department of Physics, University of Arizona, Tucson. His areas of expertise are optics, acoustics, spectroscopy, and atomic physics.

SUSAN BLACKMORE lectures in psychology at the universities of Bristol and Bath and is a writer and broadcaster. She is author of *Beyond the Body* and *The Adventures of a Parapsychologist*.

ROGER B. CULVER is an astronomer and physicist at Colorado State University and coauthor of *Astrology: True or False?*

GEOFFREY DEAN, Ph.D., is a British-born technical editor living in Perth, West Australia. He has been investigating astrological claims since 1974.

L. SPRAGUE DE CAMP has been a professional writer for more than 50 years. His works include science, history, biography, science fiction, fantasy, historical novels, and light verse. He is the author of the influential *The Ancient Engineers*, a history of technology from ancient times to the Renaissance, and *The Great Monkey Trial*, the definitive history of the Scopes evolution trial of 1925.

BILL ELLIS is associate professor of English and American Studies at Pennsylvania State University's Hazleton Campus. He is editor of *FOAFtale News*, the newsletter of the International Society for Contemporary Legend Research.

JOHN F. FISCHER is a forensic analyst with the crime laboratory of the Orange County Sheriff's Office, Orlando, Florida. An experienced crime-scene investigator, he is also president of a corporation specializing in forensic research.

ANTONY FLEW is emeritus professor of philosophy, University of Reading, England, and editor of, and contributor to, *Readings in the Philosophical Problems of Parapsychology*.

KENDRICK FRAZIER, a science writer, is the editor of the *Skeptical Inquirer* and a former editor of *Science News*. He is the author of four books, most recently *People of Chaco* and *Solar System*, and is the editor of the two previous anthologies in this series, *Paranormal Borderlands of Science* and *Science Confronts the Paranormal*.

ADRIAN FURNHAM is a reader in psychology at London University interested in psychometric assessment, the Barnum effect, and the abuse of psychology.

MARTIN GARDNER is the author of numerous books about science, mathematics, philosophy, and literature, among them *The New Ambidextrous Universe; Science: Good, Bad, and Bogus*; and *The New Age: Notes of a Fringe-Watcher*. For 25 years he wrote *Scientific American*'s Mathematical Games column.

RAY HYMAN is professor of psychology at the University of Oregon. His papers over the past two decades critically appraising parapsychological research have recently been published as *The Elusive Quarry*.

WILLIAM JARVIS, Ph.D., is a professor in the Department of Public Health and Preventive Medicine at Loma Linda University. He is president of the National Council Against Health Fraud and co-chairman of CSICOP's Paranormal Health Claims Subcommittee.

DONALD D. JENSEN received his Ph.D. from Yale in 1958, taught at Indiana University from 1960 to 1969, and since then has been at the University of Nebraska-Lincoln. His interests include research methods and conceptual styles in comparative psychology and ethology.

BARRY KARR is executive director of the Committee for the Scientific Investigation of Claims of the Paranormal.

IVAN W. KELLY is professor of educational psychology at the University of Saskatchewan in Canada and chairman of CSICOP's Astrology Subcommittee.

PHILIP J. KLASS is a technical journalist with *Aviation Week & Space Technology* magazine. For more than 20 years, Klass has investigated major UFO cases and has authored four books on the subject.

PAUL KURTZ is professor of philosophy at the State University of New York at Buffalo and founding chairman of the Committee for the Scientific Investigation of Claims of the Paranormal. He is author or editor of more than thirty books, including *The Transcendental Temptation* and *A Skeptic's Handbook of Parapsychology*.

BERNARD J. LEIKIND is a plasma physicist who worked for many years in fusion research. His more recent work has dealt with free electron lasers and fission reactors. He is with General Atomics, Inc., in San Diego.

JAMES W. LETT is professor of anthropology at Indian River Community College in Ft. Pierce, Florida, where he teaches a course in "Anthropology and the Paranormal." He is the author of *The Human Enterprise: A Critical Introduction to Anthropological Theory*.

WILLIAM J. MCCARTHY is a research psychologist in the Psychology Department at the University of California at Los Angeles.

JOE NICKELL, investigative writer and member of CSICOP's Executive Council, is author of *Inquest on the Shroud of Turin*, *Secrets of the Supernatural*, *The Magic Detectives*, and *Pen, Ink, and Evidence*.

JAMES RANDI is an internationally known conjuror and investigator of claims of alleged psychic abilities. He is the author of *The Truth About Uri Geller*, *Flim-Flam*, *The Faith Healers*, and *The Mask of Nostradamus*.

FRANK REUTER, who holds a Ph.D. in English literature, began collecting folklore while teaching at the University of Arkansas at Monticello. He now lives on a small farm in northern Arkansas and operates his own editing business.

MILTON A. ROTHMAN is professor of physics (retired), Trenton State College, a former research scientist with the Princeton Plasma Physics Laboratory, and author of *A Physicist's Guide to Skepticism* (1988) and *The Science Gap: Dispelling the Myths and Understanding the Reality of Science* (forthcoming).

JAMES ROTTON is associate professor of psychology at Florida International University and a member of CSICOP's Astrology Subcommittee.

KARL SABBAGH is an independent television producer and writer, covering scientific topics, in London.

CARL SAGAN is David Duncan Professor of Astronomy and Space Sciences and director of the Laboratory for Planetary Studies at Cornell University. He is the author of many best-selling science books, including *Cosmos, Broca's Brain*, and *The Dragons of Eden*, for which he received a Pulitzer Prize. His article in this volume is excerpted from his keynote address at the 1987 CSICOP conference in Pasadena, where he received CSICOP's In Praise of Reason Award.

NICHOLAS P. SPANOS is professor of psychology and director of the laboratory of experimental hypnosis at Carleton University, Ottawa.

PHILLIPS STEVENS, JR., is associate professor of anthropology at the State University of New York at Buffalo.

SARAH GREY THOMASON is professor of linguistics at the University of Pittsburgh and editor of *Language*, the journal of the Linguistic Society of America. Her research specialty is the study of language change, but she also enjoys investigating paranormal linguistic claims.

ARLEEN J. WATKINS is a research specialist at the University of Arizona, Tucson. Her area of expertise is applied statistics.

Aaronson, Steve, 221
Abel, E. L., 228, 232
Abell, George O., 42, 224, 227, 232
Adams, Evangeline, 308–310, 313
Adrian, E., 97, 109
Albanese, Catherine, 20, 22, 28
Alcock, James, 46, 48, 50, 52, 65, 67, 106, 108, 109, 124, 238, 239
Alexander, John, 161
Allan, Christopher, 330, 350
Allen, W. S., 197, 198
Almeder, Robert F., 92, 94
Alvarez, Luis, 340
Amin, Idi, 43
Amundson, Ron, 178
Anand, B., 97, 109
Andersen, M. L., 56, 68
Arens, W., 296
Aristarchus, 6
Arms, R. L., 228, 232
Arnold, Kenneth, 324, 337
Arnold, Larry E., 194, 198
Ashmun, J. M., 291, 294, 311
Asimov, Isaac, 229, 232
Askren, Edward, 288, 294
Aslin, R. D., 47, 52
Atanasoff, John V., 360
Atilla, 43
Atlas, R., 224, 232
Augustus, 18

Bacon, Francis, 22, 27
Bagchi, B., 97, 109
Bailly, Jean-Sylvain, 55

Baizer, Eric, 268, 270
Baker, Robert A., 58, 62, 68, 70, 71, 72, 75, 76
Bandi, Countess, 195
Barbault, André, 282
Barber, T. X., 55, 56, 59, 60, 64, 68, 69, 79, 80, 83, 112, 123
Bar-Hillel, M., 204
Barker, L. M., 114, 123
Baron, R. A., 228, 232
Barrett, Stephen, 264, 270, 367–368
Barry, T., 231, 232
Baum, A., 228, 232
Bayne, Rowan, 311
Beck, Adolph, 96
Beck, John B., 195, 199
Beck, Theodric, 195, 199
Beets, J. L., 224, 232
Bell, P. A., 228, 232
Bell, Robert, 164
Beloff, John, 165
Ben-Abba, E., 205
Benassi, Victor A., 16, 26, 27, 29, 99, 110
Bender, Hans, 316
Bennett, Gillian, 75, 76
Bennett, Thomas, 106, 109
Ben-Shaktar, G., 201, 203, 204
Benveniste, Jacques, 365–371
Berger, Hans, 107, 109
Berkeley, George, 141n
Berkner, Lloyd V., 338, 340
Berlitz, Charles, 323–326, 327, 336, 347
Bernal, John D., 366
Bernstein, M., 83

Bernstein, Robert, 170
Best, M. R., 114, 123
Best, S., 308, 312
Beyerstein, Barry, 98, 107, 123, 206
Bilin, F., 205
Birbaumer, N., 106, 109
Bishop, A., 354, 363
Biswas, S. K., 295
Bjork, Robert A., 161n
Blackmore, Susan J., 41, 48, 50, 52, 126,
    127, 128, 129, 132, 133, 134, 135, 299,
    301, 311, 312
Blair, Lawrence, 178
Blake, B. F., 226, 234
Blizin, Terry, 197, 199
Blondlot, Rene Prosper, 364
Bohr, Niels, 375
Borneman, Jay P., 273, 274
Boyle, Robert, 22
Bradshaw, Hannah, 195
Brau, J. L., 309, 312
Brazel, W. W., 324
Brazier, M., 107, 109
Brenman, M., 56, 68
Briggs, Katharine, 71, 76
Broad, C. D., 138
Broad, William J., 27, 28, 362, 363
Bronk, Detlev, 338
Brooner, R. K., 223, 234
Brown, Barbara, 107, 108, 109
Brown, F. A., 229, 232
Brown, Mark L., 265
Browne, Malcolm, 365
Bryan, J., 286, 296
Bunge, Mario, 44, 52
Burkan, Tolly, 184
Burton, J. M., 319
Bush, George, 35
Bush, Vannevar, 327, 328, 330, 331, 337,
    340, 341, 347, 349, 350

Caesar, Julius, 82
Cahn, J. P., 323
Campbell, D. E., 224, 227, 228, 232
Campbell, K., 44, 52
Cannon, M., 47, 52
Carlson, Shawn, 292, 293–294, 314, 320–
    322

Carnap, Rudolf, 378
Carnegie, Dale, 51
Carter, C. E. O., 288, 291, 294, 307, 310,
    312
Cartwright, Nancy, 378
Castro, Fidel, 35
Caton, Richard, 96, 98
Cayce, Edgar, 177
Chamberlin, T. C., 113, 123
Charry, J. M., 228, 232
Châtillon, G., 282, 294
Chaves, J. F., 80, 83
Chevalier, Maurice, 56
Chhina, G., 97, 109
Chidester, T. R., 318, 319
Churchland, P. M., 44, 52
Clamar, Aphrodite, 59
Clark, Jerome, 344–345
Clark, Vernon, 307, 312
Clements, William, 73, 76
"Clever Hans," 70
Close, Frank, 362
Clues, Mary, 195
Coe, W. C., 56, 68, 80, 83
Coetzee, C., 312
Cohen, D. B., 48, 52
Cohen, J., 232
Cole, H. W., 97, 110
Coles, E. M., 224, 232
Coles, Peter, 370
Columbus, Christopher, 43
Connor, John W., 38, 39, 304, 312
Cook, Thomas D., 161
Cooke, D. J., 224, 232
Copeland, Royal, 273
Copernicus, 5–6, 375
Coren, S., 49, 52
Corgiat, M., 223, 234
Corliss, William R., 55, 65, 68
Costello, D., 312
Crane, Stephen, 380
Crelin, Edmund, 263–264, 270
Croiset, Gerard, 36–37
Cronbach, L. J., 293
Crookes, William, 163
Crosbie, P., 294
Crow, Charles L., 28
Crown, David, 348

Culver, R., 227, 231, 232, 280, 282, 283, 294, 295, 304, 312
Cummings, M., 294
Cummings, Stephen, 368, 369
Curran, Douglas, 55, 61, 63, 68
Custer, Edith, 288, 294
Cutler, Robert, 327, 331–332, 334, 335, 345, 351, 352

Davenas, Elizabeth, 367
Davenhill, R., 225, 232
Davies, J. D., 290, 294
Davison, Gerald C., 161n
Dean, Geoffrey, 280, 283, 287, 290, 293, 294, 298, 306, 311, 312, 314, 316, 318
De Boiseon, Madame, 195
De Gaulle, Charles, 282
De Graaf, J. P., 224, 228, 230, 234
De l'Isle, Morio, 164
DeMonbreun, B. G., 230, 233
Derjaguin, Boris, 366
Derr, J. S., 282, 295
Dewar, Wilhelmina, 196
Dhlopolsky, J. G., 299, 312
Dickens, Charles, 194–195
Dickson, D. H., 299, 312
Dimond, S. J., 45, 52
Dingwall, Eric, 36, 39
Ding Wei Xin, 242–245
Dixon, Jeane, 33, 35
Dobelle, W., 50, 52
Dobyns, Zipporah, 280, 288, 294
Doherty, Jim, 188, 193
Domjan, M., 114, 123
Donnelly, F. A., 313
Douglas, Graham, 311
Downs, Hugh, 225
Doyle, Conan, 163
Drake, Frank, 7
Druckman, Daniel, 161n
Dua, M., 224, 225, 234
Dulles, Allen, 325
Dulles, John Foster, 325
Dunsany, Lord, 372
Durant, Ariel, 21, 28
Durant, Will, 21, 28
Duvall, Charles, 268, 269, 270
Dwyer, T., 292, 294

Eckert, J. Presper, 360
Edel, Leon, 71, 74, 76
Edmands, A., 312
Edmonstron, W. E., Jr., 79, 83
Einstein, Albert, 51, 352, 353, 373, 375, 379
Eisenbud, Jule, 162
Eisenhower, Dwight D., 325, 328, 330, 332, 335, 336, 337, 339, 342, 345, 347
Ellson, D., 121, 124
Elortegui, P., 231, 234
Erhard, Ludwig, 282
Erhard, Werner, 378
Ewing, R. I., 361, 363
Eysenck, Hans J., 201, 202, 205, 279, 280, 290, 294, 306, 312, 314
Eysenck, M. W., 306, 312

Falwell, Jerry, 374
Fan Yu Lin, 242
Faust, D. L., 221
Fellows, B. J., 79, 83
Festinger, Leon, 130, 135
Finke, R. A., 49
Fischer, John, 194, 199
Fisher, D. C., 315, 318
Fisher, J. D., 228, 232
Fisher, Russell S., 197, 199
Fiske, S. T., 230, 232
Flammarion, Nicolas, 354
Fleischmann, Martin, 355–362
Fleming, L. D., 319
Flew, Antony, 142
Flug, A., 205
Flugel, J. C., 290, 294
Flynn, D., 81, 84
Forrestal, James, 327, 328, 330, 338, 339, 340, 347, 351
Fourie, D. P., 307, 312
Franklin, Benjamin, 55, 136
Franks, Felix, 366
Frazier, 28n, 65, 68, 238, 239
Freireich, Emil J., 248–250, 252
Frey, J., 223, 231, 232, 233
Friedman, Stanton, 327, 330, 331, 332, 334, 336–337, 340, 343, 344, 345–346, 347, 350–351, 352
Fullam, F. A., 287, 295, 310, 312

Furnham, Adrian, 203, 205, 315, 316, 318

Gadd, Laurence D., 197, 199
Gaddis, Vincent H., 194, 197, 198
Galbreath, Robert, 21, 22, 23, 28
Galileo, 364
Gandhi, Mohandas, 207
Gardner, H., 46, 52
Gardner, Martin, 65, 67, 68, 104, 109, 165–166, 221, 277
Garzino, S. J., 222, 228, 229, 232
Gauquelin, Michel, 311, 317
Gazzaniga, M., 52
Gergen, K. J., 315, 318
Gibson, H. B., 133, 134, 135
Gieles, Mr., 293
Gilbert, G. O., 228, 232
Gill, M. M., 56, 68
Gingold, Hermione, 56
Girvin, J., 52
Glashow, Sheldon, 379
Glick, P., 303, 313, 317, 318
Godbey, J. W., 50, 52
Good, I. J., 315
Good, Timothy, 332, 334, 346
Goss, Kathleen, 368
Gottesman, D., 317, 318
Gould, Stephen Jay, 111, 112, 124
Grange, C., 293, 294, 295, 301, 313
Gray, A. L., 234
Gray, Gordon 338, 339
Greeley, A., 43, 52
Green, D. M., 119, 124
Green, Elmer, 106, 107, 108
Greene, R. L., 293, 295
Greenwood, Barry, 335
Gregory, Bruce, 377–379
Griffard, C. D., 113, 124
Grim, P., 295
Grof, Stanislav, 47, 52
Groothuis, Douglas, 180, 181
Gudjonsson, Gisle, 201, 205
Guess, George A., 275
Gunter, Barry, 202, 203
Gwynne, M. I., 81

Hagelstein, Peter, 356
Hager, J. L., 114, 124

Hahneman, Samuel, 271–272, 275
Haldane, J. B. S., 374
Haldeman, Scott, 264, 270
Halliday, D. J. X., 197, 199
Hamblin, D., 285, 295
Hand, R., 313
Handy, Rob, 297
Hansen, Chadwick, 21, 29
Hansen, R. D., 230, 233
Haraldsson, E., 46, 52
Harary, S. B., 135, 164
Harper, Edith, 163
Harris, M. E., 295
Hartley, Elda, 171, 177, 181
Hasted, John, 159, 160, 162
Hawkinshire, F. B. W., 228, 232
Haynes, B., 62, 68
Heal, L. W., 319
Hebb, Donald O., 44, 52
Heimlich, Jane, 368
Heimlich, Henry, 368
Hempel, C. G., 230, 233
Hepburn, A., 315, 318
Heslin, R., 226, 234
Hilgard, Ernest, 48, 52, 56, 57, 68, 80, 83
Hilgard, Josephine, 56, 68
Hill, Betty and Barney, 75
Hillenkoetter, R. H., 328, 332, 336, 337, 338, 339, 342, 343, 347, 348
Hines, Terence, 35
Hippocrates, 43
Hirai, T., 97, 110
Hirsch, H., 47, 52
Hirsig, Huguette, 282
Hirvonoja, R., 82, 83
Hitler, Adolf, 37, 43
Hobbes, Thomas, 18
Hoebens, Piet Hein, 37, 39, 293, 295
Hofstadter, Douglas, 208
Holton, Gerald, 26, 27, 28, 29
Home, D. D., 163–164
Homola, Samuel, 267–268, 270
Honorton, Charles, 108, 109, 127, 130, 131, 135, 144, 147, 148, 151, 161, 162
Hopkins, Budd, 54, 55, 59, 60, 65, 66, 72, 73, 75, 169
Horowitz, M., 49, 52
Horowitz, Paul, 7

Houdini, 111, 124, 167
Houston, Whitney, 33
Hubbard, G. S., 155
Hubbard, L. Ron, 46, 47, 52
Huber, B., 317
Huber, L., 317
Hufford, David J., 71–72, 74, 76
Hume, David, 136–137, 139, 141n, 142n
Humphrey, B. S., 155
Humphreys, Lloyd G., 161n
Hunsaker, Jerome, 338
Hunter, R. N., 282, 295
Hyatt, Henry Middleton, 74
Hyman, Ray, 36, 39, 41, 143–148, 152, 161, 299, 311, 312, 313

Ianna, Philip, 232, 280, 282, 283, 293, 294, 295, 304, 312
Imanishi, Kinji, 171, 181
Irving, Washington, 369
Irwin, James, 27

Jackson, Hughlings, 48
Jacobson, M., 47, 52
Jahn, Robert, 130, 151, 154, 155, 374
James, Henry, Jr., 71
James, Henry, Sr., 71, 74
James, William, 71
Janis, J., 135
Janov, Arthur, 47, 52
Jeffries, Anne, 71
Jenkins, Bill, 184
Jensen, D. D., 113, 124
Joan of Arc, 43
Johnson, F. H., 49, 52, 225, 232
Johnson, L., 102, 105, 106, 109
Johnson, Louis A., 338
Johnson-Laird, P. N., 230, 234
Jolton, J., 317, 318
Jones, Marc Edmund, 307
Jones, Scott, 161
Jones, Steven, 355, 357, 360
Jones, W., 48, 50, 53
Jones, Warren H., 65, 69, 299, 303, 313
Jung, Carl, 61, 68
Jusczyc, P., 52

Kallai, E., 301, 313

Kaminsky, Howard, 169–170
Kamiya, Joe, 98, 105, 107, 109
Kammann, Richard, 41, 50, 53, 65, 68, 233, 301, 313
Kampman, R., 82, 83
Kant, Immanuel, 377
Karnes, Edward, 206, 207
Karr, Barry, 238, 239
Kasamatsu, A., 97, 110
Katzeff, P., 228, 233
Kawai, Masao, 171, 173–174, 175, 176, 181
Kawamura, Syunzo, 171, 181
Kelly, I. W., 222, 223, 224, 225, 227, 228, 230, 231, 232, 233, 234, 280, 281, 287, 295, 299, 311, 312, 313, 316, 318
Kemp, Jack, 33
Kennedy, Caroline, 33
Kennedy, John F., 35, 282
Kepler, Johannes, 21
Kerr, Howard, 28
Keyes, Ken, Jr., 171, 176–177
Keyhoe, Donald, 323
Khrushchev, Nikita, 282
Kinder, Gary, 54, 66
Kirsch, Irving, 65, 68
Klass, Philip J., 55, 68, 76, 112, 124, 238, 239
Klein, Felix, 206, 207
Klein, Mavis, 318, 319
Klein, Robert, 72
Klimoski, Richard, 201, 205, 206, 208
Köberl, T., 316, 319
Koestler, Arthur, 164, 372, 373–374
Kolb, 45, 48, 52
Koshland, Daniel, Jr., 369
Kou, E. C. Y., 225, 233
Koval, B., 286, 295
Krakauer, Jon, 189, 193
Krier, Beth Ann, 182, 193
Krippner, Stanley, 221, 315, 319
Krogman, Wilton M., 198, 199
Krutzen, R. W., 281, 295, 311, 313
Kuhn, Thomas, 373
Kurtz, Paul, 21, 23, 29, 39, 238, 239, 241
Kusche, Larry, 112, 124
Kyler, H. J., 221

Labelle, Louise, 56, 68

Lackey, D. P., 295
Landers, Ann, 226
Landers, Daniel M., 161n
Landry, General, 341
Langmuir, Irving, 354
Larner, S. P., 286, 295
Larson, D., 113, 124
Laster, A., 284, 295
Laurence, Jean-Roch, 56, 58, 68
Lavoisier, 55
Lawrence, J., 107, 110
Lay, James, 345
Lazar, Philippe, 371
Le Doux, J. D., 45, 52
Lee, Lehman, 314
Lehman, J. L., 319
Leiman, A., 44, 53
Lenneberg, E., 47, 53
Lester, D., 201, 205, 284, 295
Levi, B. G., 359, 363
Levy, Austin, 291, 295, 311
Levy, Walter, 165, 169
Lewes, George Henry, 194–195, 199
Lieber, Arnold L., 222, 223, 224, 225, 226–227, 228, 229, 233
Liebig, Justus von, 195, 199
Lines, R., 201, 205
Linge, F., 46, 53
Lin Zixin, 235, 246
Locke, John, 22
Lodge, Olvier, 163
Loftus, Elizabeth, 48, 52, 53, 56, 57, 68
Loptson, P. J., 295
Lovick, K., 294
Lowell, Percival, 354, 367
Lu, Dr., 238–239
Luce, G. G., 229, 233
Ludwig, J., 140
Luther, Martin, 88

Maccabee, Bruce, S., 324, 343
Macharg, S. J., 307, 313
Mackay, Charles, 288, 295
MacLaine, Shirley, 23, 377, 380
MacLean, P., 48, 53
MacMahon, K., 224, 233
Macon, R. S., 295
Maddox, John, 366

Mahoney, M. J., 230, 233
Malcolm, N., 44, 53
Malik, D. C. V., 295
Mann, A. T., 316, 317, 319
Manson, Charles, 43
Marbell, N. Z., 319
Marks, David, 41, 50, 53, 65, 68, 233, 301, 313
Martens, R., 231, 233
Martin, Sandra, 231, 233
Mather, Arthur, 280, 287, 290, 293, 294, 298, 311, 312
Mathews, B., 97, 109
Matousek, Rose, 206, 207
Mauchly, John W., 360
May, Edward C., 151, 155
Mayer, M. H., 284, 295
McBain, I., 102, 110
McCall, John, 293
McCreery, C., 108, 110
McFall, R. M., 316, 317, 319
McFarland, James D., 168
McGrew, J. H., 316, 317, 319
McKellar, Peter, 64, 68
McLaughlin, S., 201, 205
Meier, Eduard, 54, 55, 61
Melzack, R., 105, 110
Menzel, Donald, 338, 343
Meriwether, Joseph, 316
Mesmer, Franz, 55
Metzger, Henry, 370
Metzner, R., 285, 295
Meyers, D. G., 226, 230, 233
Miller, G. A., 281, 295
Miller, Jon D., 25, 26, 28, 29
Millet, Madame, 195
Minors, D. S., 229, 233
Mirabile, C. S., 225, 229, 233
Miskelly, G. M., 361, 363
Mitchell, Edgar, 165
Mitchell, John, 258
Mitchinson, Avrion, 370
Mladejovsky, M., 52
Mobley, Sandra A., 161n
Modde, Peter, 265
Montague, Robert M., 338
Montague, S. P., 296
Moody, Raymond, 46, 53

Moore, K. D., 41
Moore, William L., 323–346, 347, 350, 351, 352
Morrell, F., 97, 110
Morris, Robert, 108, 110, 132, 164–165
Morrock, Richard, 47
Moss, M., 201, 205
Moss, Thelma, 221
Mozart, 43
Muhammad, 43
Mulhern, Sherrill, 76, 97
Mulholland, T., 110
Muller, Richard, 322
Murphy, Bridey, 83
Murphy, Gardner, 108, 168, 319

Nadon, Robert, 56, 58, 68
Naftulin, D. H., 301, 313
Napoleon, 207
Neher, Andrew, 48, 50, 53, 106, 110, 295, 299, 312, 313
Nelson, John, 283
Newton, Isaac, 43, 375
Nias, D. K. B., 279, 280, 290, 294, 311
Nickell, Joe, 194, 199, 350
Niehenke, Peter, 299, 313
Nisbett, R., 230, 233, 303, 305, 313
Nixon, Richard, 283
Nogrady, Heather, 58, 68
Nolen, W., 104, 110
Norman, Ruth, 61
Norstad, Lauris, 336
Nosal, G., 201, 205
Novak, A. R., 319
Nychis, William G., 274

Oakley, D., 44, 53
O'Connell, D. N., 58, 68
O'Hara, Maureen, 180, 181
Ohlemiller, Thomas J., 197, 199
Omohundro, John T., 34, 39
Orne, Martin, 48, 53, 58, 68, 102, 103, 110
Orr, Leonard, 47
Orwell, George, 377
Osborn, Albert S., 348
Osis, K., 46, 52
Ossenkopp, K. P., 228, 233

Ossenkopp, M. D., 228, 233
Ostrander, Sheila, 221
Otis, L. P., 225, 230, 233

Pace, Frank, Jr., 339
Page, Thorton, 340
Palmer, Daniel D., 263, 266
Palmer, John A., 161
Panos, Maesimund, 368
Paracelsus, 271
Parise, Felicia, 163
Park, Y. H., 232
Parker, D., 289, 295
Parmalee, A., 45, 53
Paskewitz, D., 102, 103, 110
Patrick, B., 62, 68
Paul, St., 43
Pauling, Linus, 27
Pearce, Hubert, 169
Pearson, Karl, 377
Peguy, Charles, 352
Pehek, J. O., 221
Peirce, Charles, 374, 378
Peirce, J. T., 113, 124
Pelham, B. W., 318, 319
Pell, Claiborne, 161
Perinbanayagam, R. S., 281, 295
Perkins, Maureen, 311
Perry, Campbell, 56, 58, 68
Peterson, Chase, 360
Pett, Grace, 195
Phillips, Peter, 162, 164
Pillemer, B. P., 47, 53
Pines, M., 107, 110
Pisoni, D., 52
Plato, 43
Platt, J. R., 113, 124
Plotkin, H., 44, 53
Plotkin, W., 97, 102, 105, 110
Poincaré, Henri, 4
Pitevin, Bernard, 367
Pons, B. Stanley, 355–362
Pool, R., 355, 362, 363
Popper, K., 233
Porac, C., 52
Porter, Lyman W., 161n
Posner, Michael I., 161n
Powell, J. H., 261

Pratt, J. G., 168, 169
Price, G. R., 139
Priess, Adele D., 313
Procter & Gamble, 23
Puccetti, R., 46, 53
Puthoff, Harold, 154

Quian Xue Seng, 236

Radin, Dean I., 60
Radner, Daisie, 65, 68
Radner, Michael, 65, 68
Rafael, A., 201, 205
"Ramtha," 3
Randi, James, 65, 68, 104, 110, 112, 124,
    142n, 162, 189–190, 238, 239, 241, 242,
    260, 293, 295, 316, 367, 369, 370
Rather, Dan, 33
Ray, W. J., 97, 110
Reed, Graham, 64, 65, 68
Reeser, Mary, 194, 197, 198
Reinberg, A., 229, 233
Re'is, Piri, 34
Relman, Arnold, 370
Rensberger, Boyce, 178, 180, 181
Retzlaff, P. D., 80, 83
Reuter, Frank, 261
Reverchon, J., 282–283, 295
Rhine, J. B., 108, 138, 165, 167, 168–169,
    177, 374, 375
Rhine, Louisa, 167–168
Rhodes, Richard, 27
Robertson, H. P., 340
Robbins, Tony, 182–193
Rogo, D. Scott, 165
Roll, William, 164, 319
Roman, M. B., 112, 114
Roof, N., 280, 294
Rosen, R. D., 47, 53, 104, 110
Rosenblum, Bernard, 283, 287, 289, 295,
    301, 313
Rosenthal, D., 201, 205
Rosenzweig, M., 44, 53
Ross, L. H., 230, 233, 303, 305, 307, 313
Rothman, Milton, 357, 358, 363
Rothman, T., 353, 363
Rotton, J., 222, 223, 225, 226, 227, 228,
    230, 231, 232, 233, 234

Rubin, Daniel, 221
Rudhyar, Dane, 281, 287, 288, 291, 296,
    307, 311, 317
Rukeyser, Muriel, 377
Ruppelt, Edward J., 325
Rush, Benjamin, 257–261
Russell, Bertrand, 109
Russell, D., 299, 303, 313
Russell, G. W., 224, 225, 228, 230, 232,
    234
Russell, I. S., 44, 53
Russell, R. J. H., 225, 234

Saklofske, D. H., 231, 233, 316, 318
Salamon, M. H., 362
Salmon, W. C., 230, 234
Samarin, William J., 93n, 94
Sanduleak, N., 224–225, 228, 229, 234
Santee, A., 201, 205
Sarbin, T. R., 56, 68, 80, 83
Sargant, William, 72, 73–74, 75, 77
Sargent, Carl, 127, 129, 130, 135
Schatzman, M., 49, 53
Schiaparelli, Giovanni, 354
Schmidt, Helmut, 130, 131, 151, 154, 155,
    162, 165, 374
Schneider, Walter, 161n
Schofield, S., 315, 318
Schroeder, Lynn, 221
Schultz, Ted, 178, 180, 181
Scofield, A. M., 276
Scriven, Michael, 43, 53
Scully, Frank, 323
Sechrest, L., 286, 296
Seligman, M. E. P., 114, 124
Serios, Ted, 162–163
Shandera, Jaime, 327–335, 340, 344–346,
    347, 351
Shapiro, J. L., 222, 234
Sheaffer, Robert, 55, 68
Sheldrake, Rupert, 171, 177, 181
Sherin, C. R., 223, 224, 226, 228, 233
Shivers, O., 293, 296
Shor, R. E., 58, 68
Siegel, Ronald K., 46, 49, 53, 132, 135
Sigman, M., 45, 47, 52
Silverman, P. S., 80, 83

Simkins, K., 96, 104, 105, 110
Singer, Barry, 16, 26, 27, 29, 42, 99, 110, 221
Singer, J. L., 48, 53
Singer, Jerome E., 64, 68, 161n
Singh, B., 97, 109
Smit, Rudolf, 288, 296, 311, 314, 319
Smith, M., 294
Smith, Ralph L., 264, 270
Smith, Richard Furland, 279, 296
Smith, Walter B., 338, 340
Smith, Wilbert B., 351
Smolensky, M. H., 229, 233
Snyder, M., 230, 234, 303, 313
Soal, S. G., 140, 169, 374
Soble, C., 234
Souers, Sidney W., 338, 339
Soyka, F., 234
Spanos, Nicholas P., 56, 68, 79, 80, 81, 83, 84
Spinelli, Ernesto, 127, 135
Spitz, Werner U., 197, 199
Springer, Sally P., 161n
Squire, L. R., 48, 53
Stalin, Joseph, 43
Startup, M. J., 225, 234, 280, 296, 311
Stead, W. T., 163
Stearn, J., 110, 283, 296
Steffert, B., 307, 313
Steiger, Brad, 325
Stein, Arthur, 171, 177, 181
Steiner, L. R., 286, 296
Stevens, Anthony, 289
Stevens, Phillips, Jr., 25, 29
Stevenson, Ian, 85–94
Stevenson, Thomas, 195, 199
Stewart, Gloria, 140
Stewart, Walter, 366, 367, 369, 370
Still, Andrew Taylor, 262
Story, Ronald D., 39
Stoyva, J., 48, 53
Streiner, D. L., 234
Strieber, Whitley, 54, 55, 59, 60, 61, 63–64, 66, 67, 70, 71, 72, 73–74, 75, 76, 77, 346
Suh, Dr., 264
Sullivan, Walter, 369–370
Sutcliffe, J. P., 56, 68

Swann, W. B., Jr., 230, 234, 318, 319
Swets, J. A., 119, 124, 149, 161n

Targ, Russell, 154, 164
Tart, C. T., 50, 53, 64, 69, 132, 135
Taubes, G., 362, 363
Taves, Ernest H., 70, 77
Taylor, John, 65, 69, 162
Taylor, S. E., 230, 232
Templer, D. I., 223, 234
Tenhaeff, Wilhelm, 37
Thibault, Mademoiselle, 139
Thomas, Keith, 23, 29
Thomason, Sara, 92, 94
Thompson, J. J., 377
Thompson, Richard F., 161n
Timm, U., 316, 319
Tolbert, C. R., 293, 295
Trefil, James, 380
Troscianko, T. S., 128, 135
True, R. M., 58, 69
Truman, Harry S, 327, 328, 330, 331, 332, 337, 340, 341, 344, 347, 348, 349, 350, 351
Truzzi, Marcello, 279, 296, 312
Tsumori, Atsuo, 171, 175, 181
Turner, Victor, 73, 77
Twain, Mark, 133
Twining, Nathan, 324, 327, 331, 334, 335, 338, 339, 345, 351, 352
Tyson, G. A., 292, 296, 301, 313

Ullman, Dana, 368, 369
Uttal, W., 44, 45, 49, 53
Utts, Jessica, 161

Valenstein, E., 45, 53
Vandenberg, Hoyt, 336, 338, 341
Vaughn, Freddie, 261
Veleber, D. M., 223, 234
Verny, Thomas, 47, 53
Vespasian, Emperor, 139
Vestewig, R., 201, 205
Vidmar, J. E., 307, 313
Vishveshwara, C. V., 295
Vogt, Evan Z., 41
Von Däniken, Erich, 34, 176
Von Peczely, Ignatz, 252

Walker, A. E., 46, 53
Walker, E. H., 162
Walker, Jearl, 184, 193
Wall, P., 105, 110
Wallace, L., 287, 296
Wambach, H., 78, 82, 84
Ward, L., 52
Ware, J. E., 313
Waterhouse, J. M., 229, 233
Watson, Lyall, 171–181
Watson, P., 230, 234
Weaver, H., 312
Wedow, S. M., 285, 296
Weiner, Arthur, 368
Wenger, M., 97, 109
West, L. J., 49, 53
Whishaw, 45, 48, 52
White, Rhea, 134, 135
White, S. H., 47, 53
Whorf, Benjamin Lee, 377
Wigner, Eugene, 376, 380

Wilde, Oscar, 61
Williams, N. L., 234
Williamson, Jo Ann, 334
Wilson, Harold, 282
Wilson, Ian, 69
Wilson, J. D., 52
Wilson, Sheryl C., 55, 59, 60, 64, 69
Winston, Jerold, 368
Wolf, David, 198, 199
Wolf, Kevin, 362
Wood, Robert, 364

Xu Ming Ding, 240

Yu Guangyuan, 237

Zaffuto, A., 110
Zaffuto, M., 110
Zhu Kunlong, 236
Zhu Yiyi, 236
Zusne, L., 48, 50, 53, 65, 69